Lecture Notes in Computer Science 12765

More information about this subseries at http://www.springer.com/series/7409

Sakae Yamamoto · Hirohiko Mori (Eds.)

Human Interface and the Management of Information

Information Presentation and Visualization

Thematic Area, HIMI 2021
Held as Part of the 23rd HCI International Conference, HCII 2021
Virtual Event, July 24–29, 2021
Proceedings, Part I

Springer

Editors
Sakae Yamamoto
Tokyo University of Science
Tokyo, Saitama, Japan

Hirohiko Mori
Tokyo City University
Tokyo, Japan

ISSN 0302-9743   ISSN 1611-3349 (electronic)
Lecture Notes in Computer Science
ISBN 978-3-030-78320-4   ISBN 978-3-030-78321-1   (eBook)
https://doi.org/10.1007/978-3-030-78321-1

LNCS Sublibrary: SL3 – Information Systems and Applications, incl. Internet/Web, and HCI

This Springer imprint is published by the registered company Springer Nature Switzerland AG
The registered company address is: Gewerbestrasse 11, 6330 Cham, Switzerland

Foreword

Human-Computer Interaction (HCI) is acquiring an ever-increasing scientific and industrial importance, and having more impact on people's everyday life, as an ever-growing number of human activities are progressively moving from the physical to the digital world. This process, which has been ongoing for some time now, has been dramatically accelerated by the COVID-19 pandemic. The HCI International (HCII) conference series, held yearly, aims to respond to the compelling need to advance the exchange of knowledge and research and development efforts on the human aspects of design and use of computing systems.

The 23rd International Conference on Human-Computer Interaction, HCI International 2021 (HCII 2021), was planned to be held at the Washington Hilton Hotel, Washington DC, USA, during July 24–29, 2021. Due to the COVID-19 pandemic and with everyone's health and safety in mind, HCII 2021 was organized and run as a virtual conference. It incorporated the 21 thematic areas and affiliated conferences listed on the following page.

A total of 5222 individuals from academia, research institutes, industry, and governmental agencies from 81 countries submitted contributions, and 1276 papers and 241 posters were included in the proceedings to appear just before the start of the conference. The contributions thoroughly cover the entire field of HCI, addressing major advances in knowledge and effective use of computers in a variety of application areas. These papers provide academics, researchers, engineers, scientists, practitioners, and students with state-of-the-art information on the most recent advances in HCI. The volumes constituting the set of proceedings to appear before the start of the conference are listed in the following pages.

The HCI International (HCII) conference also offers the option of 'Late Breaking Work' which applies both for papers and posters, and the corresponding volume(s) of the proceedings will appear after the conference. Full papers will be included in the 'HCII 2021 - Late Breaking Papers' volumes of the proceedings to be published in the Springer LNCS series, while 'Poster Extended Abstracts' will be included as short research papers in the 'HCII 2021 - Late Breaking Posters' volumes to be published in the Springer CCIS series.

The present volume contains papers submitted and presented in the context of the Human Interface and the Management of Information (HIMI 2021) thematic area of HCII 2021. I would like to thank the Co-chairs, Sakae Yamamoto and Hirohiko Mori, for their invaluable contribution to its organization and the preparation of the proceedings, as well as the members of the Program Board for their contributions and support. This year, the HIMI thematic area has focused on topics related to information presentation, visualization and decision-making support, information in VR and multimodal user interfaces, information-rich learning environments, and information in intelligent systems, as well as work, collaboration and design support.

I would also like to thank the Program Board Chairs and the members of the Program Boards of all thematic areas and affiliated conferences for their contribution towards the highest scientific quality and overall success of the HCI International 2021 conference.

This conference would not have been possible without the continuous and unwavering support and advice of Gavriel Salvendy, founder, General Chair Emeritus, and Scientific Advisor. For his outstanding efforts, I would like to express my appreciation to Abbas Moallem, Communications Chair and Editor of HCI International News.

July 2021   Constantine Stephanidis

HCI International 2021 Thematic Areas and Affiliated Conferences

Thematic Areas

- HCI: Human-Computer Interaction
- HIMI: Human Interface and the Management of Information

Affiliated Conferences

- EPCE: 18th International Conference on Engineering Psychology and Cognitive Ergonomics
- UAHCI: 15th International Conference on Universal Access in Human-Computer Interaction
- VAMR: 13th International Conference on Virtual, Augmented and Mixed Reality
- CCD: 13th International Conference on Cross-Cultural Design
- SCSM: 13th International Conference on Social Computing and Social Media
- AC: 15th International Conference on Augmented Cognition
- DHM: 12th International Conference on Digital Human Modeling and Applications in Health, Safety, Ergonomics and Risk Management
- DUXU: 10th International Conference on Design, User Experience, and Usability
- DAPI: 9th International Conference on Distributed, Ambient and Pervasive Interactions
- HCIBGO: 8th International Conference on HCI in Business, Government and Organizations
- LCT: 8th International Conference on Learning and Collaboration Technologies
- ITAP: 7th International Conference on Human Aspects of IT for the Aged Population
- HCI-CPT: 3rd International Conference on HCI for Cybersecurity, Privacy and Trust
- HCI-Games: 3rd International Conference on HCI in Games
- MobiTAS: 3rd International Conference on HCI in Mobility, Transport and Automotive Systems
- AIS: 3rd International Conference on Adaptive Instructional Systems
- C&C: 9th International Conference on Culture and Computing
- MOBILE: 2nd International Conference on Design, Operation and Evaluation of Mobile Communications
- AI-HCI: 2nd International Conference on Artificial Intelligence in HCI

List of Conference Proceedings Volumes Appearing Before the Conference

1. LNCS 12762, Human-Computer Interaction: Theory, Methods and Tools (Part I), edited by Masaaki Kurosu
2. LNCS 12763, Human-Computer Interaction: Interaction Techniques and Novel Applications (Part II), edited by Masaaki Kurosu
3. LNCS 12764, Human-Computer Interaction: Design and User Experience Case Studies (Part III), edited by Masaaki Kurosu
4. LNCS 12765, Human Interface and the Management of Information: Information Presentation and Visualization (Part I), edited by Sakae Yamamoto and Hirohiko Mori
5. LNCS 12766, Human Interface and the Management of Information: Information-rich and Intelligent Environments (Part II), edited by Sakae Yamamoto and Hirohiko Mori
6. LNAI 12767, Engineering Psychology and Cognitive Ergonomics, edited by Don Harris and Wen-Chin Li
7. LNCS 12768, Universal Access in Human-Computer Interaction: Design Methods and User Experience (Part I), edited by Margherita Antona and Constantine Stephanidis
8. LNCS 12769, Universal Access in Human-Computer Interaction: Access to Media, Learning and Assistive Environments (Part II), edited by Margherita Antona and Constantine Stephanidis
9. LNCS 12770, Virtual, Augmented and Mixed Reality, edited by Jessie Y. C. Chen and Gino Fragomeni
10. LNCS 12771, Cross-Cultural Design: Experience and Product Design Across Cultures (Part I), edited by P. L. Patrick Rau
11. LNCS 12772, Cross-Cultural Design: Applications in Arts, Learning, Well-being, and Social Development (Part II), edited by P. L. Patrick Rau
12. LNCS 12773, Cross-Cultural Design: Applications in Cultural Heritage, Tourism, Autonomous Vehicles, and Intelligent Agents (Part III), edited by P. L. Patrick Rau
13. LNCS 12774, Social Computing and Social Media: Experience Design and Social Network Analysis (Part I), edited by Gabriele Meiselwitz
14. LNCS 12775, Social Computing and Social Media: Applications in Marketing, Learning, and Health (Part II), edited by Gabriele Meiselwitz
15. LNAI 12776, Augmented Cognition, edited by Dylan D. Schmorrow and Cali M. Fidopiastis
16. LNCS 12777, Digital Human Modeling and Applications in Health, Safety, Ergonomics and Risk Management: Human Body, Motion and Behavior (Part I), edited by Vincent G. Duffy
17. LNCS 12778, Digital Human Modeling and Applications in Health, Safety, Ergonomics and Risk Management: AI, Product and Service (Part II), edited by Vincent G. Duffy

38. CCIS 1420, HCI International 2021 Posters - Part II, edited by Constantine Stephanidis, Margherita Antona, and Stavroula Ntoa
39. CCIS 1421, HCI International 2021 Posters - Part III, edited by Constantine Stephanidis, Margherita Antona, and Stavroula Ntoa

http://2021.hci.international/proceedings

Human Interface and the Management of Information Thematic Area (HIMI 2021)

Program Board Chairs: **Sakae Yamamoto**, *Tokyo University of Science, Japan*, **and Hirohiko Mori,** *Tokyo City University, Japan*

- Yumi Asahi, Japan
- Shin'ichi Fukuzumi, Japan
- Michitaka Hirose, Japan
- Yasushi Ikei, Japan
- Yen-Yu Kang, Taiwan
- Keiko Kasamatsu, Japan
- Daiji Kobayashi, Japan
- Kentaro Kotani, Japan
- Hiroyuki Miki, Japan
- Miwa Nakanishi, Japan
- Ryosuke Saga, Japan
- Katsunori Shimohara, Japan
- Takahito Tomoto, Japan
- Kim-Phuong L. Vu, USA
- Tomio Watanabe, Japan
- Takehiko Yamaguchi, Japan

The full list with the Program Board Chairs and the members of the Program Boards of all thematic areas and affiliated conferences is available online at:

http://www.hci.international/board-members-2021.php

HCI International 2022

The 24th International Conference on Human-Computer Interaction, HCI International 2022, will be held jointly with the affiliated conferences at the Gothia Towers Hotel and Swedish Exhibition & Congress Centre, Gothenburg, Sweden, June 26 – July 1, 2022. It will cover a broad spectrum of themes related to Human-Computer Interaction, including theoretical issues, methods, tools, processes, and case studies in HCI design, as well as novel interaction techniques, interfaces, and applications. The proceedings will be published by Springer. More information will be available on the conference website: http://2022.hci.international/:

General Chair
Prof. Constantine Stephanidis
University of Crete and ICS-FORTH
Heraklion, Crete, Greece
Email: general_chair@hcii2022.org

http://2022.hci.international/

Contents – Part I

Visualization and Decision-Making Support

Information in VR and Multimodal User Interfaces

Contents – Part II

Supporting Work, Collaboration and Design

Intelligent Information Environments

Information Presentation

The Use of New Presentation Technologies in Electronic Sales Environments and Their Influence on Product Perception

María-Jesús Agost$^{(\boxtimes)}$ ⓘ, Margarita Vergara ⓘ, and Vicente Bayarri ⓘ

Universitat Jaume I, 12071 Castellón, Spain
magost@uji.es

Abstract. New tools have appeared to improve interaction with products in online shopping: virtual reality (VR) applications, where customers are immersed in a wholly virtual environment; augmented reality (AR) applications, which allow virtual elements to be visualised in real environments; and 360° displays (360-D), which allow 3D representations to be rotated in any direction. The aim of this work is to analyse the influence of using these technologies for product presentation on the perception of different aspects of the products and also on the shopping experience, within the context of an online shopping, and compare them with traditional presentation of products through static 2D rendered images. A study has been conducted with a piece of furniture (a sideboard) and a household accessory (a gooseneck lamp). A questionnaire was developed simulating a website for selling products online, and showing the sideboard and the lamp by means of 2D images, 360-D, AR and VR techniques. The participants evaluated different aspects related to the perception of general, detailed, functional and aesthetic characteristics of the products, as well as aspects concerning the shopping experience and the use of the technologies. Results show a similar perception of product characteristics and shopping experience for the four techniques, except for the perception of details (worst for VR) and easiness of use (worst for AR and VR), with some differences depending on the type of product. In conclusion, 360-D and AR techniques are recommended for large appliances, while 360-D and 2D images are better for small ones.

Keywords: Augmented reality · Virtual reality · 360° display · Online shopping · Subjective impressions

1 Introduction

Feelings and impressions of products are crucial for purchasing decisions [1]. The number of sensory organs involved in the interaction, named as the 'affective channel width' by some authors, has an important effect on the way we perceive products [2]. After interacting with the product through the senses involved, an impression about the product is generated and from this impression the customer decides about approaching or avoiding the product. The impressions generated can be of two types: cognitive judgements about

© Springer Nature Switzerland AG 2021
S. Yamamoto and H. Mori (Eds.): HCII 2021, LNCS 12765, pp. 3–15, 2021.
https://doi.org/10.1007/978-3-030-78321-1_1

the product (also called meanings in the context of product semantics) and feelings the customer may experience (or emotions) [1]. The kind of sensory perception and the level of interaction with the product influence the user's perception [3], therefore, the impressions generated depend on the context of interaction. Real physical products allow wide interactions with products, as many senses can be involved. However, other representations of the product (such as photographs), have a lower sensory channel width, so that some impressions may not be so clear. Nevertheless, the main impressions generated from the physical interaction with the product come from the sense of sight.

Nowadays, e-commerce has increased for many reasons, including the pandemic situation we are living, and many companies are offering their products not only in physical stores, but also in online environments. As customer's purchase decisions depend on the way of presentation of the product, this way of presentation can be a key to success in online sales environments, considering the important role played by the affective factors in the purchase process [4]. Therefore, from the affective design perspective, it is essential to adapt the customer - product interaction to a new sales environment with different features. It is essential that the information offered about the product in the new virtual showcase provides confidence and security to the customer to make the purchase decision.

Beyond 2D images and photographs, new tools have progressively appeared to facilitate and improve the visualisation and perception of products in online contexts, such as 3D representations using virtual reality environments. Sometimes, 2D images are supported with texts to provide information on product details, such as materials, textures, finishes or manufacturing processes. But in the case of products that will be located in a room, such as furniture, it is necessary to contextualize the dimensions of the product to check the place where it should be located.

The online sales market has evolved and has managed to adapt the technology to its own special needs. Some of the most innovative product presentation methods are achieved through:

a) 360° display (360-D), which allow a 3D representation of the product to be rotated in any direction and viewed from different points of view.
b) virtual reality (VR) technology, in which the individual is immersed in a totally virtual environment and can look at any part of the virtual environment. Different ways of interacting and immersivity (from a simple screen like a tablet, to glasses showing different images for each eye to represent distances of the objects) are available. VR is a very established technology and has been applied for different purposes in fields such as medicine [5, 6], psychology [7], teaching [8, 9] or marketing [10, 11].
c) augmented reality (AR), which allows to visualize virtual elements (3D representations of products) in real environments. A screen and a camera are used: the screen shows the real environment captured by the camera and thanks to a printed logo or code, the virtual element is added to the physical real environment seen in the screen. AR technology has been used also in different fields such as psychology [12], health and well-being [13], or teaching [14, 15] and due to the profusion of the smart phones, their use is expected to expand to many more fields. Thus, AR can represent a great advance in online sales environments due to its immersive possibilities. In fact, some previous studies have used AR to analyse consumer response

from different types of products, in order to adjust and improve the online shopping experience [16–18].

In the field of online sales, these technologies have been previously applied to check several factors that influence perception and purchase decisions. Using 2D image presentations, the study of Yoo and Kim [4] evaluated the consumer's response to the product presentation of 2D images, combined with backgrounds and texts, concluding that product presentation with a relevant consumption background is more effective in evoking mental imagery than one with a solidwhite background.

Several studies have compared the perception of certain factors by showing the real environment vs. showing the products with some of these new technologies. Yim and Park [19] assessed through AR technique how products fit the customer bodies and analysed how body image is perceived by consumers. This study compared as well consumer responses using AR with the usual images for online sale. In this case, the opinion of consumers regarding their body image improved with the use of AR. Arbeláez and Osorio-Gómez [20] proposed the use of AR for the aesthetic assessment of products during the design process. They evaluated two products using a special type of AR vs. the real products under different lighting, obtaining similar measure of products aesthetic perception with both techniques. Galán et al. [21] analysed whether the way a product is presented (VR and real product) influences the perception of some general product characteristics such as comfortability, simplicity or originality. They concluded that physical contact affects the perception of properties such as weight, size and aesthetics. Söderman's [22] compared three different ways of presenting a car (sketch, VR and real product) and concluded that the more realistic is the presentation, the more valid is the evaluation.

In view of these studies, the evolution of the new product presentation techniques, the increase of e-commerce and the effect the impressions can have on final purchase decisions, this study aims to analyse the influence of the use of four product presentation technologies (2D rendered images, 360-D, VR and RA), on the perception of different aspects of the products and also on the shopping experience, within the context of an online shopping.

2 Materials and Methods

2.1 Development of the Study

In order to achieve our objective, an experiment was designed. A simulated web page for online sales of household products was developed, with the premise that it should be accessed from a smartphone. Two products with differentiated features were considered: a piece of furniture (a sideboard) and a household accessory (a gooseneck lamp).

Four display techniques were used:

- 2D rendering images. The product is shown by means of different 2D rendering pictures: isolated, along with decorative settings, showing some details in a zoom image, or showing detailed measures (Fig. 1). These images were next to a short descriptive text of the product (Fig. 1) and some instructions of how to go over the pictures.

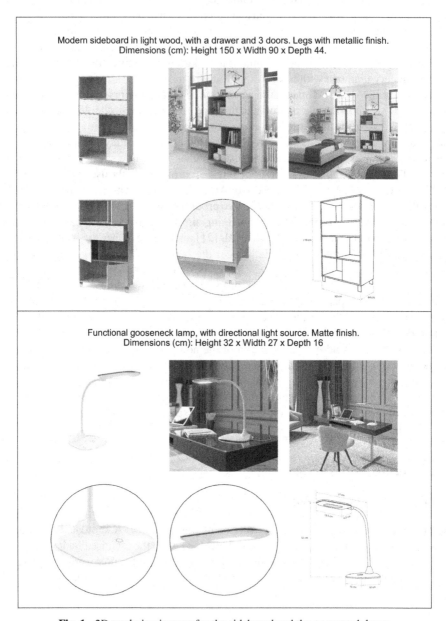

Fig. 1. 2D rendering images for the sideboard and the gooseneck lamp.

- 360-D: The product is displayed on an app (Fig. 2), allowing simple 3D interaction (rotation, zoom, and opening drawers in the case of the sideboard).

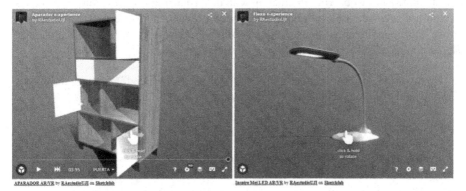

Fig. 2. 360-D display for the sideboard and the gooseneck lamp.

- AR: Detailed instructions were given for unloading an app (Sketchfab), so that a virtual representation of the product could be displayed, over any real location, chosen by the participant.
- VR: Detailed instructions were given for unloading an app (VR Media Player), so that a virtual representation of the product located in a virtual room (Fig. 3) could be displayed.

Fig. 3. Virtual representations to unload for VR.

For each product, two web versions were created. In one version, the product was shown by means of 2D images, 360-D and AR. In the other version, the display techniques were 2D images, 360-D and VR. As two products were considered, a total of 4 versions of the web page were generated, which were randomly assigned among the participants. Ninety-six students aged between 18 and 29 years were invited to participate and the questionnaire versions were assigned randomly to them.

First of all, the participants accessed a first questionnaire to give their informed consent to the study, and answer some questions such as gender, age, and previous experience in online shopping and in the use of new technologies for product presentation. Next, the participants were randomly distributed to one of the four versions of the simulated online sales web. After interacting with the product through the three techniques of their assigned version (with no pre-established order), the participants answered an online questionnaire about the perceptions of the product with each of the techniques. The questions, answered using 5-point scales are shown next:

1. Please rate (1 corresponding to "Very bad" and 5 to "Very good"), for each technique, the quality of the perception of the following product properties:

   – Dimensions and shape
   – Details and small elements
   – Reliability
   – Quality
   – Attractive

2. Please assess (1 corresponding to 'Not at all' and 5 to 'Very much'), for each technique used:

   – This technique is useful
   – This technique provides a satisfactory shopping experience
   – This technique is easy to use
   – This technique can influence on the final decision

3. Indicate the technologies that you consider optimal for displaying these types of product (more than one per product can be chosen).

   – Furniture
   – Small household products
   – Footwear
   – Sunglasses

2.2 Analysis of Results

The statistical analyses were performed with IBM SPSS statistical software (v23 for Windows). First, descriptive statistics were analysed. Next, non-parametric Friedman tests (non-parametric repeated measures) were conducted to check for differences between the techniques on the assessments of the quality of perceptions and the techniques' characteristics.

3 Results

Five participants could not perform the AR or VR technique due to technical problems with their mobiles and others were not willing to participate. Finally, valid answers from 78 participants (47% women) were obtained and analysed.

All the participants declared to have experience in online shopping, and 55% of them to buy online frequently. The products that are most bought online are fashion products, such as clothing, footwear or accessories (77%) and simple household products, such as a light bulb or small appliances (58%), followed by feeding products (20%), complex household products, such as furniture, or large appliances (17%) and technological products such as computers (10%). Figure 4 shows their previous experience with different visualization techniques.

Fig. 4. Previous experience of the participants with each technique

Figure 5 shows the mean ratings of the quality of perceptions of product properties (globally and for product type), for each technique. In general, the 360-D seems to have better ratings, while VR technique has the worst ratings, in special for the perception of details in the gooseneck lamp.

When both products are considered together, there are no major differences between the techniques on the perceptions, except for product details, where almost one point of difference is observed between the mean rating of the best-rated technique, 360-D (M = 4.05) and VR, the one with the worst rating (M = 3.08). Considering the perception for each product, this difference is enlarged for the gooseneck lamp: details are better perceived with 360-D (M = 4.1), closely followed by 2D images, while the rating falls in the case of VR (M = 2.63). In general, VR is the technique with the worst average assessment in the quality of product perceptions, except in the case of attractiveness, where it achieves a high score (M = 4.21) while 360-D is the one with the lowest score (M = 3.75). 2D images achieve good ratings for the gooseneck lamp in all cases. However, in the case of the sideboard, the 2D images are the worst valued for all perceptions, except for the dimensions, where the mean value is very slightly above that of the AR and VR. In particular, the perception of reliability through 2D images has been the worst valued. For the sideboard, 360-D is the best valued technique in all cases, except for reliability, in which it is slightly lower than AR.

Thus, it is observed that in general 360-D is well valued to perceive physical characteristics: dimensions and details. The quality of perceptions with VR and AR is rated worse in the case of the smaller product: the gooseneck lamp. In addition, for this product the perceptions are better valued through 2D images than in the case of the sideboard.

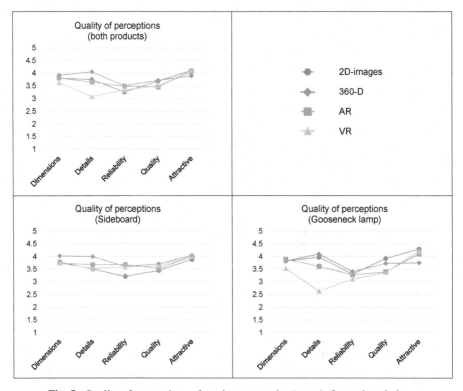

Fig. 5. Quality of perceptions of product properties (mean), for each technique.

The Friedman test showed a significant effect of the technique only on the perception of details, when the VR technique was used (Chi-square value of 10.55, p = 0.005). The post-hoc showed that specifically significant difference between 360-D and the VR technique (p = 0.012). Differentiating by product, no significant difference was detected for the sideboard. In the case of the gooseneck lamp, a significant effect of the technique was identified again, among the cases in which the VR technique had been used (Chi-square value of 10.00, p = 0.007). Specifically, the significant difference was between the 2D images and the VR technique (p = 0.045), and between 360-D and the VR technique (p = 0.045).

The mean assessment of the characteristics of the techniques is shown in Fig. 6. AR and VR are not easy to use for the participants, even they were young people with previous experience. The most useful, satisfactory and influencer techniques are AR and 360-D. Generally, the best valued is the 360-D, while the one with the worst ratings is the VR. The characteristic with the greatest differences in the ratings is 'Easy to use'. In this case, the 2D images are the technique with the highest mean rating (M = 4.22), followed by 360-D (M = 3.99), and with AR (M = 3.46) and VR (M = 3.37) being assessed as the most difficult to use. For the gooseneck lamp, these differences are bigger, reaching the 2D images better mean rating (M = 4.43) and the virtual techniques lower values: AR (M = 3.17) and VR (M = 3.26). The VR is rated low on all features, both on the gooseneck and in the sideboard. However, the AR is the best valued for the sideboard in 'Useful' (M = 4.47), 'Satisfactory' (M = 4.11) and 'Influence' (4.32). From the results it can be concluded that for the gooseneck lamp the best valued techniques are 360-D and 2D images, while AR acquires special relevance in the sideboard.

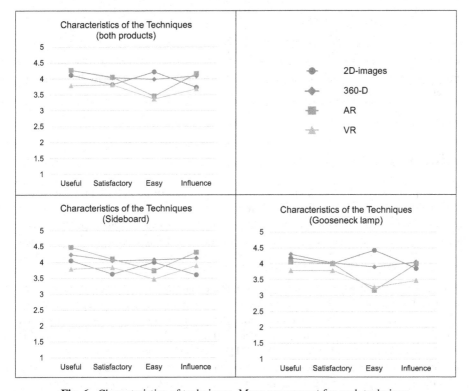

Fig. 6. Characteristics of techniques. Mean assessment for each technique.

The Friedman test showed significant difference on the assessment of 'Easy to use', both among the cases in which the AR technique had been used (Chi-square value of 28.67, p = 0.000) and in the cases in which the VR technique had been used (Chi-square value of 11.69, p = 0.003). In the first case, the specific difference was detected between the 2D images and the AR technique (p = 0.000) while in the second, a difference was detected both between the 2D images and the VR technique (p = 0.041) and between the 360-D and the VR technique (p = 0.035). By considering each product separately, significant differences between the techniques for 'Easy to use' were detected only for the gooseneck lamp: when the AR technique was used (Chi-square value of 22.36, p =

0.000), between the 2D images and the AR technique (p = 0.000), and when the VR technique was used (Chi-square value of 8.23, p = 0.016), between the 2D images and the VR technique (p = 0.036).

The technologies considered optimal for the display of the different types of product are shown in Fig. 7 (in percentage of respondents).

	Furniture	Small household products	Footwear	Sunglasses
2D-images	51%	64%	62%	64%
360-D	63%	64%	75%	56%
AR	73%	49%	24%	24%
VR	78%	31%	19%	19%

Fig. 7. Percentage of respondents that consider each technology as optimal for the display of different types of product.

4 Discussion

We have performed an experiment with young people, all of them with previous experience in online shopping, to compare different methods of presenting the products: 2D images, 360-D, AR and VR. Although most of them reported previous experience with this type of techniques, some of them were not able to complete the experience, which included the installation of specific applications on the mobile phone, as it would be necessary in the case of real shopping.

In general, the quality of the perceptions of product attractiveness and dimensions has been rated as medium-high for all techniques, with no significant differences between techniques. Previous studies have compared the perception of aesthetics through AR or VR with reality. While no difference was found for AR [20], some affection of this perception was detected in the case of VR [21]. In our study, the perception of attractiveness using AR and VR techniques have been compared with other display techniques available for online shopping, obtaining in all cases similar levels of perception. The quality of the perceptions of more functional aspects, such as quality and reliability of the products are not as well perceived as attractiveness. Finally, the perception of details has obtained the most varied scores, depending on the technique used: from average values for VR to medium-high values for 360-D.

Considering both products in the study, the assessment of each perception has similar values for all the techniques in the case of the sideboard, while in the gooseneck lamp the differences are more marked. Again, the perception of details is where the greatest differences between techniques are seen, with the VR technique below the average score. Comparing the quality of product perceptions depending on the display technique used has led to significant differences on the perception of product details, where the 360-D has been the best valued, and the VR technique the worst one. In general, it can be seen that the perceptions obtained through VR have been poorly rated, mainly when the product assessed has been the gooseneck lamp. Besides, for this product, the assessment of perceptions with 2D images have been higher to those obtained for the sideboard.

In the study, two well-known products that are commonly used at home have been chosen. They are appliances with few components and smooth surfaces, and without too much visual complexity (although in the case of the sideboard there was greater interaction with the AR technique, since the doors and drawers could be opened). The main distinguishing feature seems to be size. It must be considered that the participants have greater previous experience in purchasing small appliances (58%) than in large appliances (17%). Thus, future work is necessary to establish whether the differences obtained in the results are precisely due to the size of the products.

Regarding the assessment of the characteristics of the techniques, the 360-D has been the most accepted (the one with the highest ratings) of the 3 most innovative techniques used (360-D, AR and VR). It must be considered that it is also the one that has been previously most used by the participants. The most classical technique (2D images) is only the best assessed in easiness of use, while has similar assessment for its usefulness and satisfaction as the other techniques, but smaller influence. The AR technique has been well assessed for the sideboard (useful, satisfactory, influence on the final decision), although the participants consider it difficult to use, which can be a drawback for its application. With regard to the VR technique, it has been assessed low on all characteristics.

From the results obtained, the use of the 360-D technique is recommended to perceive the characteristics of the physical appearance of the product; i.e., dimensions, and mainly for details. In the case of large appliances, such as furniture, this technique can be accompanied by the use of AR. In the case of the gooseneck lamp (small appliances), it is recommended to use 360-D together with 2D images. VR does not currently seem an option with acceptance, neither in terms of perceptions nor in characteristics of use (although it should be pointed that it has been selected when asking for the most appropriate technique to visualize furniture). The AR technique has been highly rated for the sideboard; it has been considered useful, leading to a satisfying experience, and influencing the final decision. However, it must be considered that it is also difficult to use, compared to other techniques.

Other types of products should be considered in future work; clothes and accessories have not been included for the moment in the study. However, participants have selected 2D images and 360-D as the best techniques for the display of footwear and sunglasses. They are the most known and used techniques. Besides, it could be checked whether there are differences based on age and previous experience (since our sample is made up of young university students, all with previous experience in online shopping). Also, as mentioned, the possible influence of the size of the product on results should be analysed.

5 Conclusions

Implementing AR and VR techniques in online shopping must be considered with care, as they are difficult to use by customers, even for those with previous online shopping experience. AR technique seems to have a better acceptance than VR and presents advantages for showing large appliances and furniture (it is more useful and have more influence on the final decision). However, VR is not good for the perception of details, especially in small products. Perceptions of product attractiveness and dimensions are quite good for the four techniques (2D images, 360-D, VR and AR). In conclusion, 360-D and AR techniques are recommended for large appliances, while 360-D and 2D images are better for small ones.

Acknowledgements. This research was funded by the Universitat Jaume I, grant number 181372.

References

1. Crilly, N., Moultrie, J., Clarkson, P.J.: Seeing things: Consumer response to the visual domain in product design. Des. Stud. **25**, 547–577 (2004)
2. Schütte, S.: Engineering Emotional Values in Product Design - Kansei Engineering in Development (2005)
3. Vergara, M., Mondragón, S., Sancho-Bru, J.L., Company, P., Agost, M.J.: Perception of products by progressive multisensory integration. A study on hammers. Appl. Ergon. **42**, 652–664 (2011)
4. Yoo, J., Kim, M.: The effects of online product presentation on consumer responses: a mental imagery perspective. J. Bus. Res. **67**, 2464–2472 (2014)
5. Dubin, A.K., Julian, D., Tanaka, A., Mattingly, P., Smith, R.: A model for predicting the GEARS score from virtual reality surgical simulator metrics. Surg. Endosc. **32**(8), 3576–3581 (2018). https://doi.org/10.1007/s00464-018-6082-7
6. Huber, T., Wunderling, T., Paschold, M., Lang, H., Kneist, W., Hansen, C.: Highly immersive virtual reality laparoscopy simulation: development and future aspects. Int. J. Comput. Assist. Radiol. Surg. **13**(2), 281–290 (2017). https://doi.org/10.1007/s11548-017-1686-2
7. Formosa, N.J., Morrison, B.W., Hill, G., Stone, D.: Testing the efficacy of a virtual reality-based simulation in enhancing users' knowledge, attitudes, and empathy relating to psychosis. Aust. J. Psychol. **70**, 57–65 (2018)
8. Weibing, Y., Shijuan, L., Fei, S.: History and current state of virtual reality technology and its application in language education. J. Technol. Chinese Lang. Teach. **8**, 70–100 (2017)
9. González, N.A.A.: Development of spatial skills with virtual reality and augmented reality. Int. J. Interact. Des. Manufact. (IJIDeM) **12**(1), 133–144 (2017). https://doi.org/10.1007/s12008-017-0388-x
10. Van Kerrebroeck, H., Brengman, M., Willems, K.: When brands come to life: experimental research on the vividness effect of Virtual Reality in transformational marketing communications. Virtual Reality **21**(4), 177–191 (2017). https://doi.org/10.1007/s10055-017-0306-3
11. Grudzewski, F., Awdziej, M., Mazurek, G., Piotrowska, K.: Virtual reality in marketing communication – the impact on the message, technology and offer perception – empirical study. Econ. Bus. Rev. **4**(18), 36–50 (2018)
12. Botella, C., et al.: Treating cockroach phobia using a serious game on a mobile phone and augmented reality exposure: a single case study. Comput. Hum. Behav. **27**, 217–227 (2011)
13. Carlson, K.J., Gagnon, D.J.: Augmented reality integrated simulation education in health care. Clin. Simul. Nurs. **12**, 123–127 (2016)
14. Akçayir, M., Akçayir, G., Pektaş, H.M., Ocak, M.A.: Augmented reality in science laboratories: the effects of augmented reality on university students' laboratory skills and attitudes toward science laboratories. Comput. Hum. Behav. **57**, 334–342 (2016)

15. Martín-Gutiérrez, J., Fabiani, P., Benesova, W., Meneses, M.D., Mora, C.E.: Augmented reality to promote collaborative and autonomous learning in higher education. Comput. Hum. Behav. **51**, 752–761 (2015)

16. Huang, T.-L., Hsu Liu, F.: Formation of augmented-reality interactive technology's persuasive effects from the perspective of experiential value. Internet Res. **24**, 82–109 (2014)

17. Javornik, A.: Augmented reality: research agenda for studying the impact of its media characteristics on consumer behaviour. J. Retail. Consum. Serv. **30**, 252–261 (2016)

18. Kang, J.Y.M.: Augmented reality and motion capture apparel e-shopping values and usage intention. Int. J. Cloth. Sci. Technol. **26**, 486–499 (2014)

19. Yim, M.Y.-C., Park, S.-Y.: I am not satisfied with my body, so I like augmented reality (AR). J. Bus. Res. **100**, 1–9 (2018)

20. Arbeláez, J.C., Osorio-Gómez, G.: Crowdsourcing Augmented Reality Environment (CARE) for aesthetic evaluation of products in conceptual stage. Comput. Ind. **99**, 241–252 (2018)

21. Galán, J., Felip, F., García-García, C., Contero, M.: The influence of haptics when assessing household products presented in different means: a comparative study in real setting, flat display, and virtual reality environments with and without passive haptics. J. Comput. Des. Eng., 1–13 (2020)

22. Söderman, M.: Virtual reality in product evaluations with potential customers: an exploratory study comparing virtual reality with conventional product representations. J. Eng. Des. **16**, 311–328 (2005)

Research on Conveying User Experiences Through Digital Advertisement

Stephanie Dwiputri Suciadi[(✉)] and Miwa Nakanishi

Faculty of Science and Technology, Keio University, 3-14-1 Hiyoshi, Kohoku-ku, Yokohama 223-8522, Kanagawa, Japan
stephaniedwiputri@keio.jp, miwa_nakanishi@ae.keio.ac.jp

Abstract. Digital advertisement enables potential consumers to picture what they might experience when they use an advertised product. Creating effective user experience (UX) images in digital advertisement helps companies promote their products and attract specific audiences while allowing the designer of the product to communicate their vision to viewers. To increase the efficacy of advertising while simplifying the design processes, it is important to establish the designer's own visualization of the product to communicate to buyers. In this research, we conducted an evaluative experiment by analyzing the objective features of 80 video advertisements for home appliances to create an evaluative prediction model for the 24 defined types of UX. The results revealed that UX can be conveyed through digital media and that objective features, which are mainly related to people's appearance, visual, audio, and content, can influence this. Each UX prediction model was created with different features and might be used as a reference in creating digital content advertisements that convey the UX.

Keywords: User experience · Digital content · Human perception · Advertisement

1 Introduction

People require products with specific functions to support their daily lives, both for work and leisure. However, in recent decades, functionality is no longer the only factor in buying a product, with consumers seeking experiences that influence their senses, heart, and mind, rather than merely executing tasks [1].

The term "experience" and its importance in business developed from the field of marketing, where Schmitt described "experiential marketing" [1]. The term was also used to explain the concept of experience in human–product interactions, which became widely known as "user experience" (UX). Hassenzahl (2008) defines UX as "a momentary, primarily evaluative feeling (good or bad) while interacting with a product or service" [2]. In recent years, UX has become a primary concern in product design, along with product function.

In practice, however, UX is a challenge for product designers. Hassenzahl's (2004) model indicated that a designer's perspective is distinct to that of a user [3]. In the

© Springer Nature Switzerland AG 2021
S. Yamamoto and H. Mori (Eds.): HCII 2021, LNCS 12765, pp. 16–27, 2021.
https://doi.org/10.1007/978-3-030-78321-1_2

development of an intended product characteristic, such as "interesting" or "impressive," a designer manipulates a product's features, such as color, shape, and function; however, the unique situation that users finds themselves in can influence both the product and UX. In other words, UX is subjective and inconsistent as it is dependent on the situation of the user.

The Hassenzahl model clearly shows that there is a gap between a designer and a user's perspective of UX. Therefore, it is conceivable that to reduce the gap, designers could evaluate product UX and use it to communicate the intended product characteristic to the user.

UX evaluation is usually conducted either during or after the use of a finished product, with evaluation prior to use being less common [4].

Communicating the designer's intended product characteristic to a user could be achieved by providing product information and building consumer expectations through advertisements before using the product [5].

Companies have been using advertising for centuries to persuade consumers and promote their products. Advances in technology have enabled digital advertisement, which is exposed to wider audiences than its traditional counterparts. According to research by the Japanese Ministry of Economy, Trade and Industry, in 2018, almost half of the digital content market (including advertisement), both in Japan and globally, was in the form of video, with this proportion predicted to still be growing by 2023 [6].

As a media that contain both audio and moving images, video is more effective at persuading and is trusted in conveying information compared with other digital content [7–9]. Besides its efficacy in communicating information and educating people [10] through storytelling, video is also able to project emotion, which impacts human behavior [11–13].

There are numerous studies on the application of video in advertising and how a video's characteristics can influence viewers. For example, longer video advertisements have been shown to be more effective in terms of recall, enjoyment, and brand identification and attitude [14, 15, 17]. Other related research found that longer television advertisements exerted a greater emotional impact [16]. Moreover, it has been shown that a high level of entertainment has a positive impact on viewing interest [18, 19]. Furthermore, video advertisements that contain both an overview and specific information about a product are perceived to communicate superior product quality compared with videos with only an overview [20]. It has also been shown that campaign videos, which involve real people and product explanations, correlate with crowdfunding and product success [21, 22].

Currently, most existing research on advertising has focused on the evaluation of efficacy from the viewer's attitude toward the advertisement or brand, with none investigating the efficacy of communicating the product's intended UX.

Finding objective design standards that convey a designer's intended product characteristic or UX to viewers is an important step toward increasing the efficacy of advertisement while simplifying the design process.

In this study, we conducted an experiment using video advertisements that analyzed and modeled the objective features that convey a product's UX to the viewer.

2 Methods

2.1 Conveyance of the UX Evaluation Experiment Using Digital Advertisements

We collected 80 Japanese home appliance video advertisements from various brands that were each 30-s long and had aired after 2009. The products included air conditioners, washing machines, rice cookers, televisions, etc.

For the UX evaluation, we used 24 UX items (UX24) [23], which are presented in Table 1. The UX24 were developed as UX categories for cooking appliances by Tanaka (2013) and later adopted as essential UX classifications in home appliance design [24]. Moreover, we added two general evaluation criteria, namely, "easy to understand what product is being advertised" and "product is attractive," to establish the relationship between the perceived attractiveness and understanding of a product.

All UX24 were evaluated based on how well a participant could relate to each UX using a 7-scale score, ranging from 0 (cannot imagine the UX) to 6 (can clearly imagine the UX). The general evaluation criteria were also assessed using the same 7-scale score.

Table 1. UX24 evaluation items.

No	UX item	No	UX item
1	Having heartwarming feelings	13	Able to please someone
2	Feeling an emotional connection	14	Able to do well
3	Noticing a new self	15	Feeling visual beauty
4	Understanding what had not previously been understood	16	Experiencing auditory beauty
5	Remembering good experiences from the past	17	Feeling accomplished
6	Able to do at the right timing	18	Feeling of progress
7	Praised by someone	19	Feeling fulfilled
8	Able to trust something to someone	20	Able to concentrate
9	Able to do easily/quickly	21	Feeling comfortable/relaxed
10	Feeling tidy	22	Feeling refreshed
11	Arouses aspirations	23	Quantity and degree are just right
12	Able to imagine good experiences in the future	24	Feel unusually luxurious

We conducted our data acquisition online to observe how strongly each UX was conveyed through different advertisements. There were 19 participants (9 male and 10 female), aged between 20 and 44 years. The experiment was conducted in Japanese. The participants were able to watch each advertisement as many times as they wished during the experiment. To reduce external factors influencing their evaluations, the participants were required to use headphones and were not allowed to change the volume during the experiment. Since the duration of the experiment was 6 to 8 h, the participants were allowed to take a break between each advertisement, but had to finish the experiment within 2 weeks.

2.2 Extracting Objective Features and Categorizing Advertisement Samples

Using the data from the experiment, we compared high and low scoring advertisements to analyze objective features that corresponded with each UX. We collected 22 objective features related to the appearance of people, visual, audio, and content, including the gender and number of people present, scenes containing interactions at home, activities being performed, theme, dialog, and entertaining elements. Each advertisement was manually categorized based on these objective features.

2.3 Designing and Building Models

We used Hayashi's quantification method I to analyze the objective features data, with this data used as the explanatory variable and each UX24 evaluation score set as a response variable.

To build a qualitative model for each UX24, different features were required and selected according to the following procedure:

Step 1: Correlation ratios (between the features and UX evaluation scores) of <0.03 were ignored.
Step 2: Features with the highest correlation ratio were prioritized to be included as explanatory variables. To avoid multicollinearity, other features with a coefficient of association of >0.45 were ignored.
Step 3: A model was created using Hayashi's quantification method I and evaluated. If the explanatory variable was found to be unrelated to a UX item, another model was created by repeating Step 2 with a different combination of features.

3 Results

3.1 Analysis of UX Evaluation Items

The models were created with different objective features for each UX24. For example, each of the UX "having heartwarming feelings," "able to do easily/quickly," and "feeling visual beauty" are modeled separately against objective features to compare between the predicted and evaluated data for the three advertisements. The objective features of two sample advertisements are presented in Table 2. Sample 1 showed a wife being praised by her husband and mother-in-law for cooking better than the mother-in-law, whereas sample 2 showed a television displaying various images.

"Having Heartwarming Feelings" (UX1). The Model for UX1 is Presented in Table 3, and the Corresponding Scatter Plot is Presented in Fig. 1. The Features Used for the UX1 Model Were "number of People," "main Background," "gender of Voice," "quantitative Explanation," and "using the Product for Whom."

Table 2. Two sample features.

Feature related to	Feature	Sample 1	Sample 2
People appearance	Number of people	Three to five people	None
	Using the product for whom	All	Unknown
	Usage process	Partial	None
	User activity	Entertainment	None
Visual	Main background	Parts of a house	None
	Theme	Daily	Futuristic
	Number of scenes product appeared	One	One
	Quantitative explanation	None	None
Audio	Gender of voice	Male and female	None
	Explanation audio	Talk	None
Content	Main content	Mainly about function	Other
	Entertaining element	Yes	None
	Scenario	Introducing product as a problem solution	Introducing product function and design only

UX1 was predicted to be best conveyed by the appearance of multiple people who use the product for themselves and others within the setting of a house. Focusing on how the product is used and the users themselves, rather than in-depth explanations of the product, was most effective. The use of human voices in the advertisement, either through speech or song, was also preferable.

Our modeling showed that sample 1 was predicted to have a UX1 evaluation score of 3.89 (3.58 in the experimental data), whereas sample 2 was predicted to have a UX1 evaluation score of 1.65 (1.26 in the experimental data).

"Able to do Easily/Quickly" (UX9). The UX9 model is presented in Table 4, and the corresponding scatter plot is presented in Fig. 2. The features used were "theme," "main content," and "usage process."

Advertisements that used a realistic theme and explained the product's use and operation by showing it being used were predicted to have the highest UX9 conveyance score. Conversely, advertisements with futuristic themes where the main content was unrelated to the product's function, usage, or operation were predicted to have the lowest UX9 conveyance score.

Table 3. UX1 model.

Feature	Category	Category score	Range
Number of people	More than five people	0.0593	1.3602
	None	−0.8672	
	One person	−0.4695	
	Three to five people	0.4931	
	Two people	0.2491	
Main background	Parts of a house	0.1184	0.3644
	None	−0.2460	
Gender of voice	Both	0.0884	0.2104
	Female	−0.0916	
	Male	−0.0804	
	None	−0.1221	
Quantitative explanation	None	0.1154	0.2496
	Yes	−0.1342	
Using the product for whom	All	0.1223	0.3025
	Self	0.0487	
	Unknown	−0.1801	
Constant	2.9525		
Coefficient of determination	0.6836		

Sample 1 was predicted to have a UX9 evaluation score of 3.34 (3.26 in the experimental data), whereas sample 2 was predicted to have a UX9 evaluation score of 1.97 (2.26 in the experimental data).

"Feeling visual beauty" (UX15). The UX15 model is presented in Table 5, and the corresponding scatter plot is presented in Fig. 3. The features used were "theme," "explanation audio," "scenario," "user activity," "entertaining element," and "number of scenes product appeared."

UX15 was predicted to be conveyed best through advertising that used futuristic themes to introduce the product's function and design only, with the product's user enjoying entertainment, such as watching a movie. Neither human voices nor entertaining elements, such as joke, were preferred.

Sample 1 was predicted to have a UX15 evaluation score of 2.84 (2.74 in the experimental data), whereas sample 2 was predicted to have a UX15 evaluation score of 4.47 (5.32 in the experimental data).

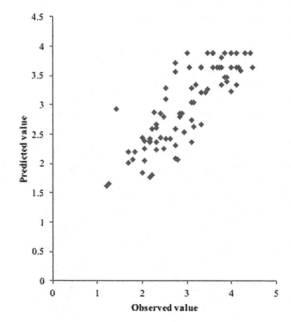

Fig. 1. UX1 model scatter plot.

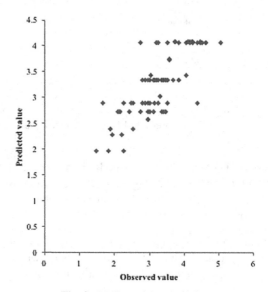

Fig. 2. UX9 model scatter plot.

Table 4. UX9 model.

Feature	Category	Category score	Range
Theme	Daily	0.0412	0.3251
	Futuristic	−0.2840	
	Traditional/historical	0.0366	
Main content	Mainly about function	0.1522	0.6332
	Mainly about usage, operation	0.1697	
	Other	−0.4635	
Usage process	Complete	0.5937	1.1498
	None	−0.5562	
	Partial	−0.1187	
Constant	3.2699		
Coefficient of determination	0.6199		

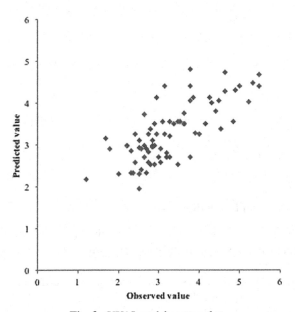

Fig. 3. UX15 model scatter plot.

Table 5. UX15 model.

Feature	Category	Category score	Range
Theme	Daily	−0.1650	0.8643
	Futuristic	0.6993	
	Traditional/historical	0.3417	
Explanation audio	None	0.0901	0.5894
	Song	−0.4993	
	Talk	0.0301	
Scenario	Introducing function and design only	0.0761	0.4061
	Introducing product as problem solution	−0.3300	
User activity	Chores	0.0710	0.9042
	Entertainment	0.3724	
	None	0.1195	
	Rest, relax	−0.5318	
	Talking	−0.1645	
Entertaining element	None	0.2515	0.5588
	Yes	−0.3074	
Number of scenes product appeared	One	−0.0361	0.4814
	Two or more	0.4453	
Constant	3.2717		
Coefficient of determination	0.5607		

3.2 Analysis of General Evaluation Criteria

The moderately positive correlation, 0.5463, between the two general evaluation criteria, "easy to understand what product is being used" and "product is attractive," is presented in Fig. 4.

It could be inferred that the ease of understanding an advertisement influences the product's perceived attractiveness; thus, if recognizing the advertised product is difficult, the viewer's attraction to the product is likely to be low.

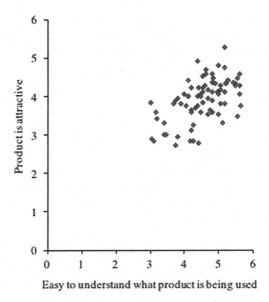

Fig. 4. General evaluation item correlation (Spearman's correlation $= 0.5463$).

4   Discussion

UX conveyance was influenced by various objective features related to people's appearance, visual, audio, and content within advertisements. However, the impact of each feature was specific to conveying different types of UX. People's appearance, for example, had an important role in communicating UX, but while it impacted positively for crowdfunding scenarios [21], each type of UX was influenced differently.

UX1 was conveyed better by more than one person compared with only one person. This can also be applied in the UXs "feeling an emotional connection," "praised by someone," and "able to please someone." This suggests that the appearance of and interaction between multiple people is necessary to provoke the viewer's empathy and relate the UX to other people.

UX15 was influenced by the activity being performed by people in the advertisement, with people enjoying entertainment having a positive influence on the evaluation.

Conversely, UX9 was not affected by people's appearance but rather by the content of the advertisement. Showing a product in full use had a positive impact on UXs related to performance, such as "able to do at the right timing," "able to do well," and "feeling accomplished."

5   Conclusion

In conclusion, UX can be conveyed to viewers through digital advertisements by implementing the objective features proposed in this paper, especially those related to people's appearance, visual, audio, and content. Communicating different types of UX can be achieved by implementing different objective features into adverts.

We hope that the objective features proposed in this study will serve as a reference for the creation of digital advertisements that convey UX and allow effective communication of a designer's UX perspective to the user. Finally, it is important to remember that the features proposed in this paper are not definitive, and there remains much creative potential in how we can convey all UXs through digital advertisement.

References

1. Schmitt, B.H.: Experiential Marketing: How to Get Customers to Sense, Feel, Think, Act, and Relate to Your Company and Brands. Free Press, New York (1999)
2. Hassenzahl, M.: User experience (UX): towards an experiential perspective on product quality. In: Proceedings of the 20th International Conference of the Association Francophone d'Interaction Homme-Machine, pp. 11–15. Association for Computing Machinery, New York (2008)
3. Hassenzahl, M.: The thing and I: understanding the relationship between user and product. In: Blythe, M.A., Overbeeke, K., Monk, A.F., Wright, P.C. (eds.) Funology. Human-Computer Interaction Series, vol. 3, pp. 31–42. Kluwer Academic Publishers, Dordrecht (2004)
4. Maia, C.L.B., Furtado, E.S.: A systematic review about user experience evaluation. In: Marcus, A. (ed.) DUXU 2016. LNCS, vol. 9746, pp. 445–455. Springer, Cham (2016). https://doi.org/10.1007/978-3-319-40409-7_42
5. Kurosu, M.: User experience and satisfaction: toward the concept and measurement method of student satisfaction (in Japanese). Journal of The Open University of Japan (28), 71–83 (2010)
6. Ministry of Economy, Trade and Industry. Overview of the global and Japanese content market (In Japanese) (2020)
7. Kleinl, R.: Creating virtual product experiences: the role of telepresence. J. Interact. Mark. 17(1), 41–55 (2003)
8. Roggeveen, A.L., Grewal, D., Townsend, C., Krishnan, R.: The impact of dynamic presentation format on consumer preferences for hedonic products and services. J. Mark. 79(6), 34–49 (2015)
9. Lee, H., Park, S.A., Lee, Y.A., Cameron, G.T.: Assessment of motion media on believability and credibility: an exploratory study. Pub. Relat. Rev. 36(3), 310–312 (2010)
10. Idriss, N.Z.: Online, video-based patient education improves melanoma awareness: a randomized controlled trial. Telemed. J. E Health 15(10), 992–997 (2009)
11. Finkler, W., Leon, B.: The power of storytelling and video: a visual rhetoric for science communication. J. Sci. Commun. 18(5), A02 (2019)
12. Lang, A., Newhagen, J., Reeves, B.: Negative video as structure: emotion, attention, capacity, and memory. J. Broadcast. Electron. Media 40(4), 460–477 (1996)
13. Barbas, T.A., Paraskevopoulos, S., Stamou, A.G.: The effect of nature documentaries on students' environmental sensitivity: a case study. Learn. Media Technol. 34(1), 61–69 (2009)
14. Newstead, K., Romaniuk, J.: Cost per second: The relative effectiveness of 15- and 30-second television advertisements. J. Advert. Res. 50(1), 68–76 (2010)
15. Varan, D., Nenycz-Thiel, M., Kennedy, R., Bellman, S.: The effects of commercial length on advertising impact: what short advertisements can and cannot deliver. J. Advert. Res. 60(1), 54–70 (2020)
16. Singh, S.N., Cole, C.A.: The effects of length, content and repetition on television commercial effectiveness. J. Mark. Res. 30(1), 91–104 (1993)
17. Pieters, R.G.M., Bijmolt., T.H.A.: Consumer memory for television advertising: a field study of duration, serial position, and competition effects. J. Consum. Res. 23(4), 362–372 (1997)

18. Hsieh, J.K., Hsieh, Y.C., Tang, Y.C.: Exploring the disseminating behaviors of eWOM marketing: persuasion in online video. Electron. Commer. Res. **12**(2), 201–224 (2012)
19. Teixeira, T., Picard, R., Kaliouby, R.: Why, when, and how much to entertain consumers in advertisements? A web-based facial tracking field study. Mark. Sci. **33**(6), 809–827 (2014)
20. Ma, R., Shao, B., Chen, J., Dai, D.: The impacts of online clothes short video display on consumers' perceived quality. Information **11**(2), 87 (2020)
21. Li, X., Shi, M., Wang, X.: Video mining: measuring visual information using automatic methods. Int. J. Res. Mark. **36**(2), 216–231 (2019)
22. Dey, S., Duff, B., Karahalios, K., Fu, W.T.: The art and science of persuasion: not all crowd-funding campaign videos are the same. In: Proceedings of the 2017 ACM Conference on Computer Supported Cooperative Work and Social Computing, pp. 755–769. Association for Computing Machinery, New York (2017)
23. Tanaka, Y.: Analysis of user experience in housework and cooking: Development of design process methodologies for home appliances (in Japanese) (Unpublished master's thesis). Keio University, Yokohama, Japan (2013)
24. Miyahara, A., Sawada, K., Yamazaki, Y., Nakanishi, M.: Quantifying user experiences for integration into a home appliance design process: a case study of canister and robotic vacuum cleaner user experiences. In: Proceedings 19th Triennial Congress of the IEA, Melbourne (2015)

Preventing Decision Fatigue with Aesthetically Engaging Information Buttons

Andrew Flangas, Alexis R. Tudor$^{(\boxtimes)}$, Frederick C. Harris Jr., and Sergiu Dascalu

Department of Computer Science and Engineering,
University of Nevada, Reno, Reno NV 89557, USA
{andrewflangas,atudor}@nevada.unr.edu, {fred.harris,dascalus}@cse.unr.edu

Abstract. In many different kinds of complex forms (financial, job applications, etc.), information button widgets are used to give context-specific information to enable users to fill out forms completely. However, in longer forms, "decision fatigue" can set in, leading to the user not absorbing these helpful tips but rather rushing through and possibly making errors. This research seeks to identify if using a more engaging version of the traditional information button widgets will reduce decision fatigue, preventing errors and increasing information retention. The research experiment involved a user study asking participants to fill out a tax form with information provided to them about a fictional person, with one group of participants given a form with traditional information widgets and another group being given a form with more engaging information button widgets. The results of this experiment show that the information button has no effect on the completion of financial forms or in reducing decision fatigue associated with filling out forms.

Keywords: Human-computer interaction · Finance · Information buttons · Decision fatigu

1 Introduction

As the internet has grown and become accessible to people of all backgrounds and education levels, information overload has become an increasingly challenging problem. When looking for information on complex topics in a domain unfamiliar to the user, it can be difficult or even impossible to find the specific information relevant to the problem at hand. That is why on many complex online forms, such as banking applications or tax documents, companies include information widgets for difficult terminology or questions. These widgets are often a useful addition to the form, allowing the user to get context-specific information right inline with difficult questions being asked. However, when forms are particularly long and complicated, decision fatigue can occur, resulting in users

ⓒ Springer Nature Switzerland AG 2021
S. Yamamoto and H. Mori (Eds.): HCII 2021, LNCS 12765, pp. 28–39, 2021.
https://doi.org/10.1007/978-3-030-78321-1_3

rushing through, making mistakes, and not using the handy information buttons. Therefore, we explore using engagement to reduce decision fatigue using human-computer interaction (HCI) and user interface design principles.

This proposed research strives to find a way to make information buttons more appealing in order to reduce errors in forms and increase the speed at which these forms can be done. At first glance, it would seem that the primary beneficiaries of this research would be large banking and financial companies that use these complicated forms as a way to get more accurate results and reduce overhead screening on form answers. However, this research also benefits the end user of these forms who will have their decision fatigue reduced, allowing for more energy to be spent on other things and to reduce errors. In this paper, we seek to answer whether or not more engaging information widgets result in a more accurate form fill out, changes in clicks for the information widgets, and more confidence in the answers given to the form.

The rest of this paper is structured as follows: in Sect. 2 we cover the background surrounding the use of aesthetics and engagement to improve performance on user interfaces and our method to undertake this work. We outline our user study setup in Sect. 3 and the results we found are presented in Sect. 4. Our work contributes to the argument that visual engagement and aesthetics of information buttons do not make a noticeable difference in the performance of tasks.

2   Background and Related Works

It is well known to anyone familiar with the field of HCI that the visual aesthetics of user interfaces is a topic with much research and debate. Designing websites, apps, and products that look good and provide good usability is a goal shared by UX designers and product developers; however there is still much debate in the community both on how to accomplish these goals and how effective increasing one to influence the other is. There is some argument that rather than the maxim of "what is beautiful is usable", the opposite is true and that usability makes things seem more attractive [11], and this is backed up by research noting that users' perception of the important part of a website changes as they use it. It is noted that before using a website, users most prioritize the visual aesthetic of it, but after use they weight aesthetic and usability equally [13].

Nonetheless, many studies have been done on the effect of aesthetic differences on users. One aspect of aesthetics that has been notably studied to mixed results is color, with some studies finding a difference depending on aesthetic [2] and others finding that aesthetic design has little to no effect on the actual message of the program [7]. Studies testing if aesthetics improve performance on tasks or in learning have been run as early as the turn of the century, and there are many claims of the advantages of better aesthetics. Researchers claim that better aesthetics can result in better information retention [6], increased readability, higher intent to purchase [10], enhanced performance on low-usability products [16], ability to find complex icons in large search environments [19], and learning more effectively [15]. Additionally, studies have found that mobile devices that look better improve the efficiency of completing tasks on the phone [20]. These studies reinforce the idea that visual aesthetics can affect the way humans interact with computers subconsciously and that better aesthetics improves performance.

The other side of aesthetics research, however, disagrees with some of those claims. Studies have found that better aesthetics have no effect on performance during experiments [5], and more to the point, some studies claim that better aesthetics only improve the perception of usability without improving user performance [12,21]. It is also worth considering that the idea of aesthetics is culturally dependent [17] and can either have a good effect if designed with the cultural target in mind or have little to no effect otherwise [18]. Also, culturally appropriate aesthetics can have an effect on trust in a website [4].

Despite the null hypothesis predicted in some studies, the general consensus is that aesthetics can have a noticeable effect on performance in HCI. A large literature survey of aesthetically-based papers [22] calculated that overall aesthetics could and have mattered in user studies. That is why we have decided to proceed with this research; because of information buttons being a widely used tool it is important to evaluate how to improve them. Many kinds of information buttons currently exist, such as information buttons that open a dialog on the screen a user is on (an example of which can be found on the Bank of America account application form [1]), information buttons that open information in another window (as seen on the Free Application for Federal Student Aid [23]), or information buttons that only pull up a tooltip (as in a Citi Bank credit card application [3]). However it is worth noting that aesthetics are not solely related to beauty and color [9], they can also relate to reflectivity, representation, meaning [14], and engagement. Our study focuses not only on the beauty of the information buttons, but their engagement and its effect on performance.

3   Methodology

The experiment involved a user study asking participants to fill out a new tax form with information provided to them about a fictional person. The information was given in the documents shown in Fig. 1 that the participant had to decipher into a complex form with niche terminology (that the participants will likely never have been exposed to prior to this point). Two groups of participants were a part of the user study. A control group was given a form using standard (non-engaging) information widgets, and an experimental group was given a form using the engaging information widgets. Then a comparison was made between the two groups using metrics such as amount of errors made, speed of completion, information widgets clicks, and confidence in the information in the form. Using these metrics we indirectly measured the effects of engaging information buttons on decision fatigue while completing complicated forms.

Fig. 1. Documents describing the person the user is filling out the tax form for, a fictional person named Olivia Smith. Original unedited image credit to the Colorado Department of Motor Vehicles and moneyinstructor.com.

3.1 Participants

For this user study, we aimed to primarily recruit participants with a limited knowledge of tax and financial information. That way, when encountering difficult terminology, the users needed to rely on provided information rather than previous knowledge. To that end, we recruited college students and young graduates (mostly within the ages of 18 to 24) from the nearby campus of the University of Nevada, Reno. In order to recruit people, we sent out emails and text messages to university students either individually using a network of known students or as a group via an email list. We expected based on the established difficulty that college students have with financial forms [8] that these participants were a good option for gauging the benefits of engagement on complex forms.

3.2 Apparatus

Due to the COVID-19 pandemic, all user studies for this research were done remotely. Using the video messaging app Zoom [24] users were able to remote access a testing desktop that had the form available to fill out. This form involved fairly complicated tax information accompanied by information buttons, either engaging or not. Figure 2 shows an example of a more engaging information button (next to the normal button) that was used to increase the involvement of the users. This button is brightly colored, and allows for both clicking and hovering. When the participants want quick information without pulling them away from the screen they are on, they can hover over the information widget in order to get a short blurb with the most important information. If they need more information, the participants can then click on the button for a pop up with detailed info.

3.3 Procedure

The user studies were performed by two different study administrators, and so to keep inconsistencies between studies a testing script was used. The script followed a procedure that performed the following steps in order:

Fig. 2. An example of what a normal information button (left) looks like, compared with what a more engaging information button looks like (right). The two buttons on the right are the same, wherein the one on the far right displays what happens when the cursor touches the orange button. The text box seen on the far right button is called a tooltip.

1. The participant filled out a pre-survey with a meaningfully generated participant ID
2. The participant joined a Zoom call with the study administrator
3. The premise and purpose of the study were briefly explained
4. The participant was asked to sign a consent form
5. The participant was given supplementary materials on Olivia Smith, the person the information needed to be filled out for
6. The testing machine screen was shared with the tax form visible on the screen
7. The participant was given remote access to the testing machine
8. The participant filled out the tax form to the best of their abilities
9. The Zoom call was ended
10. The participant filled out a post-survey

The pre-survey participants filled out contained mostly demographic information like their age or gender, as well as their level of familiarity with personal finance and tax information. The tax form included questions on banking information, identification information, and tax deductions; though it is worth reiterating that they did not fill out this form with their own information, only sample information for a fictional person. The post-survey asked the participants how confident they felt in the answers they gave to the form and how they enjoyed the feel of the information buttons.

3.4   Design

Independent variable: The information buttons changed between groups
Name: Information Button Engagement
Levels: Engaging, Standard

Dependent Variables: Speed, accuracy, information button clicks, time until first information button click, confidence in results, self-assessed confidence in results, overall assessment of information buttons, and assessed engagement of information buttons.

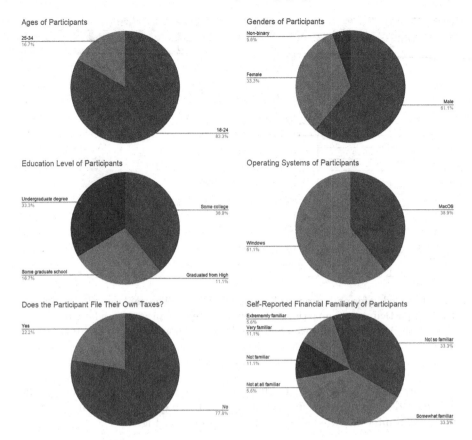

Fig. 3. Pie charts representing the demographics of the study participants based on answers to the pre-survey.

In this user study the independent variable tested was Information Button Engagement (IBE) which had two levels, engaging and standard. These levels should impact the dependent variables of speed, accuracy, information button clicks, acceleration, and retention. The conditions included a random variable as due to the remote nature of the user study, we had no control over the environment the user was in, only control over the computer they remotely accessed.

The experiment was a 10×2 between-subjects design. The amount of entry was 9 participants $\times$ 2 IBE levels = 18 phrases.

4   Experimental Results

Overall, we have managed to involve 18 participants, with nine being part of the control group and nine being given the engaging form. For each of these 18 participants information on eight dependent variables was collected and analyzed. The results were inconclusive and none of the dependent variables were

Participant	Form Completion Time (minutes)	IBE Level
1	12	Standard
2	14	Standard
3	7	Standard
4	9	Standard
5	10	Standard
6	8	Standard
7	8	Standard
8	7	Standard
9	8	Standard
10	9	Engaging
11	8	Engaging
12	11	Engaging
13	9	Engaging
14	14	Engaging
15	10	Engaging
16	7	Engaging
17	14	Engaging
18	16	Engaging
Mean	10.06	
SD	2.82	

IBE Level	Form Completion Time (minutes)	
	Mean	SD
Standard	9.22	2.39
Engaging	10.89	3.10

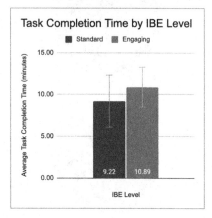

Fig. 4. Detailed information on task completion time. The table on the left shows individual participants completion times and related mean and standard deviation. The table on the right shows mean and standard deviation by level. The column chart compares the two levels.

statistically significant enough to disprove the null hypothesis. Regarding the participant demographics, as expected most of those involved were in the 18–24 range with limited knowledge of finances, as shown in Fig. 3.

4.1 Dependent Variable Analysis

We analyzed eight dependent variables, four of which were from watching the user study and four of which were from the post-survey. We performed analysis of variance (ANOVA) on each of the eight dependent variables and found none were significant. The four variables from the actual user study were time in minutes to complete the form, errors submitted in the form, amount of times the information button was clicked, and time until the first information button was clicked. The four variables from the post-survey questionnaire were the indirectly assessed confidence in results, self-assessed confidence in results, overall assessment of information buttons, and engagement assessment of information buttons.

While none of these dependent variables proved statistically significant after ANOVA analysis, we have selected four to show in more detail in this paper. The first dependent variable that is relevant is the time taken to complete the form in minutes. As shown in Fig. 4, the average time it took for all users to complete the form was 10.06 min, with IBE level Standard participants taking 9.22 min and IBE level Engaging participants taking 18% longer at 10.89 min. This difference was not statistically significant ($p > .05$, ns).

Participant	Amount of Errors	IBE Level
1	2	Standard
2	3	Standard
3	2	Standard
4	4	Standard
5	0	Standard
6	3	Standard
7	2	Standard
8	3	Standard
9	3	Standard
10	2	Engaging
11	3	Engaging
12	1	Engaging
13	5	Engaging
14	4	Engaging
15	2	Engaging
16	3	Engaging
17	6	Engaging
18	3	Engaging
Mean	2.83	
SD	1.38	

IBE Level	Amount of Errors	
	Mean	SD
Standard	2.44	1.13
Engaging	3.22	1.56

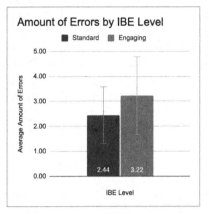

Fig. 5. Detailed information on errors submitted while completing the task. The table on the left shows individual participants errors and related mean and standard deviation. The table on the right shows mean and standard deviation by level. The column chart compares the two levels.

The second dependent variable of note is the amount of errors made and submitted on the form during the test. In Fig. 5 we see a similar pattern, with the average amount of errors made by all participants being 2.83 and IBE level Standard participants submitting an average of 2.44 errors and IBE level Engaging participants submitting 31% more errors at 3.22 on average. This difference was not statistically significant ($p > .05$, ns).

The third significant dependent variable we describe here is the amount of time until the first time a user pressed an information button in seconds. This variable was unique and could measure the ease of discovery of the information buttons. The average time until the first information button was clicked was on average 153.06 s, or a little over two minutes, as shown in Fig. 6. Interestingly, participants took slightly longer to click the engaging buttons than the standard buttons, even when the outlier participant 17 is removed from the data, though it is worth noting that this is a difference of seconds and not statistically significant. Standard level participants took an average of 91.89 s to click an information button, and Engaging level participants took 133% longer (mostly due to the outlier, participant 17) at 214.22 s on average. This difference was not statistically significant ($p > .05$, ns).

The fourth dependent variable was the indirectly assessed confidence in results, which came from the post-survey. This value is based on the answer to several questions asking participants whether they think they had completed the form accurately, whether they felt they knew more about taxes than they did before, whether they could perform better on this kind of form if they had

Participant	Time To First IB Use (seconds)	IBE Level
1	80	Standard
2	25	Standard
3	13	Standard
4	147	Standard
5	89	Standard
6	79	Standard
7	90	Standard
8	119	Standard
9	185	Standard
10	36	Engaging
11	87	Engaging
12	76	Engaging
13	31	Engaging
14	228	Engaging
15	95	Engaging
16	301	Engaging
17	737	Engaging
18	337	Engaging
Mean	153.06	
SD	171.80	

IBE Level	Time To First IB Use (seconds)	
	Mean	SD
Standard	91.89	54.15
Engaging	214.22	226.64

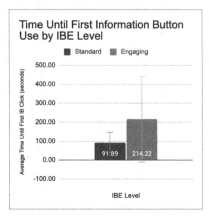

Fig. 6. Detailed information on the time it took to click an information button for the first time. The table on the left shows individual participants time until the first click and related mean and standard deviation. The table on the right shows mean and standard deviation by level. The column chart compares the two levels.

to use it again, and other similar questions assessing confidence in results. These questions were all on the five point Likert scale and thus this dependent variable is an average of the answers to those questions on a scale of one to five, where one is low confidence and five is high confidence. Figure 7 shows that the average of the indirectly assessed confidence was close to four, which on our scale was to "Agree" with statements saying they felt confidence. The average of all participants was 3.98 with Standard participants rating their confidence at 4.09, 5% higher than the Engaging participants at an average of 3.87. However, this difference was again not statistically significant ($F = 0.67$, ns).

The other variables not analyzed in depth showed the same pattern of Standard participants performing slightly better and using information buttons slightly more than Engaging participants, but not in a statistically significant manner. The average amount of information button clicks users made was 18.27 with IBE level Standard participants clicking an average of 20.11 times and IBE level engaging participants clicking an average of 16.44 times. Participants' self-assessed confidence in results on a five point scale was an average of 3.78, with Standard participants rating their confidence on average 3.89 and Engaging participants rating their confidence on average 3.67. The overall assessment of the information buttons, which was calculated using Likert scale assessments similar to those for the indirectly assessed confidence, was on average 3.93 with IBE level Standard participants ranking the buttons on average 3.78 and IBE level Engaging participants ranking the buttons on average 4.08. Lastly, we also asked participants to directly rate the engagement of the information buttons on

Participant	Indirect Confidence in Results	IBE Level
1	4.33	Standard
2	4.33	Standard
3	4.67	Standard
4	2.83	Standard
5	3.83	Standard
6	4.33	Standard
7	4.17	Standard
8	4.17	Standard
9	4.17	Standard
10	4.17	Engaging
11	4.17	Engaging
12	4	Engaging
13	4	Engaging
14	2.67	Engaging
15	3	Engaging
16	4.5	Engaging
17	4	Engaging
18	4.33	Engaging
Mean	3.98	
SD	0.57	

	Indirect Confidence in Results	
IBE Level	Mean	SD
Standard	4.09	0.52
Engaging	3.87	0.62

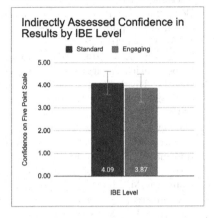

Fig. 7. Detailed information on the indirectly assessed confidence. The table on the left shows individual participants' confidence scores and related mean and standard deviation. The table on the right shows mean and standard deviation by level. The column chart compares the two levels.

a five point scale with a total average engagement score of 4.00, where both sets of IBE level participants rated the engagement 4.00 on average. None of these results were statistically significant ($p > .05$, ns).

5   Conclusion and Future Works

Our research has failed to show any difference in performance based on the style and engagement of the information buttons, however that in itself is a valuable addition to the debate on the extent aesthetics can subconsciously change how we interact with software and computers. Future work could be done to establish why this does not affect performance. Some avenues to consider include examining how small a difference can be and still change performance, or to check if modifications in aesthetics are only able to produce changes if they are part of the entire theme of the software. Additionally, it would be interesting to repeat this experiment with a within-subjects design to see if users who are aware of the differences have different opinions on the buttons. This work has laid the foundation for a deeper exploration of the aesthetics of individual widgets and their influence on improving user performance.

Acknowledgment. This material is based upon work supported by the National Science Foundation under grant number IIA-1301726. Any opinions, findings, and conclusions or recommendations expressed in this material are those of the authors and do not necessarily reflect the views of the National Science Foundation.

References

1. Bank of America Corporation: Bank of America (2020). www.bankofamerica.com/. Accessed 9 Dec 2020
2. Bonnardel, N., Piolat, A., Le Bigot, L.: The impact of colour on website appeal and users' cognitive processes. Displays **32**(2), 69–80 (2011). https://doi.org/10.1016/j.displa.2010.12.002
3. Citigroup Inc: Credit cards - apply for a new credit card online - citi.com (2020). https://citicards.citi.com/. Accessed 9 Dec 2020
4. Cyr, D., Head, M., Larios, H.: Colour appeal in website design within and across cultures: a multi-method evaluation. Int. J. Hum.-Comput. Stud. **68**(1), 1–21 (2010). https://doi.org/10.1016/j.ijhcs.2009.08.005
5. Douneva, M., Haines, R., Thielsch, M.T.: Effects of interface aesthetics on team performance in a virtual task. In: Proceedings of the 23rd European Conference on Information Systems, ECIS (2015), Research-in-Progress Papers. Paper 60 (2015). https://aisel.aisnet.org/ecis2015_rip/60
6. Douneva, M., Jaron, R., Thielsch, M.T.: Effects of different website designs on first impressions, aesthetic judgements and memory performance after short presentation. Interact. Comput. **28**(4), 552–567 (2016). https://doi.org/10.1093/iwc/iwv033
7. Duro, L., Romão, T., Karapanos, E., Campos, P., Campos, P.: How does the visual aesthetics of positively-framed messages impact their motivational capacity? In: Proceedings of the 31st European Conference on Cognitive Ergonomics, pp. 162–167. ECCE 2019. Association for Computing Machinery, New York, NY, USA (2019). https://doi.org/10.1145/3335082.3335085
8. Dynarski, S., Scott-Clayton, J., Wiederspan, M.: Simplifying tax incentives and aid for college: progress and prospects. Tax Policy Econ. **27**(1), 161–202 (2013). https://doi.org/10.1086/671247
9. Folkmann, M.N.: Exploring aesthetics in design: implications for human-computer interaction. Hum. Technol. **14**(1), 6–26 (2018). https://doi.org/10.17011/ht/urn.201805242750
10. Hall, R.H., Hanna, P.: The impact of web page text-background colour combinations on readability, retention, aesthetics and behavioural intention. Behav. Inf. Technol. **23**(3), 183–195 (2004). https://doi.org/10.1080/01449290410001669932
11. Ilmberger, W., Schrepp, M., Held, T.: Cognitive processes causing the relationship between aesthetics and usability. In: Holzinger, A. (ed.) USAB 2008. LNCS, vol. 5298, pp. 43–54. Springer, Heidelberg (2008). https://doi.org/10.1007/978-3-540-89350-9_4
12. Katz, A.: Aesthetics, usefulness and performance in user-search-engine interaction. J. Appl. Quant. Meth. **5**(3), 424–445 (2010). https://eric.ed.gov/?id=EJ919041
13. Lee, S., Koubek, R.J.: Understanding user preferences based on usability and aesthetics before and after actual use. Interact. Comput. **22**(6), 530–543 (2010). https://doi.org/10.1016/j.intcom.2010.05.002,special Issue on Inclusion and Interaction: Designing Interaction for Inclusive Populations
14. Mekler, E.D., Hornbæk, K.: A framework for the experience of meaning in human-computer interaction. In: Proceedings of the 2019 CHI Conference on Human Factors in Computing Systems, pp. 1–15. CHI 2019. Association for Computing Machinery, New York, NY, USA (2019). https://doi.org/10.1145/3290605.3300455
15. Miller, C.: Aesthetics and e-assessment: The interplay of emotional design and learner performance. Dist. Educ. **32**(3), 307–337 (2011). https://eric.ed.gov/?id=EJ953009

16. Moshagen, M., Musch, J., Göritz, A.S.: A blessing, not a curse: experimental evidence for beneficial effects of visual aesthetics on performance. Ergonomics **52**(10), 1311–1320 (2009). https://doi.org/10.1080/00140130903061717, pMID: 19787509

17. Oyibo, K., Vassileva, J.: The interplay of aesthetics, usability and credibility in mobile websites and the moderation by culture. In: Proceedings of the 15th Brazilian Symposium on Human Factors in Computing Systems. IHC 2016. Association for Computing Machinery, New York, NY, USA (2016). https://doi.org/10.1145/3033701.3033711

18. Reinecke, K., Bernstein, A.: Improving performance, perceived usability, and aesthetics with culturally adaptive user interfaces. ACM Trans. Comput.-Hum. Interact. **18**(2) (2011). https://doi.org/10.1145/1970378.1970382

19. Reppa, I., Playfoot, D., McDougall, S.J.P.: Visual aesthetic appeal speeds processing of complex but not simple icons. Proc. Hum. Factors Ergon. Soc. Ann. Meet. **52**(18), 1155–1159 (2008). https://doi.org/10.1177/154193120805201801

20. Sonderegger, A., Sauer, J.: The influence of design aesthetics in usability testing: effects on user performance and perceived usability. Appl. Ergon. **41**(3), 403–410 (2010). https://doi.org/10.1016/j.apergo.2009.09.002,special Section: Recycling centres and waste handling, a workplace for employees and users

21. Thielsch, M.T., Haines, R., Flacke, L.: Experimental investigation on the effects of website aesthetics on user performance in different virtual tasks. PeerJ **7**, e6516 (2019). https://doi.org/10.7717/peerj.6516

22. Thielsch, M.T., Scharfen, J., Masoudi, E., Reuter, M.: Visual aesthetics and performance: a first meta-analysis. In: Proceedings of Mensch Und Computer 2019, pp. 199–210. MuC 2019. Association for Computing Machinery, New York, NY, USA (2019). https://doi.org/10.1145/3340764.3340794

23. US Department of Education: Federal student aid. https://fafsa.ed.gov/. Accessed 9 Dec 2020

24. Zoom Video Communications Inc: Video conferencing, web conferencing, webinars, screen sharing - zoom. https://zoom.us/. Accessed 20 Jan 2021

A Modeling Research on How to Solve Ventilator Alarms from Behavioral and Cognitive Perspectives

Jun Hamaguchi[1]([⊠]) and Sakae Yamamoto[2]([⊠])

[1] Tohto University, Mihama, Chiba, Japan
jun.hamaguchi@tohto.ac.jp
[2] Tokyo University of Science, Katsushika, Tokyo, Japan
sakaeyam@jcom.home.ne.jp

Abstract. Objective: This paper was studied the behavior of ventilator operators from a cognitive perspective. A model was built from the behavior of the operators, especially when an alarm happened. **Introduction:** The frequency of ventilator use is increasing due to COVID-19. Thus, there is an increasing need for staff who can manipulate the ventilator appropriately. There is a need for a support system that will enable the operator to manipulate the equipment appropriately for the situation, especially when an alarm happens. **Methods:** The operator's behavior during the ventilator alarm was video-recorded. The verbal protocol data were also recorded to examine the thinking during the manipulation. After the experiment, the video recordings were reviewed with the participants and interviewed about the reasons for their speeches and behaviors. **Results:** As a result of behavioral analysis of each participant, behavioral patterns split into "Skilled Group (SG)" and "Inexperienced Group (IG)" around about 17 s after the alarm happened. From the verbal protocol data and interview data obtained from the experiment, the reasons why the behaviors divided into two groups were analyzed. Based on these results, the cognitive reasons behind the behavior were clarified. In addition, a cognitive model of the operators when a ventilator alarm happens was built. **Applications:** The results of this research can be applied not only to the manipulation of ventilator alarms, but also to the manipulation of medical equipment, for example, ECMO (Extra Corporeal Membrane Oxygenation).

Keywords: Ventilator · Cognition · Modeling · Motion study · Support system

1 Introduction

When respiratory symptoms become severe due to COVID-19, people generally need specialized medical equipment such as ventilators. The manipulations of a ventilator are directly related to life. For this reason, the operators of ventilators are required to have advanced knowledge and skills. In particular, when an alarm happens, an operator is required to clear the alarm, and the behavior for the situation is important.

However, many cases have been reported in which alarm manipulation has been the cause of patient impact [1] (see notes (1). Peters et al. describe ventilator-related deaths

© Springer Nature Switzerland AG 2021
S. Yamamoto and H. Mori (Eds.): HCII 2021, LNCS 12765, pp. 40–48, 2021.
https://doi.org/10.1007/978-3-030-78321-1_4

and injuries that include inappropriate handling of alarms [2]. Also, Xiao and Seagull described the high incidence of false alarms and inappropriate alarms [3]. From these reports, there has been a lot of research on alarms [4], including studies on improving the environment in which alarms happen and the development of new monitoring systems [5].

Currently, the alarm function of ventilators does not clearly indicate a specific problem. The main purpose is to warn of potential problems. Some information of the alarm status such as the sound and color is defined by industrial standards, and currently it is difficult to include more information in the alarm. When an alarm happens, it is the operator who find the cause of the alarm and has to handle it.

In order for an operator to be able to properly manipulate a ventilator alarm, familiarity and experience are important. Therefore, operator training is necessary. However, there is no established training method for the operation of a ventilator alarm. Therefore, a support system such as a training simulator that can simulate an alarm situation is necessary to train ventilator operators.

In order to design a support system, we focused on the operator's behavior. We measure the operator's behavior from a cognitive perspective to clarify the proper behavior when alarms happen.

2 Objective

In this study, we will clarify the proper behavior of the operator when a ventilator alarm happens. For this purpose, we will measure the behavior during manipulation from a cognitive perspective. Based on the results, we will build a cognitive model of the operator.

3 Methods

3.1 Participants and the Equipment

We measured the behavior of the participants when the ventilator alarm happened (see Fig. 1). The experimental participants (hereafter referred to as participants) were 13 nurses working in an intensive care unit. The average of years of experience in the intensive care unit of the participants was 4.8 years (SD 3.9). The equipment used in the experiment was the same type of ventilator (SIEMENS Servo 900C) that the participants usually use in their work (see Fig. 2). The purpose of this study was explained to the participants, and their consent was obtained. All participants actively cooperated with the study because it was related to their practical work.

Fig. 1. The scene of the experiment

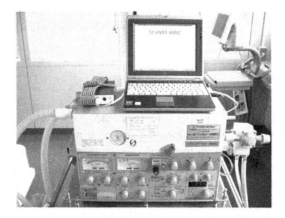

Fig. 2. The experimental equipment

3.2 Simulation Status of the Experimental Equipment

The ventilator was left running at the proper setting. A staff member with setting authority who is not the operator changes the set ventilation volume. With the change of the set ventilation volume, the lower limit of expired minute volume ("MVe") alarm was activated because it was changed to the wrong alarm setting (see notes (2)). When the ventilator alarm happened, the participants did the alarm manipulation as they usually do.

This experiment is a simulation with only a ventilator. Therefore, the vital signs, which are the bio-metric information of the patient, are not shown. Therefore, it was presented in advance to the participants that there was no change in the patient's vital signs when the alarm happened.

3.3 About Alarms that Happen

The most important thing for a patient when on a ventilator is to keep the set ventilation conditions. Therefore, in this experiment, we caused the MVe alarm, which has a serious impact on the patient's life.

There are two causes of MVe alarms. First, a change in the patient's respiratory state can cause the alarm. The other is a factor on the equipment itself. In this experiment, we focused on the factors on the device itself. Among them, we chose the MVe alarm, is caused by an "error in alarm setting" that is relatively difficult to notice.

3.4 How to Get the Data

The behavior and the speech data of the participants during the manipulation were recorded by a video camera. The time from the occurrence of the ventilator alarm to the normal state recovery was also measured.

During the manipulation, we required for the participants verbalized what they were thinking. Participants practiced their speeches before the experiment. In order to reduce individual differences in the number of speeches, after the experiment we reviewed the recorded video and interviewed the participants about their non-speaking behavior.

4 Results and Discussions

4.1 Behavioral Transitions of Participants

The behavioral and verbal protocol data of each participant were analyzed.

We found that all participants behaved the same from the time the alarm happened until an average of 17.3 s (SD., 4.03 [s]) (see (a), (b), (c), (d) and (e) in Fig. 3) ("initial behavior"). After the initial behavior, the behavior pattern was divided to two (see the star symbol in the Fig. 3 ☆). Six participants were able to clear the alarm condition in an average of 34.3 s (SD., 5.85 s) (Fig. 3 thick line). These participants were classified as the Skilled Group (SG). The average years of experience in the intensive care unit for SGs was 8.5 years (SD., 3.9 years).

On the other hand, five participants took longer to clear the alarm state (See Fig. 3 chained line). The average time was 96.3 s (SD., 26.2 s). In addition, two participants gave up trying to clear the alarm state (see α in Fig. 3). The IGs had a mean of 1.6 years (SD., 0.7 years) of experience in the intensive care unit (Table 1).

Figure 3 shows the behavioral transitions of SG and IG. Figure 4 shows the contents of the behaviors in (a) to (l) of Fig. 3.

We focused on the fact that the behavioral pattern was divided into SG and IG at the star symbol ☆ shown in Fig. 3. Therefore, we examined the cognitive thinking behind the behavior.

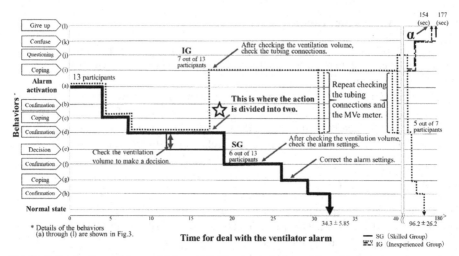

Fig. 3. Behavioral transitions of SG and IG The symbols from (a) to (h) shown on the vertical axis are arranged in the order of appropriate handling of the task. The order is such that if the participant proceeds from (i) upwards, the experiment is ended by giving up and stopping the task.

Table 1. List of behavioral contents in Fig. 3

(a) Move closer to the device
(b) Check the alarm lamp
(c) Mute the alarm
(d) Check the ventilation volume meter
(e) Decide the situation
(f) Check alarm settings and ventilation volume settings
(g) Correcting alarm settings
(h) Confirm alarm cleared
(i) Check the circuit connections
(j) Stare at the entire control panel
(k) Confused
(l) Give up

4.2 Verbal Protocol Data and Interview Analysis

We analyzed the verbal protocol data and interview results focusing on where the behavior was divided into two. From the behavioral transitions shown in Fig. 3 and the verbal protocol data, we extracted and modeled the participants' thoughts and behaviors during alarm manipulation and the information obtained from the equipment. We named this model the cognitive model for ventilator alarm manipulation ("cognitive model") (see Fig. 4).

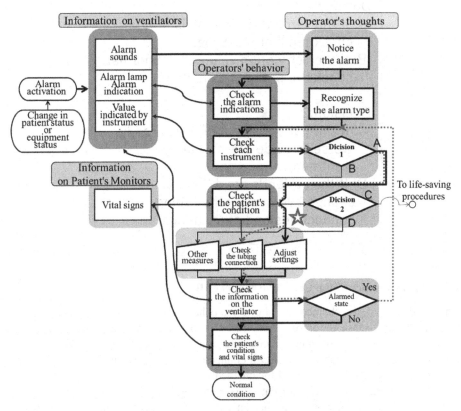

Fig. 4. A cognitive model for ventilator alarm manipulation

4.3   Characteristics of the Cognitive Model of SG

The thick line shows the process of SG's behavior and thinking (See Fig. 4). The SG participants checked the MVe meter and made decisions based on the values it showed (See Fig. 4 and Table 2, Decision 1). After that, it was clear that the participants behaved based on their decisions. The verbal protocol data was analyzed to examine this "decision". All SG participants confirmed that the patient's ventilation was maintained with the MVe meter. Afterwards, we obtained the following speech data: "The setting is wrong because the ventilation is maintained". We analyzed the interviews of this speech data. The following comments were obtained during the interview. "If I check the ventilation meter and the ventilation is not maintained, I check the patient's condition. However, if the ventilation is maintained, check the equipment settings." This shows that at "Decision 1", the SG participants were thinking about the next behavior to be done (See Fig. 4A and Table 2, Decision 1).

Table 2. List of decision details in Fig. 4

Decision 1
A: If the ventilation volume is normal, check the alarm setting.
B: If the ventilation volume is abnormal, check the patient's condition.

Decision 2
C: If the patient's condition is already worse, provide life-saving treatment.
D: If the patient's condition has not changed yet, check the tubing connections and other things.

4.4   Characteristics of the Cognitive Model of IG

On the other hand, the chained line shows the behavior and thought process of the IG participants. In this task, the alarm is not cleared by checking the connection of the circuit because it is not caused by an incorrect connection of the circuit (see (i) in Fig. 4). The participants continue to check the circuit connections repeatedly until they realize that the settings are incorrect (see Fig. 4 the chained line).

The following verbal protocol data was obtained from the IG participants. "The patient's ventilation volume is fine, but let's look at the circuit connections anyway" or "It's a ventilation volume alarm, so there must be a leak somewhere in the circuit". In the interview, we confirmed the intention of the speech data. We got the following comments; "It seemed to be ventilated, but it was an MVe alarm, so I thought I'd just check to see if the circuit was disconnected." or "It was the MVe alarm, so I thought I'd check the circuit anyway."

These data showed that the IG participants had a fixed knowledge that "If the EMV alarm goes off, there's a disconnection in the circuit." A set of decisions and actions like this "routine", that is, a script was obtained [6]. The IG participants had a script "EMV alarm has incorrect circuit connection", and we can consider that the script had an affected on the areas marked with ☆ in Fig. 3 and Fig. 4.

In the experiment, some participants repeated the same manipulation or gave up because the alarm condition did not clear (see α in Fig. 3). In addition, in the interviews, we obtained comments such as "I couldn't think of anything" and "I was panicking". In alarm manipulation, the behavior based on a script can be concise and without thinking deeply. On the other hand, we realized that if the participants do not get the expected state as a result of their actions, they may end up in a whiteout state or in a state where they cannot do anything.

Based on these results, it is necessary to prevent script-based actions from becoming a priority in emergency situations such as when an alarm happens. We considered it important to let the operator understand the information about the equipment and the patient, think carefully, and make decisions about the situation.

The cognitive model made in this study clarified not only the difference in behavior between SG and IG, but also the operator's thinking in "Decision". Building a cognitive model allowed us to clarify the proper behaviors and thoughts in manipulating the ventilator alarm.

4.5 Clinical Special Situations

If the ventilator is stopped for any cause, the patient might go into respiratory arrest. According to the Golden Hour Principle (Morley Cara's Lifesaving Curve) [7], the mortality rate is said to be 50% after 10 min of respiratory arrest. With only these figures, it may seem that there is a lot of time to spare. However, in the case of respiratory failure, where a ventilator is used, we can imagine that the situation is even more severe. In emergency situations such as when an alarm happens, it is necessary to recover the normal conditions as soon as possible. The operator needs to make a decision to choose the next action if the cause is not found. The operator should not keep seeking the cause endlessly. Considering the results of the behavioral analysis and the time constraint, we thought it was necessary to make a decision to choose the next action, such as asking for help from other staff if the alarm could not be cleared within 60 s of its happening.

For these reasons, it is important to train operators not only to be able to "become able to manipulate", but also to "think about what is the proper behavior according to the situation".

5 Application and Prospects of Cognitive Models

The operator's behavior when a ventilator alarm happens was analyzed from a cognitive perspective, and the cognitive factors behind the operator's behavior were modeled. The model we have built is adapted to the MVe alarm. We consider that we can build models for alarms of other equipment in the same way.

This study was conducted on the basic behavior of ventilator operators. This experimental method and results can be adapted to medical equipment that requires advanced technology, such as ECMO devices. In the future, we plan to study these equipment as well from a cognitive perspective.

6 Conclusions

The operator's behavior during a ventilator alarm was measured from a cognitive perspective. A cognitive model was built from the analysis of behavior and verbal protocol data. From this model, we indicated the validity of the operator's behavior and thinking during alarm manipulation.

Notes

(1) Alarm manipulation: An operation to recover from an alarm state to a normal state
(2) Expired minute volume (MVe): "Expired minute volume" is the expiratory ventilation volume per minute
   It is multiplied by the volume of each breath and the number of breaths. It can be expressed as follows; The expiratory volume [L/min] can be expressed as the number of breaths [breaths/min] multiplied by the volume of one breath [mL/breath]. For example, for a patient weighing 60 [kg], if the standard is 10 [mL/kg] per breath, the volume of each breath is 600 [mL]. If the standard respiratory frequency is 12 [breaths/min], the expiratory minute volume is 7.2 [L/min]

References

1. Japan Council for Quality Health Care, Medical Accident Prevention Center: Medical Accident Information Collection Business 3rd Report, 2004.10
2. Peters, G.A., Peters, B.J.: Medical Error and Patient Safety, pp. 114–115. CRC Press, Taylor & Francis Group (2008)
3. Xiao, Y., Seagull, F.J.: An analysis of problems with auditory alarms: defining the role of alarms in process monitoring tasks. In: Proceedings of the Human Factors and Ergonomics Society 43nd Annual Meeting, 1999, pp. 256–260 (1999)
4. Block Jr., F.E., Nuutinen, L., Ballast, B.: Optimization of alarms; A study on alarm limits, alarm sounds, and false alarms, intended to reduce annoyance. J. Clin. Monit. Comput. **15**, 75–83 (1999)
5. Phillips, J.: Clinical alarms; improving efficiency and effectiveness. Crit Care Nurs **28**(4), 317–323 (2005)
6. Abelson, R.P.: Psychological status of the script concept. Am. Psychol. **36**, 715–729 (1981)
7. Holmberg, M., Holmberg, S., Herlitz, J.: Effect of bystander cardiopulmonary resuscitation in out-of-hospital cardiac arrest patients in Sweden. Resuscitation **47**(1), 59–70 (2000)

Evaluating Digital Nudging Effectiveness Using Alternative Questionnaires Design

Andreas Mallas[(⊠)] [ID], Michalis Xenos[(⊠)] [ID], and Maria Karavasili[(⊠)] [ID]

Computer Engineering and Informatics Department, Patras University, Patras, Greece
{mallas,xenos,mkaravasili}@ceid.upatras.gr

Abstract. This paper presents a study for the effectiveness of digital nudging using two alternative questionnaire designs. Each questionnaire used the same 13 questions but included different digital nudges aiming to lead the participant to a different choice. The questionnaire topic was about students' anxiety, while 230 valid questionnaire responses were collected and used in this study. The paper presents the types of digital nudges that were used and the success or failure of each type. Results indicate that digital nudges were successful in some cases while failed in some other cases. The discussion focuses on what type of digital nudges were successful and in which cases. Based on the statistical analysis, while some digital nudges types are more effective, the topic of the question is also related to participants' resistance towards a digital nudge.

Keywords: Digital nudging · Questionnaire design · Surveys

1 Introduction

The rising use of technology resulted in significant changes to how daily life is conducted, with activities now provided in online settings. Of course, that would not have left unaffected fields that require user feedback. Such examples include inter alia marketing research which is now mostly conducted through online questionnaires instead of telephone calls, voting that is now transitioning to online environments instead of requiring a physical presence, and opinion polls that are also conducted with online surveys. Similarly, the educational sector has adopted supplementary online platforms to enhance face to face teaching that utilizes student evaluation through questionnaires. Furthermore, other fields in need of user input employ online questionnaires due to their ease of use, online presence, and low cost. What all the previous implementations have in common is the need for reliable user data.

The reliability of the data collected using questionnaires could be affected using digital nudges that guide the user behavior and choices when working in an online environment. A digital nudge is "the use of user-interface design elements to guide people's behavior in digital choice environments" [1]. This paper aims to investigate the effectiveness of digital nudging using various cases of digital nudges. Towards this goal, a questionnaire related to students' anxiety was adopted for a case study and digital nudges were used to guide responses towards either a positive answer or a negative one.

© Springer Nature Switzerland AG 2021
S. Yamamoto and H. Mori (Eds.): HCII 2021, LNCS 12765, pp. 49–60, 2021.
https://doi.org/10.1007/978-3-030-78321-1_5

The questionnaire topic was selected since anxiety is an interesting topic due to COVID-19 lockdown and distance learning being used for many cases, and we expected students to respond in the survey. That was confirmed by collecting 230 valid responses.

The research questions of this work are about the effectiveness of various types of digital nudges and about the factors that affected effectiveness, such as the type of the digital nudge and the significance of the topic of the question. The paper presents a short literature review introducing the terms used and discussing similar works in the following section. Then, Sect. 3 presents the experiment, the participants, and the tools used, while Sect. 4 outlines the results from the statistical analysis and discusses the findings. Finally, in Sect. 5 we summarize the main results and discuss limitations and future work.

2 Literature Review

The term nudge defined as "any aspect of the choice architecture that alters people's behavior in a predictable way without forbidding any option or significantly changing their economic incentive" was introduced by Thaler and Sunstein [2]. In simple terms, each individual's action is influenced by the decision environment and context, the configuration and display commonly referred to as "choice architecture" is likely to influence their decisions. Nudging has been utilized for controlling population weight gain in Japan, with institutions promoting health behaviors through screening and health record-keeping for each individual [3]. Stämpfli et al. [4] showed that placing a human-like sculpture of thin proportions near a vending machine can reduce food intake in restrained eaters by nudging users towards selecting healthier food options from the machine. Recent research shows the effectiveness of the use of nudging in communication messages in public service advertisements for public health and safety by changing the perceived threat of SARS-CoV-2 when necessitated and thus enhancing the tools of organizations and policymakers [5].

The acknowledgment of the potential of nudging for digital choice environments pointed to the introduction of digital nudging [1]. As mentioned in the introduction, a digital nudge is "the use of user-interface design elements to guide people's behavior in digital choice environments". Meske and Potthoff [6] argued that the definition of digital nudging [1] did not include the freedom to decide without necessitation or a significant variation of preferences and proposed a definition that included these terms. Specifically, they defined digital nudge as "a subtle form of using design, information and interaction elements to guide user behavior in digital environments, without restricting the individual's freedom of choice".

A recent integrative definition of digital nudging by Lembcke et al. [7] incorporates ethical restrictions to digital nudging and distinguishes it from techniques used in marketing that are not in the user's best interest. Indicatively, they specify that "nudgees must be able to easily recognize when and where they are subject to being nudged, as well as what the nudger's goals of this intervention are, in addition to how and why the nudge is working", they also mention that "increases the private welfare of the nudged individual or the social welfare in general".

Nudging on online settings was practiced successfully even before the term digital nudging was introduced. Changes in the default profile visibility settings of Facebook that set user's high school to be visible by default resulted in the doubling of users with visible high school information at Carnegie Mellon University [8]. This adoption in setting default visibility on more personal data on Facebook from 2005 to 2014 lead to an increase in user's information visibility [9]. Wang et al. [10] showed that people are influenced by their friends' ratings (social nudging) in online product ratings by rating similarly after becoming online friends. Bammert et al. [11] utilized nudges in an online experiment -that had previously proved to be effective in different fields- and applied them to a business setting leading to an improvement in customer support processes. Li et al. [12] investigated how nudging through project updates in crowdfunding could drive project followers into making pledges by adjusting the campaign update regularity and length of the messages. Dennis et al. [13] employed nudges -commonly used in offline settings- in an e-commerce website as advertisements and showed that they succeeded in influencing users' choices. Following Wikipedia declining editor retention rates, Gallus [14] conducted a field experiment to nudge new editors with randomized symbolic awards that resulted in "a sizeable effect on newcomer retention".

However, not all reports on nudging had a positive outlook. In the heath sector, a software tool that identified and nudged at-risk patients failed to decrease cardiovascular disease cases significantly [15]. Marteau et al. [16] following a literature review argued that evidence to establish the effectiveness of nudging in people's health is limited. Fellner et al. [17] reported that only one out of three nudge approaches managed to effectively increase law compliance. By conducting a systematic literature review on nudging, Hummel et al. [18] suggested "that there is evidence that it might be less effective than expected" and stated that they "identified at least twenty studies that report no or only mixed results".

A quantitative review on nudging with 100 publications showed that nudges have great variations in terms of their effectiveness [19]. A reactive effect on nudges can result in "counter-nudges" that result in deviations from the nudge's intent [20]. Caraban et al. [21] in a systematic literature review on nudging highlight 6 reasons for nudge failures: nudge results are not sustainable over time, backfiring leads to unexpected outcomes, intrusive nudges result in reactiveness, poor timing and nudge intensity, lack of educational effects, and strong preferences and established habits. On the latter point, they state that "nudges work best under uncertainty that is when individuals do not have strong preferences for particular choices, or established habits".

Following the diverse reports on nudging effectiveness in online and offline settings, we aimed at investigating nudge effectiveness in online questionnaires. The questionnaire's topic involves stress-related issues, a subject that we can expect users would have strong opinions on and constitutes an ideal scenario for evaluating nudge effectiveness.

3 Tools and Participants

For this work, a between-group design was followed. The type of questionnaire (positive or negative nudging) was randomly chosen by the system when each participant started the questionnaire, so each participant filled-in only one version of the questionnaire. The survey was carried out online and participants' responses were recorded by the system and stored for statistical analysis. When a participant failed to answer all the questionnaire questions, the system stored that the questionnaire was incomplete. Following the survey completion, the data from the incomplete questionnaires were excluded from the statistical analysis and resulted in the small inconsistency reported in the number of positive and negative questionnaire responses. Overall, 230 valid questionnaires were collected, with 123 participants filling out the questionnaire with the negative digital nudges and 107 with the positive ones. Students from various levels (B.Sc., M.Sc., Ph.D.) partake in the experiment, from two campus-based universities (involving younger students mostly under 35 years old) and one open university (involving students mostly from ages above 35 years old).

A goal of this work was to use a valid survey on a topic that the participants would find appealing enough to participate and then adjust the digital nudges to the questions and not by creating speculative questions tailored to be suitable for specific nudges. Towards this goal, the questionnaire that was used in this survey consisted of 13 items with the primary aim to record anxiety-related opinions effectively. Initially, questions were selected regarding anxiety, and then, the type of nudge was carefully considered, towards better serving the purpose of the questionnaire item, resulting in some types of nudges utilized more than others. The questionnaires were developed in the Greek language because the survey targeted Greek participants. Due to the complex nature of the topic of questionnaires (anxiety), the authors felt it was essential for the participants to answer in their native language to attain reliable results.

Consequently, two questionnaire versions were developed, each consisted of the same 13 items that incorporated digital nudges. One had the questionnaire items painted with a positive context and the other with a negative one. The aim was to influence participants towards more positive answers with the positive nudges and the contrary. Questionnaires utilized five types of digital nudges that are discussed in the following sub-sections.

3.1 Nudging Through Emotion-Evoking Photographs

Four questionnaire items (for better reference these were the questions Q1, Q3, Q9, and Q13) utilized this type of nudging and consisted of a photograph and a question with a pool of available answers. The text of the question was identical for both questionnaires as well as the available answers. In contrast, different photographs were selected to evoke the respective emotion to the participant. In the first case, the photograph's subjects were pictured in a positive context, while in the opposite case, in a negative one. For example, in the positive version of a questionnaire item, a photograph was depicting a mother having a friendly conversation with her daughter, whereas, in the negative, a mother was shouting at her, while the daughter is facing away dissatisfied.

3.2 Nudging Through Default Options

Four questionnaire items (the questions Q5, Q6, Q11, and Q12) utilized this nudge type, each version retains the same question and answers, but a different default answer was pre-selected for the participant towards the desired choice. The purpose of this nudge was that the participant would be influenced and not deviate from the default choice. Additionally, two of the four questions differed in the design of their UI. Specifically, they were two types of UI. In one type (used in Q6 and Q12), the participant chooses between a 5-point scale, while in the other type (used in Q5 and Q11), a slider with values from 0 to 100 (see Fig. 1) was used and the participant had to move the slider. In both types, a value was already selected (i.e., the slider on the left of Fig. 1 was by default at the point shown on the image, and the second option of the choices was selected as shown on the right side of Fig. 1) allowing the participant to simply press "Next".

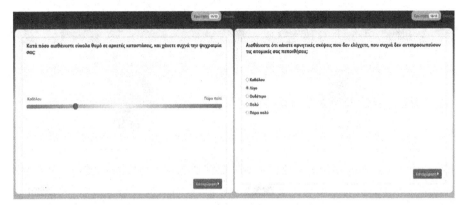

Fig. 1. A slider with values from 0 to 100 on the left and a 5-point scale on the right (Color figure online)

3.3 Nudging Through Text

Both positive and negative versions of this type of nudge (questions Q4 and Q8) had the same available answers, although the questions were carefully phrased in each version to lead towards the desired responses. The text cited information regarding the question's topic, highlighting the positive side on one version and the negative on the other while the question per se was identical.

3.4 Nudging Through Different Levels of Answers

Two questionnaire items (questions Q2 and Q10) utilized this type of nudging and consisted of a question and a pool of available answers. What differentiated the positive and the negative version was that while questions in both versions were identical, there were more available choices towards the desired responses. Specifically, there were seven available answers for both versions ranging from "definitely not" to "definitely yes", the positive version had more responses towards "definitely yes" and the negative version more responses towards "definitely not".

3.5 Nudge Based on the Vote of the Majority

One questionnaire item utilized this nudge type (in question Q7), with both versions employing the same questions and answers. In contrast, each version contained a different graph that presented the choices of other users aiming for the participant's choice not to deviate from the choice of the majority. As shown in Fig. 2, a pie chart displays different users' choices for each case; on the left side, the response with the green color is presented as having the most votes, and with the red color on the right. Also, the option presented as the most popular is showed as the first option in each case. In both cases, the goal is to sway the participant to choose what the majority has selected.

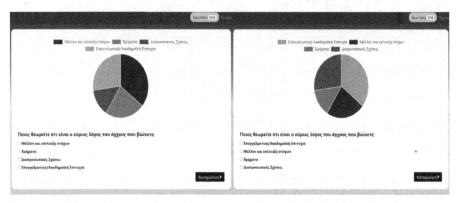

Fig. 2. A pie chart with other users' preferences and different artificial data for the majority votes were presented in each version.

4   Results and Discussion

This section presents the results of the statistical analysis of each questionnaire item. The negative and positive nudges were compared for each questionnaire item and then grouped by nudge type, each discussed in a separate subsection. The statistical analysis showed a significant departure from normality, and thus only nonparametric tests were conducted for all data.

4.1   Results from Nudging Through Emotion-Evoking Photographs

This type of nudge is aiming at influencing the participant's opinion through emotion-evoking photographs. Each questionnaire item had two versions, a positive and a negative, and utilized a different photograph to evoke contrasting emotions towards influencing the participant's choice to a positive and negative response, respectively. Questions of this nudge type were the Q1, Q3, Q9, and Q13, and the results are as follows:

In Q1, the aim was to investigate if the participant is willing to seek professional help from a mental health professional. This questionnaire item utilized a photograph that, in both versions, it portrayed a female phycologist and a student in a counseling session, with the professional depicted to take notes during their session. In the first case, the phycologist had a neutral emotion on her face, while in the second case, it had a strict and more severe expression, aiming at making her unappealing to converse.

A Mann-Whitney U test showed that there was not a significant difference between negative ($M_{rank} = 109.27$) and positive ($M_{rank} = 122.66$) nudging for Q1, $U = 7347$, $z = 1.575$, $p = 0.115$. This result can be attributed to the strong opinion of the participants on the subject due to the nature of the question (seeking professional help) or to the choice of the particular photographs for not having a strong emotional effect.

The Q3 aimed at examining the effectiveness of a recreational pursuit in coping with anxiety. This questionnaire item utilized a photograph that depicted an individual engaging in a hobby. Each version had contrasting photos, one with a more relaxing activity and the other with extreme sports. In the positive version, it depicted someone taking a photograph, while in the negative version, a man was climbing a mountain while hanging with one hand from a slope. A Mann-Whitney U test showed that there was a significant difference between negative ($M_{rank} = 107.48$) and positive ($M_{rank} = 123.57$) nudging for Q3, $U = 7444.5$, $z = 1.978$, $p = 0.048$, $r = 0.13$. The success of this nudge can be attributed to the extreme antithesis of the photographs, which effectively evoked the appropriate emotion in the participants.

In Q9, the aim was to investigate if the participant felt that their family is supportive regarding their academic performance. This questionnaire item utilized a photograph that depicted a female student having a conversation with her mother. The positive version showed a casual conversation with both family members facing each other and the mother having a supportive expression. Whereas the negative version, depicts a mother shouting at her daughter and the daughter is trying to ignore her by engaging with her mobile phone and looking away from her with dissatisfaction. A Mann-Whitney U test showed that there was a significant difference between negative ($M_{rank} = 104.7$) and positive ($M_{rank} = 127.92$) nudging for Q9, $U = 7909$, $z = 2.722$, $p < 0.01$, $r = 0.18$. Similar to Q3, the nudge was successful, and the result can be attributed to the extreme antithesis of the photographs, which effectively evoked the appropriate emotion in the participants.

Finally, the Q13 item aimed at investigating the frequency of participant's dissatisfaction and motivation loss in their academic/professional careers. This questionnaire item utilized a photograph that depicted a male in an office-like environment working on a laptop. The positive version depicted someone with neutral body language and facial emotion, whereas, in the negative version, the subject had a more concerned face with his hand resting on his head. A Mann-Whitney U test showed that there was not a significant difference between negative ($M_{rank} = 114.67$) and positive ($M_{rank} = 116.45$) nudging for Q13, $U = 6682$, $z = 0.207$, $p = 0.836$. Similarly to Q1, the ineffectiveness of the nudge can be attributed to the inability of the photographs to evoke the desired emotion. Furthermore, we cannot disregard the possibility that the nudge failed due to the participants' strong opinions on the question's particular subject.

4.2 Results from Nudging Through Default Options

This type of nudge objective was to sway the participant's opinion by setting default options. Each questionnaire item had two versions, a positive and a negative, but a different default choice was pre-selected for each version of the questionnaire. Questions of this nudge type were the Q5, Q6, Q11, and Q12 items, and the results are as follows:

In Q5, the objective was to examine if the participants felt stress symptoms during their daily lives. The participant was given a slider from 0 to 100, with 0 representing "definitely not" and 100 representing "definitely yes" as seen on the left of Fig. 1. The slider was colored with green on the one side (0) and red on the other (100), indicating that choices towards 100 were more extreme (red color). A different default choice was pre-selected for each version of the questionnaire. Instead of the slider being in the center, in the positive version, it was moved towards 0, and in the negative version more towards 100. A Mann-Whitney U test showed that there was a significant difference between negative (M_{rank} = 123.8) and positive (M_{rank} = 105.96) nudging for Q5, U = 5559.5, z = -2.029, p = 0.043, r = 0.13. The results indicate that the nudge was successful and that the slider design succeeded in influencing the participants' opinions.

In Q6, the aim was to investigate if the participants feel the need to isolate themselves from friends and family that they otherwise have close to them in periods that they are stressed. A five-level Likert item was given, with the available responses ranging from "definitely not" to "definitely yes", as seen on the right of Fig. 1. A different default choice was pre-selected for each version of the questionnaire. In the one version, the response "probably not" was selected, and "probably yes" in the other version. A Mann-Whitney U test showed that there was not a significant difference between negative (M_{rank} = 114.57) and positive (M_{rank} = 116.57) nudging for Q6, U = 6695.5, z = 0.236, p = 0.814. The results indicate that the nudge was unsuccessful since it failed to influence the participants' opinions.

The Q11 item aimed at examining if the participants regularly feel anger and fail to retain their composure. Similarly to Q5, the participant was given a slider from 0 to 100, with 0 representing "definitely not" and 100 representing "definitely yes" and a different default choice was pre-selected for each version of the questionnaire. A Mann-Whitney U test showed that there was not a significant difference between negative (M_{rank} = 120.86) and positive (M_{rank} = 109.34) nudging for Q11, U = 5921.5, z = -1.309, p = 0.19. The results indicate that the nudge was unsuccessful since it failed to influence the participants' opinions.

In Q12, the objective was to examine if the participants have negative thoughts that cannot control and do not correspond to their personal beliefs. Q12 followed an identical design to Q6 with a five-level Likert item (responses ranging from "definitely not" to "definitely yes") and a pre-selection of "probably not" in one version, and "probably yes" in the other version. A Mann-Whitney U test showed that there was not a significant difference between negative (M_{rank} = 119.78) and positive (M_{rank} = 110.57) nudging for Q12, U = 6053.5, z = -1.080, p = 0.28. The results indicate that the nudge was unsuccessful since it failed to influence the participants' opinions.

In this study we examined two different user interface (UI) designs for nudges through default options. Two digital nudges utilized a slider mechanism to select an option and two digital nudges radio buttons. Only one digital nudge using the slider design was proven effective. The nudge ineffectiveness could be attributed to the digital nudge type or to the participants' strong opinions on each question's particular subject and is a subject for further investigation.

4.3 Results from Nudging Through Text

This type of nudge utilized questions that also cited information as part of them, in an attempt to influence participant's opinion. Each version quoted different data, one focusing on the positive side and the other on the negative. Questions of this nudge type included Q4 and Q8, and the results are as follows:

The Q4 item aimed at investigating if the participants are willing to use medication for relieving anxiety symptoms with the guide of a medical professional. The text composed of two parts, the first part was quoting information about such prescriptions, and the second part was the question per se. Both versions utilized the same question (second part) but cited different information about the medication. The positive version mentioned that following the patient's treatment anxiety symptoms were significantly reduced, while the negative version focused on the side effects of such treatments aiming to discourage participants from opting for them. A Mann-Whitney U test showed that there was not a significant difference between negative ($M_{rank} = 115.01$) and positive ($M_{rank} = 116.07$) nudging for this question, $U = 6641$, $z = 0.126$, $p = 0.899$. The results indicate that the nudge was unsuccessful since it failed to influence the participants' opinions. This result can be attributed to the insufficient difference of the text in the positive and negative nudge but can also be a result of the student's strong opinion on the subject. Q4 mentions that for the participant to receive the medication, they have to visit a mental health professional, which is the exact topic of Q1, and in that case, the students were also not influenced by the nudge, further indicating the strong opinion on the subject.

In Q8, the aim was to investigate if the participants are afraid of the future and worry about the impact of their current actions on their prospects. Q8 followed an identical design to Q4, with the text composed of two parts. The positive version stated that dealing with day-to-day stressful situations can reduce anxiety about the future, while the negative version mentioned how negative news on media contributes to growth in precariousness. A Mann-Whitney U test showed that there was a significant difference between negative ($M_{rank} = 107.02$) and positive ($M_{rank} = 125.25$) nudging for this question, $U = 7623.5$, $z = 2.183$, $p = 0.029$, $r = 0.14$. The results indicate that the nudge was successful and that the text cited succeeded in influencing the participants' opinions.

4.4 Results from Nudging Through Different Levels of Answers

Questionnaire items that utilized nudging through different levels of answers, attempted to influence the participant's opinion by offering more responses towards the positive or negative scale depending on the version. Questions of this nudge type included Q2 and Q10, and the results are as follows:

The objective of Q2 was to examine if participants believed that their negative thinking adversely influences their relationships with others. Both versions utilized the same question and the same range of responses (from "definitely not" to "definitely yes") but differed in the in-between responses. Specifically, the positive version had more positive answers than negative whereas the opposite was true for the negative version. A chi-square test showed that there was no significant association between the type of nudging and the participants' answers for this question, $\chi^2(2, N = 230) = 4.55$, $p = 0.103$. Based on the results, the nudge failed to sway the participants' opinions which can be likely attributed to the participants' strong opinions on the subject.

In Q10, the aim was to investigate if the participants had feelings of fear, concern, or internal tension without a distinct cause. Q10 followed an identical design to Q2, with responses ranging from "definitely not" to "definitely yes" while the positive version had more positive answers and the reverse. A chi-square test was performed to examine the relation between the type of nudging and the participants' answers to this question. The relation between these variables was significant, $\chi^2(2, N = 230) = 18.84, p < 0.01$. The results reveal that the nudge was successful.

4.5   Results from Nudge Based on the Vote of the Majority

This nudge type was aimed at influencing the participant's opinion by presenting a graph beside the question that displayed what others had selected. There was one question of this nudge type and the results are as follows:

The aim of Q7 was for the participants to select their principal cause of anxiety. Both versions used identical questions along with a pool of answers that included four available choices: future/goal achievement, money, interpersonal relationships, and academic/professional career. The one version used a fictitious pie chart that presented what other participants had supposedly selected, and the other a similar chart where users had made different choices. Additionally, the choice that was promoted as the most popular was offered as the first option in each case. A chi-square test was performed to examine the relation between the type of nudging and the users' answers to this question. The relation between these variables was significant, $\chi^2(3, N = 230) = 10.85, p < 0.01$. The results show that the nudge was successful.

5   Summary and Future Work

The findings of this study suggest that some nudges were successful, while others failed to lead the participant to specific choices. Specifically, all types of nudges (nudging through emotion-evoking photographs, nudging through different levels of answers, nudging through text, nudging through default options, and nudge based on the vote of the majority) had at least one question that influenced participants' decisions. This result suggests that all types of digital nudges considered had the potential to sway participants' opinions. The difference in effectiveness can be attributed to diverse factors.

An apparent determinant could be the choice of the nudge implementation, e.g., in the emotional nudging through a photograph, the selection of photographs might represent an essential factor for the nudging effectiveness. For instance, if the photographs fail to convey the positive or negative feeling that the nudge should infer. Another example is in the nudge through text description, where the difference of the text in the positive and negative nudging might be inadequate to affect the participant's opinion. Alternative text with wording more towards the positive or negative side could have different effectiveness.

A divergent factor could also depend on the choice of questions: a participant's strong view on a specific question could leave their choices unaffected by the nudge. For example, when voting in an election, an individual that is yet undecided between two candidates might be more easily influenced towards either one by a nudge, while someone with a strong view that leans heavily towards a candidate might not. Seemingly questions that the participant has possibly a strong opinion, e.g., the participant's willingness to visit a psychiatrist, were unaffected by the digital nudging. Considering the topic of the questionnaire that focused on anxiety symptoms, an individual that is aware of their mental condition might exhibit higher resistance to nudging, partially explaining the difference in nudging effectiveness.

This study highlights that although all types of nudges had at least a successful case, there might be more to the question of nudging effectiveness than the UI design, and future research should focus more on underlying factors impacting nudge effectiveness. Such factors are related to personal profiles and opinions and not the UI design.

Limitations of this work are related to the use of a small number of digital nudges since we focused on having a valid and meaningful questionnaire on an interesting topic. Therefore, we tailored the digital nudges to a short questionnaire aiming to stimulate natural responses from the participants rather than having a predetermined set of digital nudges and craft the questions to fit into the requirements of the digital nudges. This helped towards the ecological validity of our survey but offered uneven numbers of the digital nudges that were analyzed. Future work will be investigating more types of digital nudges using more questionnaires. Another limitation is the nature of the sample since we collected responses from students. To broaden the sample, we surveyed students from various fields, various institutes, and a broad variety of ages since the sample included younger students, as well as older students from the open university. Nevertheless, a future goal is to conduct more surveys using topics for a broader population. Finally, another future goal is to investigate the effectiveness of the digital nudge in relation to the opinion strength of the participants, which will require using interviews after the questionnaires.

References

1. Weinmann, M., Schneider, C., Vom Brocke, J.: Digital nudging. Bus. Inf. Syst. Eng. **58**, 433–436 (2016)
2. Thaler, R.H., Sunstein, C.R.: Nudge: Improving Decisions About Health, Wealth, and Happiness. Yale University Press, New Haven (2008)
3. Borovoy, A., Roberto, C.A.: Japanese and American public health approaches to preventing population weight gain: a role for paternalism? Soc. Sci. Med. **143**, 62–70 (2015)
4. Stämpfli, A.E., Stöckli, S., Brunner, T.A.: A nudge in a healthier direction: how environmental cues help restrained eaters pursue their weight-control goal. Appetite **110**, 94–102 (2017)
5. Kim, J., et al.: Nudging to reduce the perceived threat of Coronavirus and stockpiling intention. J. Advert., 1–15 (2020)
6. Meske, C., Potthoff, T.: The DINU-model–a process model for the design of nudges. In: Proceedings of the 25th European Conference on Information Systems, Guimarães, Portugal, pp. 2587–2597 (2017)
7. Lembcke, T.-B., Engelbrecht, N., Brendel, A.B., Herrenkind, B., Kolbe, L.M.: Towards a unified understanding of digital nudging by addressing its analog roots. In: PACIS, p. 123 (2019)
8. Stutzman, F.D., Gross, R., Acquisti, A.: Silent listeners: the evolution of privacy and disclosure on Facebook. J. Priv. Confid. **4**(2) (2013)

9. Acquisti, A., Brandimarte, L., Loewenstein, G.: Privacy and human behavior in the age of information. Science **347**, 509–514 (2015)
10. Wang, C., Zhang, X., Hann, I.-H.: Socially nudged: a quasi-experimental study of friends' social influence in online product ratings. Inf. Syst. Res. **29**, 641–655 (2018)
11. Bammert, S., König, U.M., Roeglinger, M., Wruck, T.: Exploring potentials of digital nudging for business processes. Bus. Process Manage. J. **26**(6), 1329–1347 (2020)
12. Li, Y., et al.: Exploring the nudging and counter-nudging effects of campaign updates in crowdfunding. In: PACIS, p. 281 (2018)
13. Dennis, A.R., Yuan, L., Feng, X., Webb, E., Hsieh, C.J.: Digital nudging: numeric and semantic priming in E-commerce. J. Manage. Inf. Syst. **37**, 39–65 (2020)
14. Gallus, J.: Fostering public good contributions with symbolic awards: a large-scale natural field experiment at Wikipedia. Manage. Sci. **63**, 3999–4015 (2016)
15. Holt, T.A., Thorogood, M., Griffiths, F., Munday, S., Friede, T., Stables, D.: Automated electronic reminders to facilitate primary cardiovascular disease prevention: randomised controlled trial. Br. J. Gen. Pract. **60**, e137–e143 (2010)
16. Marteau, T.M., Ogilvie, D., Roland, M., Suhrcke, M., Kelly, M.P.: Judging nudging: can nudging improve population health? Bmj, 342 (2011)
17. Fellner, G., Sausgruber, R., Traxler, C.: Testing enforcement strategies in the field: threat, moral appeal and social information. J. Eur. Econo. Assoc. **11**, 634–660 (2013)
18. Hummel, D., Toreini, P., Maedche, A.: Improving digital nudging using attentive user interfaces: theory development and experiment design. In: DESRIST 2018 Proceedings, pp. 1–37 (2018)
19. Hummel, D., Maedche, A.: How effective is nudging? A quantitative review on the effect sizes and limits of empirical nudging studies. J. Behav. Exp. Econ. **80**, 47–58 (2019)
20. Sunstein, C.R.: Nudges that fail. Behav. Public Policy **1**, 4–25 (2017)
21. Caraban, A., Karapanos, E., Gonçalves, D., Campos, P.: 23 ways to nudge: a review of technology-mediated nudging in human-computer interaction. In: Proceedings of the 2019 CHI Conference on Human Factors in Computing Systems, pp. 1–15. Association for Computing Machinery, New York (2019)

Cultivation System of Search-Query-Setting Skill by Visualizing Search Results

Chonfua Mano[✉] and Tomoko Kojiri

Kansai University, 3-3-35 Yamate-cho, Suita-shi, Osaka, Japan
{k587558,kojiri}@kansai-u.ac.jp

Abstract. In using a search engine, we propose words as a search query that we believe are likely to locate various kinds of desired information. When a search result fails to include the desired information or does not answer our questions, we often change the search query based on the difference between the contents of the search result and the desired information. Unless the change is effective, we will continue to get unsatisfactory information even if we repeatedly change our search query. In this research, we define a search-query-setting skill, which we define as the ability to effectively change a search query, and propose a system that improves such skills by helping users find effective words as search queries. Our system shows the distributions of search results before and after changing search queries. By comparing them, users can evaluate their changes.

Keywords: Search-query-setting skill · Skill cultivation · Visualization

1 Introduction

We usually use such search engines as Google to find needed information or answer questions. In search engines, we enter a word or a phrase as a search query that we believe is likely to locate the desired information. When the search result fails to provide the desired information, we change the search query based on the differences between the search result's contents and the desired information. If the query is not effectively changed, we will fail to get our desired information regardless how many times we change the search query. Our research defines the ability to change search queries as a search-query-setting skill and supports its cultivation.

As support for finding desired information, Nguyen et al. proposed a search result visualization system that arranges search results that reflect user's interests [1]. Their system presents the words in the web pages of the search results as choices to users, and makes the user select interesting words. It calculates the weight of each searched page based on the selected words and rearranges them by their weights to simplify the identification of interesting pages. With this system, users can easily find fruitful information from many search results. However, if the required information is not included in the search results, users have no awareness of it. To get the desired information in a list of search results, appropriate search queries are required.

© Springer Nature Switzerland AG 2021
S. Yamamoto and H. Mori (Eds.): HCII 2021, LNCS 12765, pp. 61–75, 2021.
https://doi.org/10.1007/978-3-030-78321-1_6

Since search queries are often inadequate at the first stage, they must be modified until the desired information is obtained. Some research has targeted dynamic modifications of search queries. Yoshida et al. proposed a system that automatically chooses the next search query based on the user's interest in the searched topics [2] by distributing of words in the search results as a graph structure. Their graphs have three nodes: a search query, a topic word that represents the topic of the search result, and a co-occurrence word that appears at the same time as the topic word. The search query and the topic word are connected by a link, and the co-occurrence words are arranged around the topic word. Their system displays the graph, and users can move nodes of interest closer to the query node and those of no interest farther from it. The system infers the user's interest based on the distance of each node and automatically adds or deletes words in the search query. Although this research determines search queries that more closely reflect the interests of users, they themselves do not understand how the appropriate search query was derived. The system has little transparency. Thus, users cannot learn how to appropriately establish search queries for themselves.

To support exploratory searches, Ma et al. proposed a search query recommendation method that increases user awareness of new topics that are related to their search query [3]. As an example of exploratory search, suppose him search for French favorite birthday gifts. Since flower is popular as a birthday gift, he used "flower like French" as a search query, and found that flower is not popular with French and wine is popular. So, he used "wine like French" as a search query and found hopeful gift. In their research, his search goal shifts from the "flowers" to "wine" is called a search goal shift. Their system has offline and online phases. In the offline phase, the system uses machine learning to identify a search goal shift upon which it builds a graph using the search goal shift's search query. In the online phase, the system identifies whether the search goal was shifted; if it has, it recommends a query for the next search goal based on the search goal shift graph using a random walk algorithm. Although this method can recommend candidate words for search queries, users do not know why the words are recommended. Since users cannot set search query without a system, they cannot improve their search-query-setting skills.

For cultivating search-query-setting skills, Saito et al. proposed a method that promotes reflection of the user's search activity by presenting their search processes and asking questions and offering guidance to reflect on their search process [4]. The system's questions, which compel users to explain their own search process, include such examples as "how did you combine the keywords so far?" and "how many links did you follow from one page?" This activity promotes reflection about the decision to change the search query or follow the results' links. However, it provides no support for recognizing the words that the user should input as a search query. Therefore, in our research, we propose a system that helps users identify appropriate words as search queries and improve their search-query-setting skills.

Nishiyama et al. concluded that visualizing the effect of activities is beneficial for reflecting on them [5]. The effect of changing a search query is judged by comparing the new search results with the previous ones. Therefore, understanding the change of the searched for topics based on the modified search query leads to appropriate evaluation. In this paper, we propose a system that shows the distributions of the search results before and after search-query changes. By comparing them, the modified search queries can be evaluated.

2 Approach for Cultivating Search–Query-Setting Skill

2.1 Information Retrieval Process

The information retrieval process consists of the steps shown in Fig. 1. When using a search engine, we first enter words as a search query that we believe will locate the desired information. After getting the search results, we judge whether they provide the desired information. If they do not, we change the search query based on the search result's contents. This process is repeated until we get the desired information. For example, suppose you want to create the program of uploading a file and set "file upload" as a search query, you may not get the desired information. In this case, you need to modify the search query to "file upload programming".

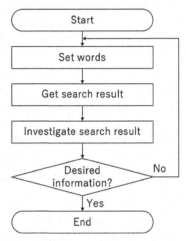

Fig. 1. Information retrieval process

2.2 Contents of Search Results Page

When we use a search engine, the search results are shown on web pages. Figure 2 shows the composition of a search result page of Google that consists of the following: (1) the approximate number of search hits, (2) a list of searched web pages, (3) the search query candidates related to the input search query, and (4) the number of pages of search results.

The following is the information of the searched web page: (2-1) URL, (2-2) title, and (2-3) snippet. The title often encapsulates the entire web page. The snippet connects sentences in the web page that contain the words in the search query. It does not always represent the exact topic of the web page. Therefore, when we judge whether a web page contains the desired information, we mainly utilize its title.

2.3 Search-Query-Setting Skill

Broder [6] classified queries into three types based on the search's intention: navigational queries whose objective is to reach a specific web page that contains the required information; informational queries whose objective is to obtain specific information; and transactional queries that seek the service provider's web page that can provide the required information. A transactional query is defined as a navigational query when the purpose is to reach a specific web page that provides a service.

Fig. 2. Search results page of Google

When the search results do not contain the appropriate information, a subsequent search query is made based the differences between the search results and the desired information. For a navigational query, the search results are investigated to determine whether their contents are close to a specific web page. In an informational query, however, they are investigated to determine whether their contents are close to the desired information. In both queries, the state of the search results must be accurately recognized.

The words in the title of the searched web pages represent the topics derived by the search query. The differences of the topics between the desired information and the search results shape the strategy for making the next search query. Here we assume various situations of the topics in the results (search-query-topic space) and the desired information (desired-topic space) (Fig. 3). Red circles represent the topics. When the searched topics do not contain the desired information (Fig. 3(a)), every word in the current search query indicate undesired information. Therefore, every word in the current search query should be deleted and new words that more accurately represent the desired information should be added. If the search query contains both desired and undesired topics (Fig. 3(b)), words that may belong to the undesired information should be deleted and those that may represent the desired information (but are not included in the search query) should be added. When the search query is part of the desired information (Fig. 3(c)), words should be added that may represent the desired information that is not included in the search query. In Fig. 3(d), the desired information was successfully acquired. In this research, this ability to judge which words to add or delete from a search query is called search-query-setting skill.

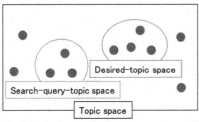

a) Search query does not overlap with desired-topic space

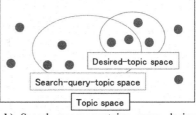

b) Search query contains some desired-topic space

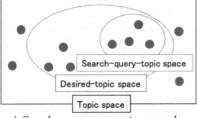

c) Search query represents some desired information

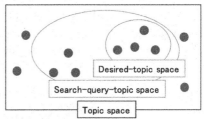

d) Search query covers all desired information

Fig. 3. Relations of topics of search query and searched results (Color figure online)

2.4 Support System for Cultivating Search-Query-Setting Skill

To acquire search-query-setting skills, users must change the search query by themselves and determine whether the subsequent search query improves the search results. To support the recognition of the state of the search results, this research proposes a system that visualizes the distribution of desired/undesired information in search results and allows users to compare the distribution of such information before and after changing a search query. By comparing two visualizations, the users can evaluate whether their modifications were effective.

Figure 4 shows the interaction between the system and the user. When the user inputs a search query, the system displays the search results and the words included in the titles of the searched web pages. The user judges whether the displayed words reflect the information she desires and inputs the judgment results to the system, which visualizes the state of the search results based on her word judgment. She changes the search query based on the visualized search results and obtains the visualizations of both the original and the modified search queries. By comparing these two visualizations, she can evaluate the appropriateness of the search query's modifications.

Figure 5 shows the configuration of our system that achieves interactions of Fig. 4. When a search query is entered through an interface, the search function inputs the words in the search query into Google and gets search results. The topic extraction function extracts words that represent topics from the titles of the web pages of the search results and displays them by a search interface. A judgment that shows whether the words represent the desired information is sent (with the words) to the search-result-visualization function, which calculates the frequency of the occurrence of the words

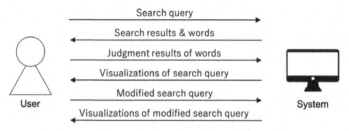

Fig. 4. Interaction between system and user

and their degree of similarity and visualizes them in two-dimensional space. When the modified search query is input, the search function is also used to acquire the search results. The distribution of the words in the new search results is visualized based on the original user's word judgment.

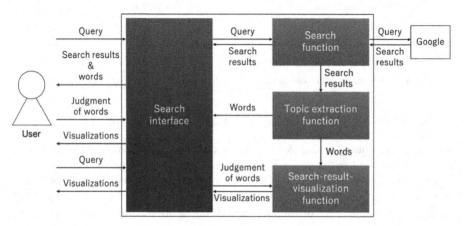

Fig. 5. System configuration

3 Topic Extraction Function

The topic extraction function extracts words that represent topics from the titles of the searched web pages. Since a topic is often expressed by nouns, the system outputs all the nouns from the titles of the top ten web pages obtained as search results.

A morphological analyzer extracts the nouns. Morphological analysis divides a sentence written in natural language into morphemes, which are the smallest language units, and judges their parts of speeches. In our system, when the same words are found in the titles multiple times, only one is output to avoid duplicates. As an example, Table 1 shows the morphological analysis results of the title "20 selections of delicious izakaya in Fukushima (Osaka) - From popular shops to hidden spots - Retty." It also shows the extracted words that express the topics of the nouns: "20, selection, izakaya, Fukushima, Osaka, popular shops, hidden spots, Retty."

Table 1. Results of morphological analysis

Morphemes	Part of speech
20	noun
selections	noun
of	preposition
delicious	adjective
izakaya	noun
in	preposition
Fukushima	noun
(	symbol
Osaka	noun
)	symbol
–	symbol
from	preposition
popular shops	noun
to	preposition
hidden spots	noun
Retty	noun

4   Search-Result-Visualization Function

The search-result-visualization function displays the distributions of the extracted words that represent topics derived by the search query. The distribution of words should show how many desired and undesired topics are included in the search query as well as their topics. Therefore, the extracted

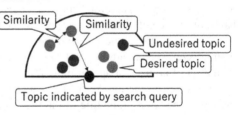

Fig. 6. Visualization method of search results

words are distributed around the current query. Figure 6 shows their distribution. The semicircle's center represents the topic indicated by the current search query, and the circles in it show the topics of the extracted words. The distance between the words represents their similarities. In addition, the color distinguishes the desired/undesired topics: desired topics in red and undesired in blue. We can instantly recognize how many desired topics were derived by the current query.

Since determining the locations of all the words is difficult on a two-dimensional plane that reflects their similarity, this research decides the location words by calculating the following two elements shown in Fig. 7:

- Distance from center: co-occurrence of extracted words with the words in the search query.
- Angle from right radius: conceptual similarities of extracted words with the words in the search query.

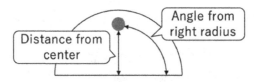

Fig. 7. Method of similarity calculation

If a word represents the topic indicated by the search query, it is assumed to be included in all the titles obtained as the search result. Therefore, the co-occurrence ratio of an extracted word is derived by Eq. 1 by calculating the number of titles including the words out of the top ten titles in the search results. The distance of the circles from the center of the semicircle is calculated by Eq. 2:

$$\text{Co-occurrence} = \frac{\text{Titles that contain word}}{10} \tag{1}$$

$$\text{Distance from center} = \text{radius} \times (1 - \text{co-occurrence}). \tag{2}$$

On the other hand, the word's concept is inferred by the search result whose search query is the word. Therefore, the conceptual similarities of the words are calculated based on the search results whose search query is the target words.

Assuming search query C and the word in search result $W = \{W_1, W_2, \ldots, W_n\}$, their concept vectors are V_C, and V_{W_n}. The elements of the concept vectors are created by the set of nouns that appear more than twice in the title of each search result. Their values are set by the number of their appearances. For example, here we assume that the nouns included in the title of the C search result are (a, a, b, b, c, c, c, c), and the nouns included in the W_1 search result are (a, a, c, c, d, d, e). In this example, V_C and V_{W_1} elements are (a, b, c, d). Using the number of occurrences of each noun, V_C is set to (2, 2, 4, 0) and V_{W_1} is (2, 0, 2, 2).

The position of W_n is an angle that corresponds to the cosine similarity between V_c and V_{W_n}. Equation 3 calculates the cosine similarity of two vectors. The angle from the right radius of W_n corresponds to the angle of the cosine similarity between V_c and V_{W_n}. However, since V_c and V_{W_n} take only positive values, the cosine similarity takes a value of 0 to 1. To derive the angle from 0 to 180°, the angle derived by the cosine similarity is doubled:

$$\cos\left(\overrightarrow{V_C}, \overrightarrow{V_{W_n}}\right) = \frac{\overrightarrow{V_C} \cdot \overrightarrow{V_{W_n}}}{\left|\overrightarrow{V_C}\right|\left|\overrightarrow{V_{W_n}}\right|}. \tag{3}$$

For example, if $V_c = (2, 2, 4, 0)$ and $V_{w1} = (2, 0, 2, 2)$, the cosine similarity is 0.707106781, and the angle corresponding to this cosine similarity is 45°. The angle is doubled, and W_1 is placed 90° from the right radius. If W_1 is included in five of the top ten titles, the co-occurrence is 0.5. When the radius is 10, the distance from the center of W_1 is 5 (Fig. 8).

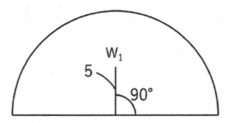

Fig. 8. Example of word placement

Since the usual results of search engines do not represent the state of the topic of the search results, we cannot recognize how the state of the topic of the search results has changed even if we compare the search results. Therefore, the user cannot evaluate the modification of the search query. On the other hand, in the visualization of the proposed system, the user can evaluate the effect of modifying the search query. For example, if the desired word becomes closer to the center by modifying a search query, she evaluates that her modification was good. If the undesired word becomes closer to the center, she evaluates that her modification was inappropriate.

5 Prototype System

Figure 9 shows the interface of our prototype system that was developed using HTML, CSS, Python, Node.js, and JavaScript. When a user inputs a search query in the text box on Screen 1 and pushes the search button, a list of search results is displayed on Screen 1 and the nouns included in their titles are displayed on Screen 2. In Screen 2, the user selects the radio button for the necessary words in red, unnecessary words in blue, and unknown words in gray, and clicks the visualize button to display the distribution of the words shown on Screen 3. Screen 4 checks the effect of changing the search query. The user enters the modified search query in the text box and push the search button. Then the distribution of the words appears for the modified search query. At this time, the words on Screen 4 are displayed in the same color as Screen 3, and new words are depicted in green.

In this system, Custom Search Engine displays the Google search results. Custom Search JSON API obtains the top ten titles in them, and mecab-ipadic-NEologd [7] is used as a morphological analyzer to extract the nouns.

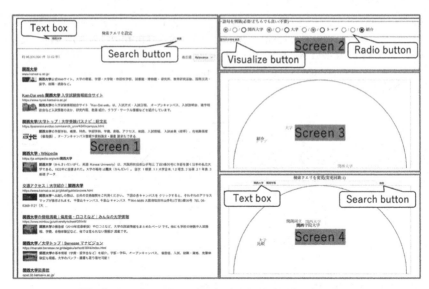

Fig. 9. Interface of prototype system (Color figure online)

6 Evaluation Experiment

6.1 Experimental Setting

We experimentally evaluated the validity of the visualization manner (*Experiment 1*) and the effectiveness of the support system for cultivating search-query-setting skills (*Experiment 2*). Our participants were 11 undergraduate students of our university.

Before the experiment, we checked their search-query-setting skills (*pre-task*). They were given target information to acquire and a current search query, which they were asked to change to get the target information. Table 2 shows the target information and the current search query given to them and the words in desirable search query. We identified the participants (students) who set all the necessary words as a search query as having search-query-setting skills and deemed them to be outside of our system's target users because they do not have to change the search query. From this task, we labeled all the students as having search-query-setting skills except four (*A* to *D*).

Experiment 1. Experiment 1 evaluates whether our proposed visualization manner can communicate the similarities among words to users. Our participants were given the titles of the search results for a search query, "Fukushima Izakaya" (a Japanese bar), the words in the titles, and their visualization with all words in black. Table 3 shows the titles and the words, and Fig. 10 shows the visualization for the search query: "Fukushima Izakaya." Since "Fukushima" and "Tabelog" appeared twice, they are situated closer to the center. "Osaka-shi," "Osaka," "Noda," "Fukushima-ku," and "Fukushima" are words related to Fukushima, and so they are also situated close.

Next, they were given the titles and the words of another search query, "Earphones Bluetooth," and asked to draw the distribution of the words in a semicircle based on

Table 2. Given information as pre-task

Task ID	Target information	Current search query	Words in desirable search query
1	Third and fifth highest mountains in Japan	Japan mountains	"Japan," "mountain," "rank," and "high"
2	Popular games currently available	Popularity latest game	"popularity," "game," and "available"
3	Name of back part of a tail plane	Airplane tail plane	"airplane," "tail plane," and "back"

Table 3. Title and words of search query "Fukushima Izakaya"

(a) Titles of search results	
1	Fukushima/Noda popularity ranking of izakaya [Tabelog]
2	Izakaya of recommendation in Fukushima-ku, Osaka-shi [Tabelog]
3	Around Fukushima (Osaka) Reservation of izakaya, Hot pepper gourmet

(b) Words in titles	
1	Fukushima, Noda, Izakaya, Popularity, Ranking, Tabelog
2	Osaka-shi, Fukushima-ku, Recommendation, Izakaya, Tabelog
3	Fukushima, Osaka, Around, Izakaya, Reservation, Hot pepper, Gourmet

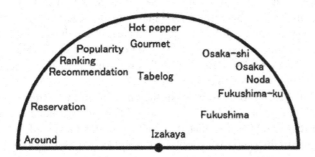

Fig. 10. Visualization for "Fukushima Izakaya"

what they learned from the visualization. Table 4 shows the titles and the words given to them. Figure 11 is the ideal visualization. Since "Bluetooth," "Selections," and "Recommendation" appeared twice, they are situated nearer the center. If the students correctly understood the meaning of the visualization, they drew the distribution of the words that resembled the ideal visualization.

Table 4. Title and words of search query "Earphones Bluetooth"

(a) Titles of search results	
1	Recommendation 10 selections of Bluetooth earphones
2	Bluetooth earphones popular selling ranking
3	Recommendation 20 selections of completeness wireless earphones
(b) Words in titles	
1	Bluetooth, Earphones, Recommendation, 10, Selections
2	Bluetooth, Earphones, Popularity, Selling, Ranking
3	Completeness, Wireless, Earphones, Recommendation, 20, Selections

Fig. 11. Visualization for "Earphones Bluetooth"

Figure 12 shows the distribution of the words that were drawn by the students, all of whom placed the search query "Earphones Bluetooth" in the center. Regarding the distance from the center, for "Recommendation" and "selections," no student put both words closer to the center. This result shows that they understood that the words related to the search query should be put nearer to the center, although they did not know how to derive the similarity. Concerning the similarity between words, since everyone placed "Ranking" and "Popularity" or "10" and "20" close to each other, they understood that similar words should be close together.

They were also asked to describe what they read from the visualization and drew the word distributions. Table 5 shows the results. Every student noticed that the search query and their related topics are situated around the center. Only student C commented about the meaning of the nearer words: "I think the place words have a similar meaning, which is why they are close to each other."

Experiment 2. In Experiment 2, we evaluated whether the students could change a search query to a more effective one based on experience with our system. The system's effectiveness was evaluated by the change of the pre-task answers after using it.

Students were given pictures and asked to identify objects (Fig. 13(a)) and places (Fig. 13(b)) using our system.

After using the system, they were allowed to change the search query from the pre-task. If they did change it, they explained why in the post-task.

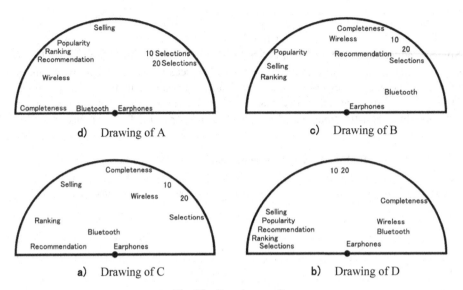

d) Drawing of A

c) Drawing of B

a) Drawing of C

b) Drawing of D

Fig. 12. Drawing results

Table 5. What participants read from visualization

Student	Answers
A	Words in search query are placed in center Before and after words in title are placed nearer to it
B	Words are placed based on number of appearances Before and after words in title were also placed nearer
C	Words related to search query are placed in the center Words that appear less often are on the edge of the circle Words with similar meanings are placed close together
D	Search query is placed in the center Words are placed based on word derivation relationships

(a)   (b)

Fig. 13. Search task

Tables 6 summarizes the results of the pre- and post-tasks. Students changed the search queries more than once, but not for all tasks. All the changes were effective. Students added desirable words or deleted undesirable words. For example, student A changed "ranking" to "highest ranking" for task 1 because he got a different ranking other than "height" at the pre-task and wanted the height ranking in the post-task. Student D deleted "construction" and added "part" for task 3 because he thought "part" was more common than "construction." He also added "name" for task 3 for easily acquiring it.

Table 6. Result of changing search queries in post-tasks

Student	Task	Changed words	Added desirable words	Deleted undesirable words
A	1	Y	Y	–
	2	N	–	–
	3	N	–	–
B	1	N	–	–
	2	N	–	–
	3	N	–	–
C	1	Y	Y	–
	2	N	–	–
	3	N	–	–
D	1	N	–	–
	2	Y	N	Y
	3	Y	N	Y

These results suggest that our system is less effective when often changing the types of search queries. However, with it, the students can change the search query to a more appropriate one at least once. Therefore, our system might encourage search-query-setting skills. We need further experiments to verify our result.

7 Conclusion

We proposed a system for cultivating search-query-setting skills by comparing the states of search results before and after changing queries. To understand a search result's state, our system has a function that shows a word that represents the topics of a search result and visualizes its state based on their similarities. In addition, to understand the effect of changing a search query, an interface is provided where the visualization of the states of the search results before and after changing the search query can be compared. Using this system, our participants repeatedly changed the search query to a more effective one. We need further experiment to verify the result.

Since this experiment's participants were university students who routinely search using their smartphones, they have probably already acquired search-query-setting skills to some extent. Therefore, we must experiment with different groups who search much less often, for example, senior citizens.

References

1. Nguyen, T.N., Zhang, J.: A novel visualization model for web search results. IEEE Trans. Vis. Comput. Graph. **12**(5), 981–988 (2006)
2. Yoshida, T., Oyama, S., Nakamura, S., Tanaka, K.: Query transformation by visualizing and utilizing information about what users are or are not searching. In: Proceedings of the 11th International Conference on Asia-Pacific Digital Libraries (ICADL 2008), vol. 5362, pp. 124–133 (2008)
3. Ma, C., Zhang, B.: A new query recommendation method supporting exploratory search based on search goal shift graphs. IEEE Trans. Knowl. Data Eng. **30**(11), 2024–2036 (2018)
4. Saito, H., Miwa, K.: Construction of a learning environment supporting learners' reflection: a case of information seeking on the Web. Comput. Educ. **49**(2), 214–229 (2007)
5. Nishiyama, T., Suwa, M.: Visualization of posture changes for encouraging meta-cognitive exploration of sports skill. Int. J. Comput. Sci. Sport **9**(3), 42–52 (2010)
6. Broder, A.: A taxonomy of Web search. ACM SIGIR Forum **36**(2), 3–10 (2002)
7. Satou, T., Hashimoto, T., Okumura, M.: Implementation of a word segmentation dictionary called mecab-ipadic-NEologd and study on how to use it effectively for information retrieval. In: Proceedings of the Twenty-third Annual Meeting of the Association for Natural Language Processing, pp. 875–878 (2017). (in Japanese)

A Support Interface for Remembering Events in Novels by Visualizing Time-Series Information of Characters and Their Existing Places

Yoko Nishihara[1(✉)], Jiaxiu Ma[1], and Ryosuke Yamanishi[2]

[1] College of Information Science and Engineering, Ritsumeikan University, Shiga 5258577, Japan
nisihara@fc.ritsumei.ac.jp
[2] Faculty of Informatics, Kansai University, Osaka 5691095, Japan
ryama@kansai-u.ac.jp

Abstract. The style of reading books has shifted. People use electronic devices to read e-books. People often read books in a short time because it is difficult to have enough time for reading books at one time. If there is a time gap between readings, they may forget what has happened in the already read parts of books. Especially if people read novels where many characters appear, it is more difficult to remember what has happened for each character. Reading several stories in parallel should increase the number of rereads. The more times they reread books, the more difficult they enjoy reading. This paper proposes a support interface for remembering events in novels by visualizing time-series information of characters and their existing places. The proposed interface consists of two parts: (a) displaying of a novel text and (b) visualizing of time-series information of character names and their existing places. The proposed interface uses a historical chart as a metaphor for the visualization. If a novel text is given, the proposed interface extracts character names and places with the Conditional Random Fields (CRFs). The proposed interface sets the horizontal axis as the sentence's number and the vertical axis as character names. The existing places are mapped as labelled dots on the plane given by the two axes. We experimented the proposed interface with participants and verified that the proposed interface could support them to remember events in novels.

Keywords: Support for resuming reading · Remembering events · Visualization like a historical chart

1 Introduction

The style of reading books has shifted. People use electronic devices to read e-books. They use smartphones, tablets and special devices like Amazon Kindle. The weight of the e-books is not heavy while the weight of the print books increases in proportion to the number. People can bring several e-books in their

S. Yamamoto and H. Mori (Eds.): HCII 2021, LNCS 12765, pp. 76–87, 2021.
https://doi.org/10.1007/978-3-030-78321-1_7

devices without caring the weight. It is expected that the style of reading several e-books in parallel will spread widely.

People often read books in a short time because it is difficult to have enough time for reading books at one time. If there is a time gap between readings, they may forget what has happened in the already read parts of each book. Especially if people read novels where many characters appear, it is more difficult to remember what has happened for each character. Reading several stories in parallel often make people forget events in the already read parts, which may increase the number of rereads. The more times they reread books, the more difficult they enjoy reading.

A novel is consist of events related to characters in time-series. An event is information about when, where and what the character did, what the character thought, and so on. If the events are visualized, people can grasp the outline of the already read parts of a novel at a glance. The visualization may remove rereads and help people in resuming reading novels.

This paper proposes a support interface for remembering events in novels by visualizing time-series information of characters and their existing places. The target users of the proposed interface are people who read several novels in parallel. We believe that the visualized information can support the users to remember when and where each character did something and what he/she thought in the already read parts. We hope to support the users to resume reading novels smoothly. We design our original quizzes about events in novels to evaluate the effect of the proposed interface. The quizzes can be answered if people remember events in the already read parts.

We focus on a historical chart as a metaphor which makes people remember the outline of events. Visualization methods often use the metaphors to make the users understand easily [5,6]. The benefit of the historical chart is to show the conditions of several objects at the same time. We utilize the historical chart as a metaphor to visualize the events in the already read parts of novels. The proposed interface sets the horizontal axis as the sentence's number and the vertical axis as characters. The existing places are mapped on the plane given by the two axes. The users watch the visualization on the interface in the same way of watching a historical chart. They can remember events in novels easily with the proposed interface.

2 Related Work

2.1 Remembering Support System

The proposed interface is one of the remembering support systems. Most of them support for remembering schedules and plans in the near future and memories in the past [3,8]. The proposed interface will support for remembering events in novels that are one of the memories in the nearer past.

The systems to support for remembering in the nearer past have been studied in the field of learning support systems. The learning support systems estimate the degree of learner's memory. If the degree drops down to a threshold, the systems prompt the learner to remember again [2]. The learner can retain what he/she has remembered. However, in reading books, readers do not need to retain contents in the already read parts. They need to remember them just before resuming their readings. It is redundant to estimate the degrees of their memories and prompt to remember again. Therefore, we design the proposed interface to be used just before resuming readings.

2.2   Support for Reading Long Novels

To support reading long novels, methods on visualizing characters features and relations among characters have been studied [1,4]. In reading novels, if users choose a character name on the visualization interface, the interface shows the scenes of the character. If users forget the events about the character, the users reread the scenes to remember. The users can continue to read novels without wasting much time for rereading. However, the rereading wastes time certainly. Therefore, the proposed interface will support users to resume readings in less time than rereading.

We assume that a novel story is driven by events arranged continuously. An event can be represented as the information of "when," "where," "who," "what," and "how." The proposed interface extracts "where" and "who," and visualizes them in the order of sentences. The order of sentences corresponds to "when." The information is visualized like a historical chart, that is our original point.

3   Proposed Interface

We describe the outline of the proposed interface. The proposed interface consists of two parts: (a) displaying of novel texts and (b) visualizing of time-series information of character names and their existing places. The proposed interface works on a Web browser and uses D3.js as a JavaScript library to visualize the information.

Figure 1 shows the part (a) of the proposed interface that displays a novel text. If a user selects a novel from the left-side menu, the proposed interface shows the novel texts in the right-side window.

Figure 2 shows the part (b) of the proposed interface, which visualizes time-series information of character names and their existing places. The proposed interface shows the part (b) if a user clicks a hyperlink of a novel title in the part (a). The proposed interface sets the horizontal axis as the sentence's number and the vertical axis as character names. The existing places are mapped as labelled dots on the plane given by the two axes.

Figure 3 shows a flowchart of visualization. The proposed interface is given a novel text. A processing module in the proposed interface extracts sentences from the text. It extracts phrases for character names and places from each sentence. If a sentence does not have both the character names and places, the existing places of each character and the ranges are estimated based on the extracted information. Then, the triplet data (a sentence number, an appeared character name and its existing place) is made. The proposed interface uses the set of the

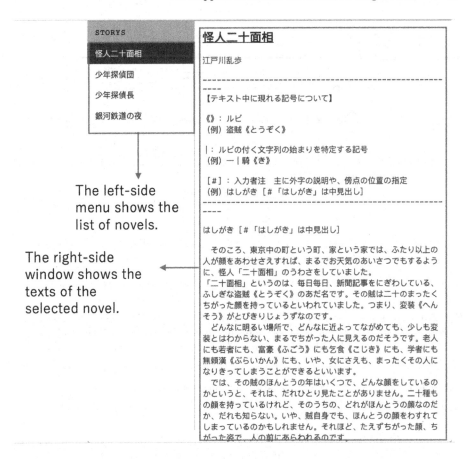

Fig. 1. Part (a) of the proposed interface. It displays novel titles (on the left-sided) and a novel text selected by a user (on the right-sided).

triplet data for the visualization of time-series information. If a user would like to resume reading, he/she uses the proposed interface and can remember the events in the already read parts.

3.1 Extraction of Phrases for Character Names and Places

The processing module of the proposed interface extracts phrases for character names and places. The module uses the Conditional Random Fields (CRFs) that is one of the Named-entity recognition methods.

The module divides a text into sentences. Then, the module extracts phrases and obtains information of their part of speeches by a morphological analyzer. We use MeCab[1] as the analyzer and NEologd[2] as a dictionary.

---

[1] https://taku910.github.io/mecab/.

[2] https://github.com/neologd/mecab-ipadic-neologd.

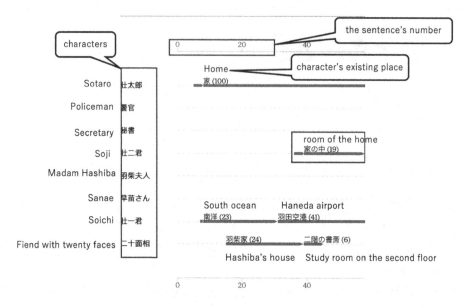

Fig. 2. Part (b) of the proposed interface. It visualizes time-series information of character names and their existing places. Notes in balloons explain the visualized information. The figure is based on Japanese story. The authors gave translations for character names and their existing places.

Table 1. BIO2 tags for giving annotations on phrases for character names and places.

Tag	Definition
B-CHAR	The beginning of a phrase for a character name
I-CHAR	The intermediate of a phrase for a character name
B-POS	The beginning of a phrase for a place
I-POS	The intermediate of a phrase for a place
O	Others

Next, an annotator (the 2nd author) gives tags for character names and places on phrases in the sentences. We use BIO2 tags for the annotation. The BIO2 tags are used for named-entity recognition, which gives a B-tag to the first alphabet of a named-entity, an I-tag to another alphabet except for the first, and O-tag to non-named-entity. We use five types of tags shown in Table 1 for annotation. After annotations, a data set for training of the CRFs will be obtained. The character names (the places) are limited to phrases that explicitly identify the characters (the places), which means pronouns should be omitted.

If the CRFs learns the feature of i_{th} phrase, it uses each three phrases located in front and back. The information phrases, their part of speeches and BIO2 tags for seven phrases are used for machine learning. The processing module obtains a named-entity extraction machine by the CRFs to extract phrases for character names and places.

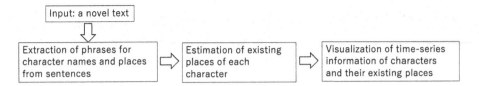

Fig. 3. Flowchart of visualization time-series information about events in the already read parts.

Table 2. Experimental results about relations between character names and places in three novels. Each number denotes the number of sentences which satisfied each condition.

Novel	Only character name	Only place	Both	Neither	Total
Momo Taro	34	12	12	24	126
Little Red Riding Hood	53	17	32	24	82
Night on the Galactic Railroad	251	44	55	274	624
Total	338	73	99	322	832

3.2   Estimation of Existing Places for Each Character

The processing module estimates existing places for each character. We assume if one sentence includes phrases for both a character name and a place, the character name should be related to the place, which means the character may exist in the place. We conducted preliminary experiments to show the validity of the assumption. We used three novels "Momo Taro," "Little Red Riding Hood," and "Night on the Galactic Railroad" in the experiments.

Table 2 shows the experimental results. There were 338 sentences which included only phrases for character names while there were 73 sentences which included only phrases for places. 99 sentences included both of them, while 322 sentences did not include either. We investigated for the 99 sentences whether or not phrases for character names and places were related to each other. As a result, we found that 71% sentences satisfied the assumption. That means if one sentence includes both character names and places, the character existed in the place. The experimental results showed the validity of the assumption. Therefore, if one sentence includes both a character name and a place, the processing module estimates that the character exists in the place in the sentence.

The characters in novels would stay in the same place for a certain period. They rarely move places every sentence. If a sentence $s(j-1)$ include only a phrase for a character name A, his/her existing place is estimated by the sentence $s(i)$ which includes both phrases for the character name A and a place B $(i < j)$. Figure 4 shows the outline of the estimation. The module estimates the character A continues to stay in the same place B between the i_{th} and the $j-1_{th}$ sentences [3]. According to the estimation, the processing module obtains

---

[3] The assumption must be investigated experimentally. That will be our future work.

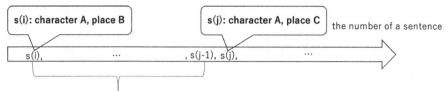

The processing module estimates that character A exists in place B between s(i) to s(j-1).

Fig. 4. Estimation of a character's existing place.

a set of triplet data (a sentence number, an appeared character name and its existing place) for visualization.

3.3 Visualization of Time-Series Information of Characters and Their Existing Places

The proposed interface uses the set of triplet data to visualize the time-series information of characters and their existing places. Figure 2 shows an example of the visualization. The visualization is based on a historical chart as a visualization metaphor. The visualization makes people find where each character existed in the read parts. They use the information about characters and places as clues and may remember what each character did in each place. The effects of the visualization will be evaluated in the next section of evaluation experiments.

4   Evaluation Experiments

We evaluated the proposed interface. We conducted experiments with participants who read several novels in parallel. We evaluated whether or not the proposed interface could support the participants to remember events in the already read parts of novels.

4.1 Experimental Procedures

We conducted the experiments as follows.

1. The experimenter (the 2nd author) divides participants into three groups named Group A, B, and C.
2. Each participant reads four novels in a day with an assigned interface. He/she reads a part of the novel. The experimenter gives the beginning and the ending points to read.
3. After reading the novels, each participant waits for one week. After one week, he/she answers quizzes about events in the novels.
4. The experimenter compares the correct rates of the quizzes among groups and evaluates the proposed interface.

Table 3. Used novels in the experiments. # denotes the numbers.

Title	Author	# of Japanese alphabets to read
Boy Detectives Club	Ranpo Edogawa	9,979
Shonen Tantei Cho	Jyuza Unno	6,101
Fiend with Twenty Faces	Ranpo Edogawa	8,907
Night on the Galactic Railroad	Kenji Miyazawa	8,903

Participants. The participants were university students in their 20s. There were 21 students (nine Japanese and 12 Chinese). The Chinese students had JLPT N1. They could read Japanese novels as well as Japanese students. Each group had three Japanese and four Chinese.

Used Interface for Reading Novels. The experimenter asked the participants to use our proposed interface to read the novels. They used the part (a) of the proposed interface shown in Fig. 1.

Used Novels. We chose novels from "AOZORA BUNKO" that was a Web site providing with book texts with expired copyrights [4]. People can read short novels at one time. Therefore, we chose long novels that required several times to stop reading. Table 3 shows information about the novels. We used four novels "Boy Detectives Club," "Shonen Tantei Cho," "Fiend with Twenty Faces," and "Night on the Galactic Railroad." We suppose that the target users of the proposed interface would read novels in a short period. The period was set to be 20 min for one reading. Since the average reading speed was 400 Japanese alphabets every minute, we set the ending point of each novel which the participants would read 8,000 Japanese alphabets for a novel (= 20 min × 400 Japanese alphabets). The participants read 33,890 Japanese alphabets of the four novels in total.

Quizzes About Events in the Novels. We designed quizzes about events in the novels. If the proposed interface supported the participants, they could remember the events in the already read parts. They would mark higher scores in answering quizzes.

Table 4 shows the example of the quizzes. All quizzes were related to the events in the novels. For example, a quiz "what did Giovanni buy before leaving?" was related to the events of shopping by a character in the novel "Night on the Galactic Railroad." There were quizzes about *a character, about characters, about a place,* and *about places.* For example, "who was in place B with character A?" was a quiz about characters. "Where did character A move from place B?" was an example of quizzes about places. The quizzes about characters and places were difficult to answer because the participants needed to remember more information about events. There were five quizzes for one novel, 20 quizzes

---

[4] https://www.aozora.gr.jp/.

Table 4. Examples of quizzes used in the evaluation experiments. The numbers in parentheses are the number of quizzes answered by choosing options.

Quiz type	Number of quizzes	Example of quizzes (Novel title)
A character	9(5)	What did Giovanni buy before leaving? (Night on the Galactic Railroad)
Characters	3(1)	Who did clime up Kan-nuki mountain with Haruki? (Shonen Tantei Cho)
A place	5(0)	Where was Hashiba's jewel? (Fiend with Twenty Faces)
Places	3(2)	Where did Kei run to bright town from? (Boy Detectives Club)

in total. The participants answered 12 of 20 quizzes by writing words and sentences while they answered the rest by choosing options. They had five minutes to answer 20 quizzes.

In answering quizzes, each group had different conditions in remembering the events. The experimenter asked the participants in Group A not to reread novels or use the proposed interface. They answered the quizzes without any references. The experimenter asked those in Group B to reread novels in three minutes with the part (a) of the proposed interface before answering the quizzes. The experimenter asked those in Group C to use the part (b) of the proposed interface in three minutes to remember the events in the novels before answering the quizzes.

4.2 Evaluation Method

We used the correct rates of quizzes to evaluate the proposed interface. By comparing the correct rates between Group A and B, we can find the effect of rereading for remembering events in the already read parts of a novel. Then, by comparing those between Group B and C, we can find the effect of visualizing the time-series information of character names and places for remembering events.

4.3 Experimental Results

Table 5 shows the correct rates of the quizzes. The correct rates were average in each group. The correct rate of Group C that used the part (b) of the proposed interface was the highest among them.

Table 6 shows the correct rates of each quiz type. Those correct rates of Group C for all quiz types were also the highest among them.

5   Discussion

Table 5 shows the difference between Group A and B was +10%. The difference was positive. Therefore, rereading novels could support the participants to

Table 5. Correct rates of the quizzes for each group. The values in parentheses are the differences between Group A and B, and between Group A and C.

Group	Correct rates
A	26%
B	36%(+10%)
C	**54%(+28%)**

Table 6. Correct rates of each quiz type. The values in parentheses are the differences between Group A and B, and between Group A and C.

Group	A character	Characters	A place	Places
A	40%	14%	6%	33%
B	38% (−2%)	43% (+29%)	20% (+14%)	52% (+19%)
C	**57%** (+17%)	**62%** (+48%)	**40%** (+34%)	**62%** (+29%)
Average	45%	40%	22%	49%

remember the events in the novels. The difference between Group A and C was +28%. The difference was also positive. Therefore, visualization of time-series information also could support the participants to remember the events. The difference between Group A and C was bigger than that between Group A and B. The result indicates that visualization of time-series information could support more for remembering the events than rereading.

Then, we discuss the correct rates of the quizzes about a character. The correct rate of characters was lower than that of a character. We guessed there were two reasons. One reason was the complexity of events about characters. In general, an event about a character is less complex than that about characters because people need to remember more items about characters. Rereading and visualization could support the participants to grasp the relations of characters in each event. Therefore, the correct rates of Group B and C were higher than that of Group A. Another reason was the rates of quizzes to answer by choosing options. There were five of nine quizzes about a character answered by choosing options while there were one of three quizzes about characters. The rate of quizzes answered by choosing options about a character was bigger. In general, a quiz answered by choosing options is easier than that answered by writing words and sentences. Therefore, the correct rate about characters of Group A was lower relatively, and the differences were bigger.

Next, we discuss the correct rates of quizzes about a place and places. The correct rate about a place was lower than that about places. We guessed the reason was the rate of quizzes answered by choosing options. The quizzes about a place were all descriptive type, which means the participants must write words and sentences to answer them. The correct rate of all descriptive type quizzes was 40% in Group C. The result might indicate that the participants with the part (b) of the proposed interface could remember 40% events at the highest. In future work, we improve the quizzes to evaluate the proposed interface more precisely. Our research group is studying methods to generate quizzes automatically for evaluating people's understood levels [7]. By using the quizzes, we may be able to estimate which parts people often forget. We would improve the proposed interface to show only the parts users forget by using the quiz generation method.

6   Conclusion and Future Work

This paper proposed a support interface for remembering events in novels by visualizing time-series information of characters and their existing places. The proposed interface consists of two parts: (a) displaying of novel texts and (b) visualizing of time-series information of character names and their existing places. The processing module of the proposed interface extracts phrases for character names and places from each sentence in a novel text. The processing module makes triplet data (a sentence number, an appeared character name and its existing place). The set of the triplet data is used for the visualization. We use a historical chart as a metaphor for the visualization, which is our original point. The proposed interface sets the horizontal axis as the sentence's number and the vertical axis as characters. The existing places are mapped as labelled dots on the plane given by the two axes. The proposed interface can support users for remembering events in the already read part for resuming the readings smoothly.

We experimented the proposed interface with participants. The experimenter asked the participants to read four novels and answer quizzes about events in the novels after one week. The experimenter divided the participants into three groups: Group A, B, and C. Each group had different conditions in answering the quizzes. The participants in Group A used nothing to remember events. Those in Group B reread novels in three minutes in total. Those in Group C used part (b) of the proposed interface to remember events. We compared the correct rates of answering the quizzes among the groups and found that the correct rate of Group C was the highest. The results indicated that the proposed interface could support the participants to remember the events.

As the future work, we improve the quizzes to evaluate the proposed interface and find the limitation of the proposed interface. We would like to examine the proposed interface with many participants. We will refer more literatures of remember support system to design experiments for testing hypothesis.

Acknowledgement. This work was partly supported by JSPS grant (20K12130). We show our great appreciates.

References

1. Agarwal, A., Kotalwar, A., Rambow, O.: Automatic extraction of social networks from literary text: a case study on alice in wonderland. In: Proceedings of the Sixth International Joint Conference on Natural Language Processing, pp. 1202–1208. Asian Federation of Natural Language Processing, Nagoya, Japan (2013). https://www.aclweb.org/anthology/I13-1171
2. Asai, H., Yamana, H.: Detecting learner's to-be-forgotten items using online hand-written data. In: Proceedings of the 15th New Zealand Conference on Human-Computer Interaction, p. 1720. CHINZ 2015. Association for Computing Machinery, New York, NY, USA (2015). https://doi.org/10.1145/2808047.2808049
3. Brewer, R.N., Morris, M.R., Lindley, S.E.: How to remember what to remember: Exploring possibilities for digital reminder systems. Proc. ACM Interact. Mob. Wearable Ubiquitous Technol. **1**(3), 1–20 (2017). https://doi.org/10.1145/3130903
4. Elson, D., Dames, N., McKeown, K.: Extracting social networks from literary fiction. In: Proceedings of the 48th Annual Meeting of the Association for Computational Linguistics, pp. 138–147. Association for Computational Linguistics, Uppsala, Sweden, July 2010. https://www.aclweb.org/anthology/P10-1015
5. Li, Y.N., Li, D.J., Zhang, K.: Metaphoric transfer effect in information visualization using glyphs. In: Proceedings of the 8th International Symposium on Visual Information Communication and Interaction, p. 121130. VINCI 2015. Association for Computing Machinery, New York, NY, USA (2015). https://doi.org/10.1145/2801040.2801062
6. Pang, P.C.I., Biuk-Aghai, R.P., Yang, M.: What makes you think this is a map? suggestions for creating map-like visualisations. In: Proceedings of the 9th International Symposium on Visual Information Communication and Interaction, p. 7582. VINCI 2016. Association for Computing Machinery, New York, NY, USA (2016). https://doi.org/10.1145/2968220.2968239
7. Shan, J., Nishihara, Y., Maeda, A., Yamanishi, R.: Extraction of question-related sentences for reading comprehension tests via attention mechanism. In: Proceedings of the 2020 International Conference on Technologies and Applications of Artificial Intelligence, p. 2328 (2020)
8. Stawarz, K., Cox, A.L., Blandford, A.: Don't forget your pill! designing effective medication reminder apps that support users' daily routines. In: Proceedings of the SIGCHI Conference on Human Factors in Computing Systems, p. 22692278. CHI 2014. Association for Computing Machinery, New York, NY, USA (2014). https://doi.org/10.1145/2556288.2557079

Experimental Evaluation of Auditory Human Interface for Radiation Awareness Based on Different Acoustic Features

Dingming Xue$^{(\boxtimes)}$ (ID), Daisuke Shinma (ID), Yuki Harazono (ID), Hirotake Ishii (ID), and Hiroshi Shimoda (ID)

Graduate School of Energy Science, Kyoto University, Kyoto, Japan
{dingming,shimoda,harazono,hirotake,shimoda}@ei.energy.kyoto-u.ac.jp

Abstract. During the maintenance of a nuclear power plants (NPP), being aware of radioactive strength at one's position is an important issue to ensure workers' safety. For that purpose, a human interface for radiation awareness is useful. Auditory sensation is a reliable sensation widely used for different kinds of warning signals. Different acoustic features such as loudness, frequency and interval between beeps can give different acoustic stimuli which represent radioactive intensity information to the users. In this study, experiments were conducted to find out which acoustic feature perform the best in radiation awareness and avoidance of radiation in NPP field work. Three different kind of alarms were designed based on different acoustic features. In order to evaluate these alarms, a virtual experimental environment was developed where virtual radiation distribution was simulated. Participants were required to move around and do calculation tasks in the environment. Meanwhile, they were also required to avoid the high radioactive area and reduce their radiation exposure listening to the alarm as the human interface. Their traces were recorded and their total radiation exposure was calculated and compared among three kinds of alarms. The result showed that loudness was the best acoustic feature for radiation awareness and avoidance comparing with the two acoustic features. Besides, staying time in radioactive environment has strong influence on radiation exposure. Although human interface is effective in radiation awareness, reduction of staying time in radioactive environment is very important to keep workers' safety.

Keywords: Acoustic features · Radiation awareness · Virtual experimental environment

1 Introduction

Nowadays nuclear power plants are built in large scale for massive energy supply. In a nuclear power plant, maintenance work is carried out to keep the plant operating safely. It is important to keep workers from being exposed to long-term radiation dose during the daily maintenance work. According to ALARA

S. Yamamoto and H. Mori (Eds.): HCII 2021, LNCS 12765, pp. 88–100, 2021.
https://doi.org/10.1007/978-3-030-78321-1_8

Principle [1], which is 'as low as reasonably achievable', it is important to put effort in minimizing the harm of radiation as we can. Therefore, enhancing the protection such as radiation awareness system has vital significance. Currently different measurements are taken to protect workers safety in NPPs. The most commonly used is shields and covers on the radiation contaminated materials and equipment. Besides, workers should strictly comply with the safety protocols to reduce the radiation exposure. Personal dosimeter is a kind of human interface [2], used to detect the current dose rate at user's position in NPP. It uses an audible alarm to warn the operator that a specific dose is exceeded. However, little study has been conducted on which kind of alarm is more effective in radiation awareness. Human beings have different kinds of sensations and auditory sensation is one of the most widely used for perceiving dangerous signals [3]. When designing an alarm, different acoustic features can carry information in the change of loudness, frequency or even the interval between the beeps. The purpose of this study is therefore to figure out which acoustic feature is suitable for radiation awareness and avoidance by conducting an evaluation experiment.

2 Design of Alarm with Acoustic Features

2.1 Selection of Sensation

Multiple sensations can be used to perceive the physical stimulation, which carries radioactive intensity information. Table 1 demonstrates several physical stimuli and the corresponding types of common human body sensations. Among these sensations, the ones which are easy to be applied in NPPs for radiation awareness are vision, audition and touch. With the radiation intensity changes at user's position, the human interface should have precise changes correspondingly with no delay. It is hard to control the concentration of scent or taste to show these changes and the spread of the scent requires time. Therefore, the low feasibility of smell and taste makes them hard to be a suitable sensation for radiation awareness. Audition has many advantages. Equipped with auditory human interface, user can capture the changes in alarm with almost no time delay. Although NPP is an environment filled with obstacles such as pipes, tanks and steam, using auditory human interface does not cause any interference in his movements. Vision, audition and touch are three feasible options for human interface design and auditory human interface is focused at current stage.

Table 1. Human sensations.

Sensation category	Physical stimulus	Sensory organ	Feasibility in NPPs
Vision	Light	Eyes	Feasible
Hearing (audition)	Sound	Ears	Feasible
Smell	Chemical substance	Nose	Low feasibility
Taste	Chemical substance	Tongue	Low feasibility
Touching	Position, motion, Temperature	Skin	Feasible

2.2 Acoustic Features

Auditory human interfaces are expected to present radioactive information to the users, helping them to reduce unnecessary radiation exposure in NPPs when conducting various kind of tasks. Sound has different acoustic features which present information in different ways. The audio signal used in alarm can be explained by the amplitude, frequency and phase factors. There are infinite combinations of these features and therefore different sounds can be created. Different acoustic features can be used for radiation awareness in NPPs, for example, radioactive intensity can be described by the loudness of the alarm, which is the alarm's amplitude feature. When the user approaches to the source of the radiation, the loudness of the alarm increases and the user would notice that it is dangerous to move forward. Totally, loudness (amplitude feature), interval (amplitude feature) and frequency were used for radiation awareness human interface design in this study.

2.3 Auditory Alarm Design

Overall Design of Alarm. Several principles are followed when designing the alarms in this study. The principles are from the perspective of pulse rate, frequency, audibility.

Pulse rate is a frequency which represents how many times a pulse is repeated in one second. According to a research on warning tone selection for a reverse parking aid system [4], pulse rate between 2 Hz to 8 Hz was urgent to nearly all of the participants and the urgency increases with pulse rate.

Frequency describes the frequency of an alarm. Different frequency of an alarm is associated with the perception of the danger, therefore selection of frequency is of vital significance. ISO 7731 [5] and ISO 24500 [6] suggests that danger signal should include components from 500 Hz to 1500 Hz. In another experiment [4], alarms with different frequency (500 Hz, 666 Hz, 750 Hz, 1000 Hz, 1200 Hz, 1500 Hz) were tested in both quiet environment and noisy environment and gathered participants' opinions. 500–1000 Hz tones were mostly liked by the participants in quiet environment and the tone pitch was considered 'about right' by the most participants. However, the effectiveness of the alarm tone changed when it came to noisy environment. Effectiveness of 500 Hz–666 Hz dropped significantly and 1000–1200 Hz was considered as the most desirable frequency. 1500 Hz constantly ranked least desirable in both environments. Considering NPP is a very noisy working place which contains huge noise of heavy machinery, 1000 Hz and 1200 Hz would be the suitable choice for the alarm design.

Danger signal should be clearly audible. ISO 7731 [5] gives criteria which shall be met for the alarm design: The difference between the two A-weighted sound-pressure levels of the signal and the ambient noise shall be greater than 15 dB. However, in NPPs the ambient noise is around 59.5 dB–90.4 dB and the noise caused by maintenance work is around 79.1 dB–93.5 dB [7]. In a study of Israeli industrial workers who were exposed to industrial noise, the workers' blood pressure and heart rate were significantly higher in over 85 dB (A) environment than in under 85 dB (A) environment [8]. In order not to do much damage to the workers' health, the sound-pressure level should not be higher than the environment noise but similar to it.

Loudness Condition. Radioactive intensity information can be presented to the user by the change in alarm loudness. Therefore, loudness is a candidate for the experiment. For the loudness condition, we designed an alarm whose loudness changes as the radioactive intensity changes at the user's position. By recognizing the loudness change, user could effectively avoid the dangerous radioactive area in working environment.

Interval Condition. Alarms usually consist of on and off periods, which are known as 'beeps'. In the interval condition, the length of the interval between beeps represents the radioactive intensity at user's position. The period of the beep is set to 0.1 s, which is a strong signal pattern for caution [6]. The maximum interval between the beeps is 1.0 s, which is not urgent and disturbing. As it gets dangerous, the interval decreases to almost zero. The alarm itself changes to be a strong caution signal and would sound critically urgent, warning the user to leave the dangerous area as soon as possible.

Frequency Condition. Human can hear the sound ranging from 12 Hz to 20000 Hz and be sensitive to the frequency change below 5000 Hz. In the frequency condition, the alarm's frequency is from 3 Hz (lowest frequency that can be achieved by the programming) to 1500 Hz in the highest radioactive area. The maximum frequency is selected from the highest limitation of the recommended frequency by ISO standard. The change in frequency makes the alarm sounds a little weird but it is very distinct for the user to detect the change.

Mapping Between Acoustic Features and Radioactive Intensity. In this study, we used a virtual environment simulating the real working environment in NPP and we applied various radiation distributions for repeated experiments. When the participant moved in the virtual environment, the radioactive intensity changed in different positions. Therefore, the alarm will make the user feel the difference in radioactive intensity by changing the loudness, interval or frequency correspondingly. According to Steven's Power Law [9], the relationship between the increased intensity of acoustic feature's stimulation and the perceived magnitude is not linear. Therefore, after repeated modifications and experiencing in the preliminary experiments, the mapping between radiation intensity and the parameters for the acoustic features we applied was described as follows. In (1), x represents the radioactive intensity at participant's position and f(x) is the intensity of acoustic feature's stimulation.

$$f(x) = \frac{e^{5x} - 1}{e^5 - 1} \tag{1}$$

The reason for this mapping is to make it easier to distinguish between low radiation area and high radiation area. Therefore, users can identify the real dangerous area faster and make quick reactions to it.

3 Experiment

3.1 Experiment Outline

In this study, we tried to find out which acoustic feature had best performance in radiation awareness and avoidance by conducting an evaluation experiment. The experiment had been planned to be conducted in a real NPP with real workers. However, in order to establish a mature experiment system including experiment method, analysis method and alarm design, the experiment was supposed to be conducted repeatedly to narrow down the candidates. In the past ten months, nine preliminary experiments were conducted for this purpose. Besides, due to the impact of COVID-19, conducting experiments with less face contact was better. Therefore, virtual experiment environment was built to solve this problem. We simulated the working environment in the real NPPs by using the real scenes as wall paper and generating noisy ambient sound which was recorded in NPPs in advance. Experimental participants were instructed to move and do some tasks in the virtual environment to simulate working scenarios in NPPs. Virtual experimental environment contained virtual radiation distributions with different number of radiation sources and they should try to avoid radiation exposure under four conditions of auditory alarms which were: loudness, interval, frequency and no alarm. In the no alarm condition, no sound alarm was given to support participants to avoid the dangerous radioactive area. After repeated 6 different radiation distribution patterns for each condition, their radiation exposure data was calculated and compared as primary index to evaluate which acoustic feature is the best.

3.2 Experiment Period and Participants

The experiment was conducted from Dec. 24th, 2020–Jan. 10th, 2021. The participants were 12 university graduate students of Kyoto University, Japan. The participants were aging from 22–25 with no visual and auditory disability.

3.3 Experimental Procedure

Figure 1 shows the procedure of this experiment. The duration of this experiment was 1 h and 45 min in total. In the preparation and practice period, participants were firstly required to check if the experiment application which realized the virtual experimental environment could be successfully executed and they were also required to adjust the loudness of their personal computer to a comfortable level. After that, they conducted 4 practice experiments under 4 different conditions, respectively. The only difference between the practice experiment

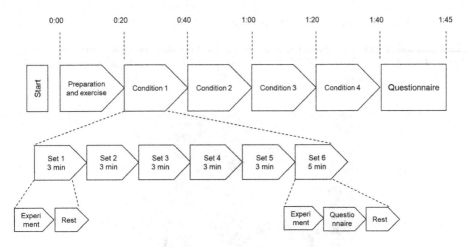

Fig. 1. Experimental procedure.

and the formal experiment was that there was a health bar in practice experiment, which represented participants' virtual exposure damage condition. The health bar decreased it's length as participants' radiation exposure increased. This aimed to let participants feel the severity of the radiation and understand how the alarm changes as the radioactive intensity changes. After the practice period, participants could be more familiar with the experiment method. There were four conditions in this experiment and participants experienced human interfaces based on the three different acoustic feature conditions and a no alarm condition. For each condition, experiments were conducted repeatedly for 6 times (set 1–6). In each set, the virtual experimental environment contains different distributions from other sets in the same condition to avoid the learning effect. Besides, the order of the conditions and the order of the distributions in each condition were randomized to eliminate the ordering effect. At the end of each condition, a questionnaire was give to the participants to acquire their impressions about the human interface. At the end of experiment, a final questionnaire was given to acquire participants' impression on the whole experiment and which human interface is most satisfied from their perspective. The questionnaire items are shown in Table 2.

Table 2. Final questionnaire.

No.	Questionnaire content
1	Which human interface do you think is the best?
2	Please state your reasons for question 1
3	Do you have any suggestions for improving the human interfaces?
4	Do you have any suggestions for improving the experiment method?

3.4 Virtual Radiation Distribution

The virtual radiation distributions used in the experiment were generated by P.H.I.T.S code [10], which is used for professional radioactive calculation. Figure 2 shows an example of a virtual radiation distribution.

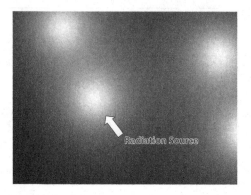

Fig. 2. Virtual radiation distribution.

3.5 Experiment Flow

The experiment flow is shown in Fig. 3. In this experiment, participants were required to do calculation tasks as a simulated task in work field in a virtual experimental environment where virtual radiation distribution existed. Under three experimental conditions alarm sound was given to the participants corresponding to the radiation intensity at their position while no alarm sound was give under no alarm sound condition. Figure 4 shows an example of the virtual experimental environment. When a participant started the set, a sector appeared in the experimental environment. Entering this sector, 2 calculation questions were displayed in the lower left of screen. Meanwhile, the sector will disappear and another sector will appear in a random position in the environment. Each calculation task has a time limit and will disappear when the time limit past. The set will be completed when participant answered 16 calculation tasks in total. They were required to finish the calculation tasks as accurately as possible and by using the human interfaces they should try to avoid radiation exposure as much as they can.

4 Experiment Result

4.1 Radiation Exposure

Average radiation exposure in each condition is shown in Fig. 5. The radiation exposure for loudness condition is 2872.48 (S.D. = 461.23), interval condition

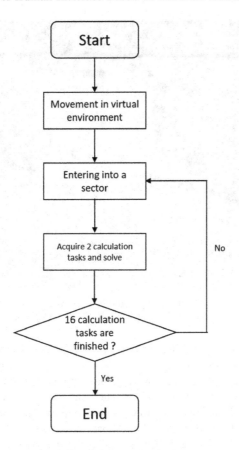

Fig. 3. Experiment flow.

is 2982.87 (S.D. = 547.34), frequency condition is 3108.32 (S.D. = 599.01) and no alarm condition is 2654.36 (S.D. = 469.92). Loudness is the best among three acoustic features in radiation awareness since participants have the lowest radiation exposure under loudness condition while frequency condition has the worst performance in radiation avoidance. This suggests that frequency feature may not be suitable option for radiation warning signals. Besides, participant's radiation exposure is the least without using any alarms. This is because of the limitation of virtual experimental environment. In the experiment application, the movement is not smooth and flexible as it is in

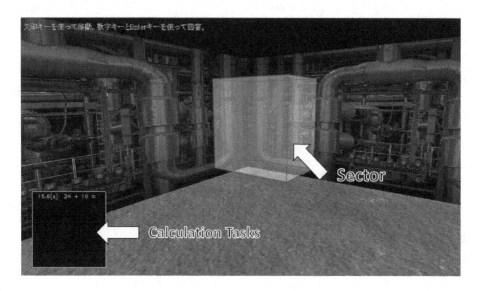

Fig. 4. Virtual experimental environment.

reality. Therefore, every time they tried to avoid the radioactive area, a certain amount of time would be wasted, which led to more radiation exposure. Under no alarm condition, however, participants did not have any intention to avoid the radioactive areas and they moved directly between sectors. The accumulation of the unnecessary radiation exposure caused the circumstance that they performed better than those under other conditions. One way repeated ANOVA analysis was conducted to verify whether there is a significant difference in radiation exposure between the experimental conditions. The result of the ANOVa analysis is shown in Table 3. The data of four conditions is normally distributed and passed the homogeneity of variance test. However, the p value of all four conditions is larger than 0.05, showing there is no significant difference between each conditions. With more participants involved, significant difference may appear.

Table 3. ANOVA analysis between experimental conditions.

Source	SS	df	MS	F	Prob > F
Groups	1.3358e+06	3	444933.1	1.63	0.1961
Error	1.20114e+07	44	272987.2		
Total	1.33462e+07	47			

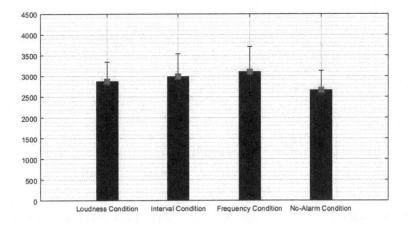

Fig. 5. Average of radiation exposure under experimental conditions.

4.2 Time Consumption

The average time consumption under each condition is shown in Fig. 6. The average time consumption for loudness condition is 96.00 s (S.D. = 19.69 s), interval condition is 97.63 s (S.D. = 18.51 s), frequency condition is 111.78 s (S.D. = 26.84 s) and no alarm condition is 81.50 s (S.D. = 16.65 s). Among the four conditions, frequency condition has the longest time consumption and no alarm condition consumed the shortest time. A correlation test was conducted to check the relationship between radiation exposure and time consumption using participants' data under all conditions and the result is as shown in Fig. 7. The huge R and F-statistic value indicates that there is a significant correlation relation between radiation exposure and time consumption. Therefore it can be concluded that even when using an alarm to avoid radiation exposure, it

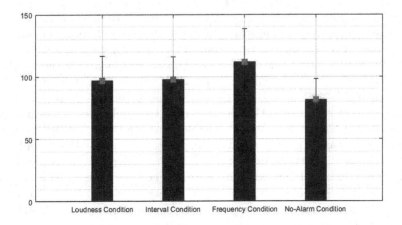

Fig. 6. Average of time consumption under experimental conditions.

```
Dependent Variable: DOSE
Method: Least Squares
Date: 02/03/21  Time: 01:50
Sample: 1 48
Included observations: 48
DOSE=C(1)+C(2)*TIME
```

	Coefficient	Std. Error	t-Statistic	Prob.
C(1)	1041.382	194.0452	5.366699	0.0000
C(2)	0.019221	0.001950	9.858689	0.0000

R-squared	0.678757	Mean dependent var	2904.435
Adjusted R-squared	0.671773	S.D. dependent var	532.8813
S.E. of regression	305.2934	Akaike info criterion	14.32120
Sum squared resid	4287386.	Schwarz criterion	14.39916
Log likelihood	-341.7087	Hannan-Quinn criter.	14.35066
F-statistic	97.19376	Durbin-Watson stat	2.132181
Prob(F-statistic)	0.000000		

Fig. 7. Regression result between radiation exposure and time consumption under all conditions.

is important to focus on both radiation intensity awareness and avoidance as well as staying time in the radiation exposed area. Besides, without any extra movement such as avoiding radioactive areas, they consumed least time in no alarm condition. Thus, no alarm condition has better performance than other conditions. It took significantly more time for them to complete the experiment using the frequency alarm than the two other conditions. This is the root cause that under frequency condition they have the highest radiation exposure.

4.3   Trace Analysis

In this experiment, participants' position data was recorded every time they made a movement in the virtual experimental environment. And their traces could be generated and analyzed as in Fig. 8. In the trace figure, the green line represents participant's trace while the red spot is the position where they finished a calculation task. As the radioactive intensity changes, it is more easy to distinguish the difference in frequency change compared to the other two acoustic features according to their impression. However, the result of frequency condition was negative. After comparing their trace under frequency condition and other conditions, it was found that they showed more intention to avoid radiation under frequency condition. The reason is that they were not accustomed to the experimental environment and they spent much time and attention to the avoidance. Besides, because of the limitation of the application, movement was not smooth especially when avoiding the dangerous area by turning around to search for a safer area. The increased staying duration caused a significantly high radiation exposure. With the promotion of the virtual experimental environment or conducting in a real environment as well as getting more familiar with the application, it is supposed that their performance becomes better under the three alarm conditions, and there may be a significant improvement under the frequency condition.

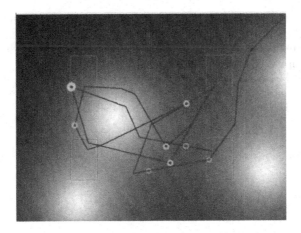

Fig. 8. An example of participant's trace.

5   Conclusion

In this study, an evaluation experiment in the virtual experimental environment was conducted to find out which acoustic feature among loudness, interval and frequency performs the best in radiation awareness and avoidance. As the result of the experiment, participants had least radiation exposure under loudness condition, which indicates that loudness feature can be a suitable option for radiation awareness human interface used in radioactive working environment such as NPPs field work. Although participants are sensitive to frequency feature, but it may not be suitable for radiation avoidance. More important, it is necessary to reduce the staying time in radioactive area no matter how we avoid the radiation.

References

1. Prasad, K.: Radiation protection in humans: extending the concept of As Low As Reasonably Achievable (ALARA) from dose to biological damage. Br. J. Radiol. **77**(914), 97–99 (2004)
2. Zera, J.: Preferred levels of auditory danger signals. Int. J. Occup. Saf. Ergon. **6**(sup1), 111–117 (2000)
3. Lee, B.: Performance analysis of Electronic Personal Dosimeter (EPD) for external radiation dosimetry. J. Radiat. Prot. Res. **40**(4), 261–266 (2015)
4. Zobel, P.: Warning tone selection for a reverse parking aid system. In: Proceedings of the Human Factors and Ergonomics Society, vol. 42, no. 17, pp. 1242–1246 (1998)
5. ISO 7731.: Ergonomics – danger signals for public and work areas – auditory danger signals (2003)
6. ISO 24500.: Ergonomics – accessible design – auditory signals for consumer products (2010)

7. Hikono, M.: The effective way of Shisa-Kosho (calling loudly while pointing at object) in plants (4). J. Jpn. Ergon. **38**, 512–513 (2002)
8. Green, M.: Industrial noise exposure and ambulatory blood pressure and heart rate. J. Occup. Med. **33**(8), 879–883 (1991)
9. Stevens, S.: The psychophysics of sensory function. Am. Sci. **48**(2), 226–253 (1960)
10. Sato, T.: Features of Particle and Heavy Ion Transport Code System (PHITS) version 3.02. J. Nucl. Sci. Technol. **55**(6), 684–690 (2017)

Comprehending Research Article in Minutes: A User Study of Reading Computer Generated Summary for Young Researchers

Shintaro Yamamoto[1(✉)], Ryota Suzuki[2], Hitokatsu Kataoka[2], and Shigeo Morishima[3]

[1] Waseda University, Tokyo, Japan
s.yamamoto@fuji.waseda.jp
[2] National Institute of Advanced Industrial Science and Technology (AIST), Tsukuba, Japan
{ryota.suzuki,hirokatsu.kataoka}@aist.go.jp
[3] Waseda Research Institute for Science and Engineering, Tokyo, Japan
shigeo@waseda.jp

Abstract. The automatic summarization of scientific papers, to assist researchers in conducting literature surveys, has garnered significant attention because of the rapid increase in the number of scientific articles published each year. However, whether and how these summaries actually help readers in comprehending scientific papers has not been examined yet. In this work, we study the effectiveness of automatically generated summaries of scientific papers for students who do not have sufficient knowledge in research. We asked six students, enrolled in bachelor's and master's programs in Japan, to prepare a presentation on a scientific paper by providing them either the article alone, or the article and its summary generated by an automatic summarization system, after 15 min of reading time. The comprehension of an article was judged by four evaluators based on the participant's presentation. The experimental results show that the completeness of the comprehension of the four participants was higher overall when the summary of the paper was provided. In addition, four participants, including the two whose completeness score reduced when the summary was provided, mentioned that the summary is helpful to comprehend a research article within a limited time span.

Keywords: User study · Scientific paper summarization · Scientific paper comprehension

1 Introduction

The number of scientific articles has been increasing exponentially [2]. For example, in recent years, there have been several conferences in computer science that have accepted more than 1,000 papers in (Fig. 1). Consequently, researchers need to spend more time reading research articles. A literature survey is an important

© Springer Nature Switzerland AG 2021
S. Yamamoto and H. Mori (Eds.): HCII 2021, LNCS 12765, pp. 101–112, 2021.
https://doi.org/10.1007/978-3-030-78321-1_9

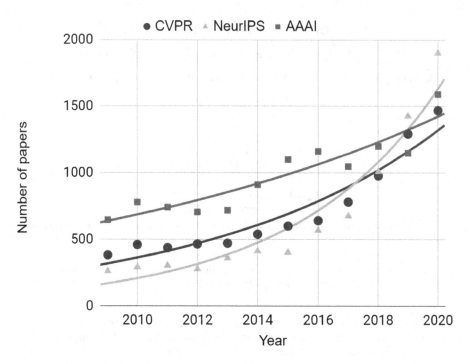

Fig. 1. Number of papers accepted in computer science conferences.

step in research activity that enables researchers to learn about the latest findings or clarify the position of their own research. However, limited time is available to read research articles because researchers also need to perform other tasks, including conducting experiments and analyzing results. Nevertheless, university students involved in research need to read more papers due to their lack of research experience. Thus, understanding the content of research articles within a short period of time is necessary, especially for students.

To cope with the rapid increase in the number of scientific publications, several Japanese groups perform activities to archive brief summaries of conference proceedings from domains such as computer vision [11] and human–computer interaction [15]. These summaries help researchers to comprehend trends in the field quickly without needing to read the original papers, which are much longer than the summaries. While the abstract in a scientific paper is written from the author's point of view [8], summaries written through such activities reflect the reader's opinion. However, limited summaries are available because writing a summary of scientific paper requires enormous time and a good understanding of the research.

Automatic text summarization is an active research topic in the field of natural language processing, and to address the aforementioned challenges, several studies have focused on automatic scientific paper summarization [1,5–7,16,18,19,22]. An automatic summarization system is expected to provide a brief description of the research overview without human effort, and to enable users comprehend the research in a short time. Manual evaluation of such text summarization systems is challenging, as it would require human evaluators to read a lot of text, and also possess sufficient background knowledge on the topic. Instead, an automatic evaluation metric such as ROUGE [13] has been used to evaluate the system's utility in most automatic summarization research. ROUGE measures the content overlap between the system generated summary and a reference summary written by a human, using available datasets of manually written summaries. However, most studies ignore whether and how system generated summaries actually help readers comprehend scientific research. In this work, we aim to investigate the effectiveness of automatically generated summaries of scientific publications for students who are inexperienced in research and need to obtain knowledge of the research field, to help them comprehend research articles within limited period of time.

To conduct a user study, we asked six students in Japan, who were enrolled in either bachelor's or master's programs, to join a workshop for reading scientific papers. The participants were given two comprehension tasks. In one task, they were provided a paper alone (a), and in the other task, they were provided a paper along with its automatically generated summary (b), as illustrated in Fig. 2. In both tasks, they were given 15 min to read the material provided. We then asked the participants to present an overview of the research that they had read to an audience comprising four evaluators, who evaluated the participant's comprehension of the research. Based on the results of our user study, we discuss the effectiveness of the current approach toward automatic summarization of scientific papers in facilitating students' comprehension of the overview of a research work.

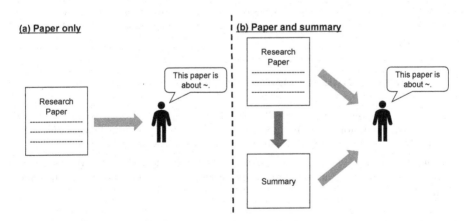

Fig. 2. Experimental outline of summary reading. Participants are asked to read either (a) a paper alone, or (b) the paper and its summary generated by computer. After reading, they are asked to present an overview of the research in the paper.

2 Related Work

Automatic text summarization is a well-researched topic in the field of natural language processing. To alleviate the burden of reading scientific publications, several researchers have focused on the automatic summarization of scientific papers [1,5–7,16,18,19,22]. However, to manually evaluate an automatic summarization system is costly; therefore, automatic evaluation metrics are widely used instead. For example, ROUGE [13] evaluates a system-generated summary by measuring its similarity to a reference summary (generally written by a human).

Although ROUGE evaluate the system without reading system generated summaries, several studies have reported that automatic evaluation metrics for natural language generation, including ROUGE, do not accurately reflect human judgment [4,12,14,20]. For example, Kryscinski et al. [12] asked annotators to rate summaries based on four aspects: relevance, consistency, fluency, and coherence. The experimental results showed that ROUGE does not correlate well with human judgment on these four aspects. Therefore, the prior studies on automatic summarization of scientific papers relying on automatic evaluation metrics do not properly reflect the actual effectiveness of scientific paper summarization in real use.

Some research has adopted alternative methods to evaluate automatic summarization systems. For example, Huang et al. [10] defined eight errors in summaries, such as omission (missing a key point) and duplication (unnecessary repetition) to evaluate automatic summarization systems. They revealed the characteristics of several algorithms based on the errors identified by humans. However, these works studied the human evaluation of summaries of news articles [9]. Several researchers have pointed out that scientific paper summarization is different from news article summarization [6,7]. For example, scientific papers are much longer than news articles and follow a normalized discourse structure [24]. In this paper, we investigate the effectiveness of scientific paper summarization.

3 Automatic Summarization System

To generate summaries of research articles, we implement an automatic summarization system for scientific papers based on our previous work [25] with some updated components. The summarization system consists of two primary components: textual summarization and figure selection (Fig. 3). We generate a single page power point file that contains the title, author, textual summary, and a selected figure.

Linguistic Component. We apply an automatic text summarization method to generate a textual description of an overview of the research. To extract the body of paper, we utilize the ScienceParse library[1], which converts a PDF file into JSON format. Among several methods of automatic text summarization,

---

[1] https://github.com/allenai/science-parse.

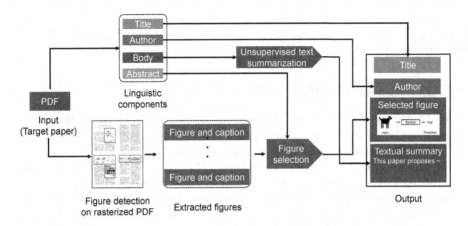

Fig. 3. Overview of the paper summarization system used in this study. Linguistic components, including title, author, body and abstract are extracted from the input paper. An unsupervised text summarization method is used to generate textual summary from body of the article. Figures are extracted from the rasterized PDF of the paper and one figure is selected based on content similarity between figure caption and abstract.

we consider that the ability to summarize arbitrary papers is important in a system for aiding scientific paper comprehension. We employ an unsupervised text summarization method, TextRank [17], as it only requires text and does not rely on training data. In TextRank, sentence importance is calculated according to the connection between sentences. We extract important sentences until the output summary reaches 250 words.

Visual Component. In scientific papers, figures are used in addition to text to convey information. According to findings in psychology [3], visual information helps readers understand text. To further facilitate comprehension, therefore, We attach one figure to the output summary along with the textual summary.

We first extract figures and their captions from the PDF file using the image-based method proposed by Siegel et al. [23]. To minimize the space allocated for the figure, we select only one figure for the summary. Following our previous work [25], we measure the similarity between abstract and figure caption. Specifically, the figure is selected based on the cosine similarity of the tf-idf vectors of the abstract and figure captions. The figure whose caption is most similar to the abstract is used for the summary.

4 User Study

To investigate the effectiveness of computer-generated summaries, we conducted a user study wherein students read (a) a paper alone, or (b) a paper and its summary generated by the method described in Sect. 3. After reading the material provided, the students were asked to present a brief overview of the research that they read to an audience.

4.1 Experimental Setting

Subjects. We asked six students of computer science to join the scientific paper-reading workshop for beginners. Four students (B_A B_B B_C B_D) were 4th-year undergraduate students who were working on their bachelor's thesis, and the other two were 1st-year master's students (M_A M_B). We focus on students because they do not have sufficient research experience and knowledge. The native language of participants were Japanese; therefore, we conducted the experiments in Japanese; however, the reading material was written in English.

We also asked an audience of fours to evaluate the participants' comprehension of the research based on the presentation. Two members of the audience were researchers in computer science who had a Ph.D. degree in the field. The others were computer science Ph.D. students with a Master's degree. All of them were native Japanese, similar to the participants.

Material. We used papers from computer vision research as the reading material. Specifically, papers accepted to either the Conference on Computer Vision and Pattern Recognition (CVPR) or the European Conference on Computer Vision (ECCV) in 2018, which are considered as top-tier conferences in computer vision, were used in the study. We selected one paper for demonstration and the other six for experiments that were not relevant to the participants' research projects to ensure that the participants had not read the papers before.

Methodology. We explained to the participants that the study was designed as a scientific paper-reading workshop for beginners and that experimental data would be collected. Participants first read one paper for practice, and then two papers for which they were required to present overviews. Each of six papers chosen for the experiment was read by two participants under different conditions. Thus each participant was assigned two different papers: (a) a paper alone, and (b) a paper with its summary. The reading time was limited to 15 min because we assumed that summaries help users to comprehend the research in a short span of time. During the workshop, all reading materials were printed. Participants were not allowed to connect to the Internet or use a dictionary, to prevent them from using external means to comprehend the paper.

The workshop was conducted according to following steps (Fig. 4):

1. As a demonstration, we first presented an overview of a computer vision paper using a single page slide. We told the participants that this was an example of the presentation that they were required to prepare.
2. As a preparation, the participants were then asked to read the paper used for the demonstration and prepare a presentation of the overview of research from the paper in 10 min. The aim of this step was to ensure that the participants

Table 1. Evaluation criteria.

Score	Criteria
4	Satisfied
3	Somewhat satisfied
2	Somewhat unsatisfied
1	Unsatisfied

correctly understand their task. We told them their presentation material from this step would be collected to duplicate the experimental situation that presentation would be evaluated later.

3. We again showed the participants our original presentation material from Step 1. Specifically, subjects were told that the overviews of the "objective," "proposed method," and "experimental result" were required in their presentation.

4. The participants were told to read the first paper provided to them. They were assigned to either (a) original paper only, or (b) original paper and its summary. Those provided with the paper only, without the summary, read the article for 15 min. Those provided both the paper and the summary, were allowed to read only the summary for the first 5 min first, and could then read both the summary and the original paper for the next 10 min. At the end of the 15 min reading session, the participants were given 5 min to prepare a single-page presentation. Three papers were used for this step. Each paper was read by two participants based on the two conditions, that is, each paper was used once for condition (a) and once for condition (b). This was to ensure that our procedure was free from bias between the two conditions.

5. The participants read different papers and prepared presentation of them. The condition to which they were assigned was swapped to ensure that they were assigned to a different condition from that in the previous step. The other procedures were the same as in Step 4. The remaining three papers not used in the previous step were used as reading materials in this step.

6. The participants gave presentations to an audience of four evaluators. The evaluators were not aware of the experimental condition of each subject (with or without summary). They evaluated the presentation based on the following criteria: (1) comprehensiveness, which measures the three aspects mentioned in Step 3 were included in the presentation. (2) completeness, which measures the comprehensiveness of the presentation and the accuracy of the content. The evaluations were conducted using the four-point Likert scale in Table 1. The evaluators were able to read the original papers whenever they wanted to confirm the correctness of the content. After presentation, the participants were asked to submit their presentation materials.

7. We asked the participants to answer a few questions, including those on the usability of the summary and their English skills.

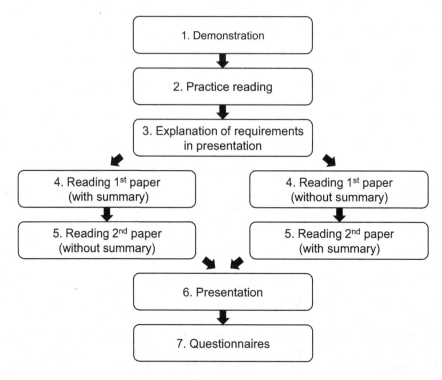

Fig. 4. Process of the user study.

4.2 Result

The scores provided by the evaluators are listed in Table 2. These results reveal that when the summary was provided, the overall score in completeness increased in terms of completeness, but decreased slightly in terms of comprehensiveness. Regarding the scores of individual participants, two undergraduate students (B_A and B_B) obtained higher scores in both metrics whereas only completeness increased for the two graduate students (M_A and M_B). By contrast, the other two participants (B_C and B_D) received higher scores in both metrics without the summary.

The four participants, including B_C and B_D, who obtained lower comprehensiveness scores with the summary, mentioned that the summary was helpful to understand the research. For example, M_A said, "the summary provided me with an overview of the research, which helped me to read the original paper after five minutes." According to Ozuru et al. [21], the comprehension of a scientific text is highly correlated with prior knowledge. Therefore, the summary is useful not only to understand the overview of a research paper in a short span of time, but also to obtain prior knowledge before the reading the body of the paper. Moreover, B_C mentioned, "this workshop was a good opportunity for me to think about my reading strategy". Generally, bachelor's and master's degree

Table 2. Experimental results. The scores indicate the average of four evaluators. The condition with better score is highlighted in bold.

Subject	Condition	Paper ID	Comprehensiveness	Completeness
B_A	w/ summary	5	**1.75**	**1.75**
	w/o summary	1	1.50	1.25
B_B	w/ summary	3	**3.00**	**2.75**
	w/o summary	4	2.00	2.00
B_C	w/ summary	4	2.75	2.75
	w/o summary	5	**3.75**	**3.75**
B_D	w/ summary	2	2.00	2.00
	w/o summary	6	**3.00**	**2.75**
M_A	w/ summary	6	3.00	**3.75**
	w/o summary	3	3.00	3.00
M_B	w/ summary	1	2.50	**2.25**
	w/o summary	2	2.50	2.00
Overall	w/ summary	–	2.50	**2.63**
	w/o summary	–	**2.54**	2.46

students do not have sufficient research experience, and therefore are not used to reading research articles. This comment suggests that summaries of scientific papers might be helpful into acquiring the abilities required to read research articles.

The participants also mentioned some shortcomings of the summaries used in the study. B_C said, "some difficult words should be translated into Japanese." Indeed, the summary that B_C read (paper ID 4) included some specialized terminology such as "albedo" and "geometry" that less experienced Japanese students might not know. Although the participant's score indicates a negative effect of the summary, B_C mentioned that the summary was helpful. We can infer that drop in B_C's score when the summary was provided probably occurred because the topic was difficult to comprehend due to their lack of knowledge. We also observed that the summary read by B_D (paper ID 2) used specialized vocabulary such as "normal maps" and "wrinkles," which might have caused the decrease in their score. Moreover, all participants mentioned that they were not confident reading text written in English. Research articles are generally written in English, but they hinder non-native students' comprehension. Therefore, language is an important aspect to consider for aiding non-native students in comprehending a research article.

Two participants (B_A and B_D) mentioned that they would have preferred more figures. As our summarization system [25] aims to generate a single page summary, only one figure was extracted for the readers. Figures convey several aspects of research including methods, data, and experimental results and therefore multiple figures can provide more information. In particular, computer vision focuses on image data; therefore, the importance of figures is further heightened. Most prior studies have focused primarily on text summarization [1,5–7,16,18,19,22], and have ignored the role of figures. Nevertheless, further research is important to help students better comprehend scientific papers within short time spans.

5   Conclusion

In this paper, we conducted a user study to investigate the effectiveness of a computer-generated summary for students with low research experience. The participants were asked to read either (a) a paper only, or (b) a paper with a summary, in 15 min, and to present an overview to an audiences four people. We observed that four participants obtained higher scores, as rated by an expert audience, when they read a summary before reading the original paper. The comments from the participants suggest that a summary is helpful in acquiring prior knowledge and thinking about the reading strategy. The participants also highlighted several challenges and insights. For example, the lack of fluency in English is a significant hurdle toward understanding for non-native students, and measures could be adopted to address this, such as by providing a translation of difficult terms. Further, some participants would have preferred more figures in the summary.

In future work, we intend to study users' request for features in summaries of scientific papers. As mentioned in Sect. 2, scientific papers possess unique characteristics in terms of text length and structure, which might necessitate features that are specific to research article summarization. We will intend to investigate what users consider an ideal summary for a scientific publications. Another research direction is to investigate the effectiveness of figures in scientific paper summarization. We observed that two participants requested more figures in the summary, which indicates that they found figures useful. Therefore, investigating the role and selection of figures in summaries, such as the kind of figures that would be useful and the relationship between the textual summary and figures, can be valuable to enhance the effectiveness of summaries in aiding the comprehension of research articles.

Acknowledgment. This work was supported by JST ACCEL (JPMJAC1602). We would like to thank the members of cvpaper.challenge for their help in the user study.

References

1. Abu-Jbara, A., Radev, D.: Coherent citation-based summarization of scientific papers. In: Proceedings of the 49th Annual Meeting of the Association for Computational Linguistics: Human Language Technologies, pp. 500–509 (2011)
2. Bornmann, L., Mutz, R.: Growth rates of modern science: a bibliometric analysis based on the number of publications and cited references. J. Assoc. Inf. Sci. Technol. **66**(11), 2215–2222 (2015)

3. Carney, R.N., Levin, J.R.: Pictorial illustrations still improve students' learning from text. Educ. Psychol. Rev. **14**(1), 5–26 (2002)
4. Chaganty, A., Mussmann, S., Liang, P.: The price of debiasing automatic metrics in natural language evalaution. In: Proceedings of the 56th Annual Meeting of the Association for Computational Linguistics (Volume 1: Long Papers), pp. 643–653 (2018)
5. Cohan, A., et al.: A discourse-aware attention model for abstractive summarization of long documents. In: Proceedings of the 2018 Conference of the North American Chapter of the Association for Computational Linguistics: Human Language Technologies, Volume 2 (Short Papers), pp. 615–621 (2018)
6. Cohan, A., Goharian, N.: Scientific article summarization using citation-context and article's discourse structure. In: Proceedings of the 2015 Conference on Empirical Methods in Natural Language Processing, pp. 390–400 (2015)
7. Cohan, A., Goharian, N.: Scientific document summarization via citation contextualization and scientific discourse. Int. J. Digit. Libr. **19**(2–3), 287–303 (2018)
8. Elkiss, A., Shen, S., Fader, A., Erkan, G., States, D., Radev, D.R.: Blind men and elephants: what do citation summaries tell us about a research article? J. Am. Soc. Inf. Sci. Technol. **59**, 51–62 (2008)
9. Hermann, K.M., et al.: Teaching machines to read and comprehend. In: Advances in Neural Information Processing Systems, vol. 28, pp. 1693–1701 (2015)
10. Huang, D., et al.: What have we achieved on text summarization? In: Proceedings of the 2020 Conference on Empirical Methods in Natural Language Processing (EMNLP), pp. 446–469 (2020)
11. Kataoka, H., et al.: cvpaper.challenge in 2016: futuristic computer vision through 1,600 papers survey. arXiv preprint arXiv:1707.06436 (2017)
12. Kryscinski, W., Keskar, N.S., McCann, B., Xiong, C., Socher, R.: Neural text summarization: a critical evaluation. In: Proceedings of the 2019 Conference on Empirical Methods in Natural Language Processing and the 9th International Joint Conference on Natural Language Processing (EMNLP-IJCNLP), pp. 540–551 (2019)
13. Lin, C.Y.: ROUGE: a package for automatic evaluation of summaries. In: Text Summarization Branches out, pp. 74–81 (2004)
14. Liu, C.W., Lowe, R., Serban, I., Noseworthy, M., Charlin, L., Pineau, J.: How NOT to evaluate your dialogue system: an empirical study of unsupervised evaluation metrics for dialogue response generation. In: Proceedings of the 2016 Conference on Empirical Methods in Natural Language Processing, pp. 2122–2132 (2016)
15. Matsumura, K., et al.: Chi benkyokai 2017: nettowâku renkei shita benkyokai to sono shien shisutemu. [chi study meeting 2017: Networked study meeting and support system.] (in japanese, title translated by the author of this article.). IPSJ SIG Technical Reports 2017-HCI-174(13), pp. 1–8 (2017)
16. Mei, Q., Zhai, C.: Generating impact-based summaries for scientific literature. In: Proceedings of the ACL-08: HLT, pp. 816–824 (2008)
17. Mihalcea, R., Tarau, P.: Textrank: bringing order into text. In: Proceedings of the 2004 Conference on Empirical Methods in Natural Language Processing, pp. 404–411 (2004)
18. Nanba, H., Kando, N., Okumura, M.: Classification of research papers using citation links and citation types: towards automatic review article generation. Adv. Classif. Res. Online **11**(1), 117–134 (2000)
19. Nanba, H., Okumura, M.: Towards multi-paper summarization reference information. In: Proceedings of the 16th International Joint Conference on Artificial Intelligence, vol. 2, pp. 926–931 (1999)

20. Novikova, J., Dušek, O., Cercas Curry, A., Rieser, V.: Why we need new evaluation metrics for NLG. In: Proceedings of the 2017 Conference on Empirical Methods in Natural Language Processing, pp. 2241–2252 (2017)
21. Ozuru, Y., Dempsey, K., McNamara, D.S.: Prior knowledge, reading skill, and text cohesion in the comprehension of science texts. Learn. Instr. **19**(3), 228–242 (2009)
22. Qazvinian, V., Radev, D.R.: Scientific paper summarization using citation summary networks. In: Proceedings of the 22nd International Conference on Computational Linguistics (Coling 2008), pp. 689–696 (2008)
23. Siegel, N., Lourie, N., Power, R., Ammar, W.: Extracting scientific figures with distantly supervised neural networks. In: Proceedings of the 18th ACM/IEEE Joint Conference on Digital Libraries, pp. 223–232 (2018)
24. Suppe, F.: The structure of a scientific paper. Philos. Sci. **65**(3), 381–405 (1998)
25. Yamamoto, S., Fukuhara, Y., Suzuki, R., Morishima, S., Kataoka, H.: Automatic paper summary generation from visual and textual information. In: Eleventh International Conference on Machine Vision (ICMV 2018), vol. 11041, pp. 214–221 (2019)

Possibility of Reading Notes as Media to Enrich Communications Between Reader and Book

Satoko Yoshida[1]([✉]), Madoka Takahara[2], Ivan Tanev[1], and Katsunori Shimohara[1]

[1] Doshisha University, 1-3 Tatara Miyakodani, Kyotonabe-shi, Kyoto-fu, Japan
yoshida2016@sil.doshisha.ac.jp
[2] Muroran Institute of Technology, 27-1 Mizumoto-cho, Muroran-shi, Hokkaido, Japan

Abstract. Reading gives readers intellectual stimuli, elicits their imagination and creativity, and expands their inner world. If reading has such importance, in Japan, many people don't understand the importance of reading. In this research, we introduce the concept to postulate that reading is communication between a reader and a book, even though a book is not reactive but non-reactive entity. An idea is to change a book into a reactive entity, and to achieve interactions between a reader and a book. Especially, we introduce a mechanism to feedback a reading note as response from a book to a reader's gaze information that the reader generates during reading as stimulus to the book. A reading note is a record that a reader makes about a book, based on the reader's intention and interest. Reader's gaze information generated during reading shows his/her mind about the book. Thus, we regard that a reading note is useful as response from a book, and employ reader's gaze information is useful stimuli to a book. The technological goal of this study is to build an automatic reading notes creation system that automatically generate reading notes based on gaze information that a reader naturally generates during reading. For that goal, we conducted experiments to measure a subject's gaze coordinate data, and to verify whether we could identify eye movements for specific four reading patterns through the analysis of the acquired data. As the result, we have confirmed it possible to identify a reader's reading patterns from the change of gaze information.

Keywords: Reading · Reading notes · Communication · Media · Gaze information

1 Introduction

People have considered that "reading" has many advantages for a long time. The recent research found that the childhood reading experiences has a good influence on adult consciousness, motivations, and activities [1]. On the other hand, in recently, the increase of young people who don't read books has become a serious problem in Japan. According to the latest survey about college students which is conducted by National Federation of University Co-operative Associations (NFUCA) every year, the percentage of students who spend no time reading books in a day is 48.0% [2].

© Springer Nature Switzerland AG 2021
S. Yamamoto and H. Mori (Eds.): HCII 2021, LNCS 12765, pp. 113–124, 2021.
https://doi.org/10.1007/978-3-030-78321-1_10

Reading gives readers surprise, knowledge, and intellectual stimuli, excites imagination and creativity, and expands their inner world. If reading has such importance, the survey result shows that many people don't understand the importance of reading.

The research concept we introduced for that purpose is to postulate that reading is communication between a person and a book. Communication between people is the activity or process of expressing ideas and feelings or of giving information each other so that they can share idea, feelings, and/or information, and can get mutual understanding. Visual and auditory information is in general used as media of communication between people [3, 4]. Reading is the activity or process in which a person as reader tries to share and/or understand the contents of a given book. Accordingly, we can postulate that reading is communication between a reader and a book, even though a book is not a reactive but non-reactive entity.

An idea is to change a book as non-reactive entity into a reactive one, and to achieve interactions between a reader and a book. Here, we introduce a mechanism to feedback a reading note as response from a book to a reader's gaze information that the reader generates during reading as stimulus to the book. A reading note is a record that a reader makes about a book, based on the reader's interest, intention, and/or sense of value. The contents of a reading note include many information; for example, the title of the book, the author, favorite sentences and expressions, ideas and impressions the reader comes up with, and so on. A reader records his/her feelings and thoughts about the book in a reading note. Thus, we can regard that a reading note is rich and useful as response from a book to a person.

To enable people to re-recognize and rediscover the importance of reading, we propose reading notes as media that intermediates interactions between a person and a book. However, to take reading notes, people have to take considerable time and effort. In this research, we use gaze information generated through his/her reading behaviors. Readers change reading patterns depending on the difference of interest in and difficulty of the contents of books. Therefore, gaze information shows reader's mind about the book. We think it is useful to create reading notes. Researches on gaze information and reading have been conducted so far. However, these studies don't aim to create reading notes that reflect reader's mind.

The research objective is to build the system that create automatic reading notes by gaze information during reading. This paper reports an outcome of the basic research toward the goal. Concretely, we set four reading patterns that readers often apt to have during their reading as the experiment conditions, and acquired coordinate data of a reader's gaze with an eye-tracking system. After conducting smoothing on the acquired data, we confirmed that they indicate characteristics inherent to the four reading patterns.

This paper is organized as follows: Sect. 2 explains reading note as media and an automatic reading note creation system, and Sect. 3 describes the originality of this research comparing with related research. Section 4 reports the details of experiments, and Sect. 5 investigates and discusses the experimental result. Finally, our conclusion is given in Sect. 6.

2   Creation of Reading Notes as Media

2.1   Importance and Application of Communication

In this study, we postulate that reading is communication between a reader and a book.

There is an idea that the importance of the communication between people is defined as the expansion of one's inner world by eliciting imagination and creativity each other [5]. We cannot completely understand and predict someone to communicate with. However, we feel various emotion and get unexpected information by communication with unpredictable people.

Even in reading where there is no explicit interaction, readers can get intellectual stimuli, elicit their imagination and creativity, and then expand their inner world. Postulating reading where communication between a reader and a book could be accomplished, however, we could envision possibilities to extract new advantages of reading different from one with one-directional communication as well as enhancing elicitation of imagination and creativity.

2.2   Problem of Reading as "Communication Between a Reader and a Book"

In reading, a book has no stimulus to receive from the reader, let alone no reaction to respond to the reader. That is, a book itself is a non-reactive entity by nature, and interactive communication between a book and a reader is not basically presumed. To achieve interactions between them, we need to change a non-reactive book into an entity reactive to a reader.

To solve this problem, we use reading notes as a kind of feedback from a book to a reader's gaze information during reading as a kind of stimuli from the reader, as shown in Fig. 1. The reason why we employ reading notes as feedback from a reader was already described in Sect. 1, and the reason why we utilize a reader's gaze information to create a reading note is explained in Sect. 2.4 and 2.5.

Fig. 1. Communication between a reader and a book

2.3   Reading Notes as Media

In this study, reading notes not only helps to communicate between a reader and a book but also plays a role as media which mediates interaction between them. Media denote, here, some entity to intermediate something such as ideas, thoughts, feelings, and so on, between people, and sometimes to provoke changes of people's behavior, way of thinking and sense of value.

Reading notes record words and sentences which extends reader's inner world: unknown knowledge, favorite expression, idea, and so on. Reading the notes again after time pass, people remember various emotion, thought, and memory, and get new discoveries about themselves and feel their changes. Reading the notes possibly makes people to re-read the book. And then, a new reading note should be created and new interactions are naturally generated. Through comparing the new reading note with the old one, a reader could realize some change and/or his/her own growth in way of thinking. Through comparing his/her own reading one with the other's one, a reader could find difference in way of thinking and feelings to the same book, and it means that reading notes can function as media between people as well as between people and books.

2.4 Gaze Information During Reading

We use gaze information generated through his/her reading behaviors to create the reading notes. People change reading patterns depending on the difference of interest in and difficulty of the contents of books. For example, they repeatedly read a part difficult to easily understand, and re-read a part with back tracking. We have a Japanese proverb saying "Read a hundred times over, and the meaning will become clear of itself." It means that people could see the meaning naturally by repeatedly reading many times. In the same way, we could see the same functionality of repeatedly reading not only a book itself but also a part of a book; if the same part of a book is repeatedly gazed, it would be difficult for a reader to understand the meaning of the part. If some parts of a book are skipped over, the parts would not be important and interesting to a reader.

In the case that we study a thing unknown so far, we sometimes read and study several books that contain the same unknown thing. Those books frequently have the common description about basic ideas, a reader who knows the basic ideas should skip the corresponding parts in those books. If some parts and/or pages of a book are intentionally skipped in reading, those parts and/or pages should not be attractive to the reader.

Hence, gaze is rich in useful information to show reader's will, interest, intention and mind about a given book. We employ, thus, reader's gaze information as the useful stimuli from a reader to a book to create reading notes as the feedback from a book to a reader.

2.5 Automatic Reading Notes Creating System

To identify a part of book to interest a reader, other modalities such as line maker and voice utterance are available. We employed, however, the gaze information of a reader for that purpose from the following two reasons.

The first reason is that the gaze information doesn't pose any action and operation to extract sentences from a book to a reader. In the cases of using line marker and/or voice utterance to note, for example, the reader has to trace it with his/her finger and/or make his/her voice. Those action and operation should give inconvenient to the reader under some situations in which he/she feels difficulty to do so.

The second reason is that gaze information often implies a reader's unconscious mind and will on the content of book. In the case of creating reading notes with line maker and/or voice utterance, the reader should intentionally select words and sentences to take it. The reader can do the same thing too with eye movement defined for such action, for example, blinking 3 times, at the starting and finishing point to record. However, we emphasize again that gaze information implies the reader's unconscious mind and will. The reader might read repeatedly not only an interesting part but also a difficult part to understand. In this research right now, we investigate a possibility of gaze coordinate only, but there should be a possibility for a system to be able to judge that the coordinate would be selected whether consciously or unconsciously by introducing other parameters such as the speed of eye movement and the change of pupil diameter into a system.

We postulate that a reading note involving parts that a reader doesn't select should make some difference as media from one involving only part that the reader intentionally selects for record.

2.6   Automatic Reading Notes Creating System

We have 3 purposes to enable people to re-recognize and rediscover the importance of reading as follows:

1. Suggestion on automatic reading notes creating system using gaze information during reading

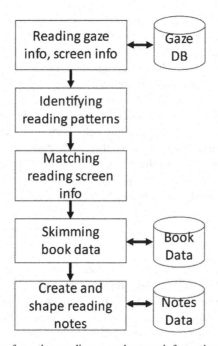

Fig. 2. Process of creating reading notes by gaze information during reading

2. Investigation the influence of reading notes on the cognitive characteristics
3. Propose a new type of reading notes as media that intermediates interactions between the reader and the book

We aim to build the system shown in the first purpose to achieve the second and third ones. Figure 2 shows the process to create reading notes in the proposed system. Automatic creation of reading notes is not executed in real time during reading but after a reader finishes reading. Gaze information including coordination data of eye movement, diameter of the pupil, and time stamp during reading is stored Gaze Data Base (GDB).

To guess reader's mind about the contents of a book, specific reading patterns should be identified by analyzing gaze information stored in GDB. Reading patterns that should be identified are not specified yet because gaze information is now examined.

Through matching analysis to identify some text parts in reading screen that correspond to specific reading patterns, the text information of the book is skimmed accordingly. The text information is extracted on the basis of one sentence or multiple ones to keep readability for a reader.

Reading notes are automatically generated and shaped based on the skimmed text data. As the first step, right now, only the skimmed text data in a book is contained, and any memorandum of a reader's is not done. In that sense, the current version of reading notes is a sort of document to summarize pieces of sentences in a book.

3 Related Research

Research about gaze information and reading have conducted since before. For example, there is the research that use eye features to detect reading behaviors [4]. The final goal of this research is to create reading life log automatically. The reading life log is the long-term digital recording that when, what, and how people read in a daily life [5]. In this related research, the experiment to detect reading behavior in the daily activities with mobile eye tracker was conducted. The researchers compared the accuracy of 4 methods: Bulling et al. [6], Yoshimura et al. [7], Kunze et al. [8], and all these methods. The purpose of this study is detecting reading behaviors, not creating reading notes.

In the support of reading by eye tracking, there is the research to track eye movement during reading an e-book and show the point at where people leave off reading [9]. The point is shown by changing the color of part that people finish reading.

These studies don't aim to create reading notes that reflects reader's mind.

4 Experiment

The aim of this research is to build the automatic reading notes creating system introduced in the Sect. 2.6. As the first step for that purpose, we conducted experiments to acquire a subject's gaze coordinate data with an eye-tracking system, and to verify whether we could identify eye movements for specific reading patterns through the analysis of the acquired data. A number of subjects was one.

4.1 Eye Tracker

We used Tobii Pro X3-120 which is the eye tracker provided by Tobii Technology Company [10]. It gets coordinates of left eye and right eye per 1/120 s: the sampling rate is 120 Hz. The axis of coordinate puts (0, 0) on the upper-left of the screen and (1, 1) on the lower-right. In the document area for this experiment, the range of X coordinate is 0.20–0.78, the range of Y coordinate is 0.17–0.93. Figure 3 shows the reading screen and the coordinates.

Fig. 3. Coordinate of reading screen

4.2 Reading Patterns

We measured gaze information of 4 reading patterns as follow:

1. Along sentences: a subject reads the document from the beginning to the end.
2. Repeat: a subject reads the direction sentence twice repeatedly.
3. Return: after reading until the direction sentence of the end side, a subject return
4. to that of the beginning side.
5. Skip: after reading until the direction sentence of the beginning side, a subject
6. skip to the next sentence of that of the end side.

The experimental conditions are set based on the reasons described in Sect. 2.4 and the author's reading experience.

The direction sentences for conditions 2–4 are changed to red.

4.3 Document for Experiment

To create a document for this experiment, we referenced Soseki Natsume "I am a cat" published on the Internet [11]. Table 1 shows the layout detail of the document. A participant read only the first page displayed on the screen. Table 2 shows the coordinates of the direction sentences for conditions 2–4.

Table 1. Layout details

Paper orientation	Lateral
Character orientation	Vertical writing
The number of lines	21 lines
The number of letters per line	33 characters
Margin sizes	Upper, Under, Left, Right: 12.7 mm

Table 2. Coordinate of sentences changed red

	X	Y
Condition 2	0.44–0.46	0.32–0.79
Conditions 3 and 4	Begging side: 0.39–0.44 Tail side: 0.56–0.61	0.21–0.89

5   Results and Discussion

Figures 4, 5, 6, 7 and 8 show results of data averaged and processed by smoothing with sliding window. The sampling rate of the eye tracker is 120 Hz, but the raw data is too many for analyzing it. We took the average every 12 data: per 1/10 s. When people stare a point voluntarily, eyes vibrate involuntarily. It is called involuntary eye movement during fixation. It causes the experiment result to include many vibrations. Therefore, we performed the smoothing processing by sliding window. Sliding window needs a window width and a sliding width. We decide that the window width is 6 and the sliding width is 2. It means that these values are calculated by averaging data for 600 ms with every 200 ms difference. In the figures, the ranges of red and green indicate the coordinates of the direction sentences shown on Table 1. In the conditions that have 2 direction sentences, the green is the beginning side, the red is the end side. The range of yellow indicates the section in which the characteristic eye movement of each case occurs.

Figure 4 and 5 show the results of condition 1 An experiment participant moves eyes from upper to lower and from right to left in Fig. 3. The expected eye movement of X coordinate is from (1, y) to (0, y). The one of Y coordinate is the oscillation between (x, 0) and (x, 1). Figure 4 and 5 records the explained motion.

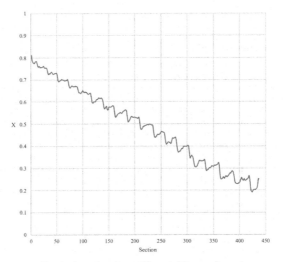

Fig. 4. Result of condition 1 (X coordinate)

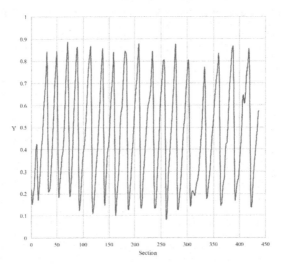

Fig. 5. Result of condition 1 (Y coordinate)

The result of Y coordinate of condition 2 is shown in Fig. 6. In the second experiment, a participant read the direction sentence twice repeatedly. In Fig. 6, the motion that eyes move from the end of the direction sentence to its beginning is observed.

The result of X coordinate of the third experiment is shown in Fig. 7. The X-axis motion is expected that eyes move from (0, y) to (1, y). In Fig. 7, the expected movement is observed.

The result of X coordinate of condition 4 is shown in Fig. 8. The motion is expected that eyes move from (1, y) to (0, y). In Fig. 8, the expected movement is observed.

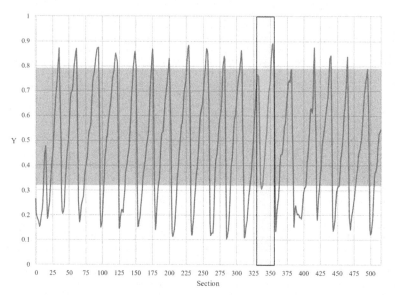

Fig. 6. Result of condition 2 (Y coordinate/R: y = 0.32–0.79)

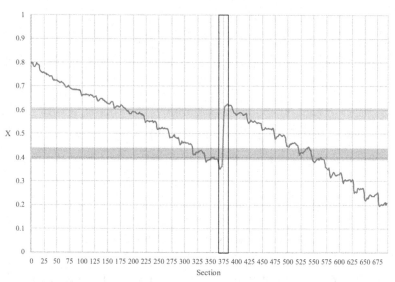

Fig. 7. Result of condition 3 (X coordinate/R: x = 0.39–0.44, G: x = 0.56–0.61)

The results show that the characteristic eye movement differs depending on the reading patterns. Therefore, it is possible to identify the reader's reading patterns from the change of gaze coordinate. Especially, in vertical sentences, the analysis of the eye movement of X coordinate is useful for identifying the 4 patterns measured in this experiment. In addition, the analysis of Y direction is useful for identifying where the eye movement started and ended within a line.

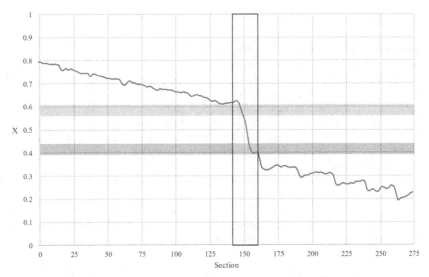

Fig. 8. Result of condition 4 (X coordinate/R: x = 0.39–0.44, G: x = 0.56–0.61)

6 Conclusion

The aim of this study is to enable people to re-recognize and rediscover the importance of reading. For that purpose, we introduced the concept to postulate that reading is communication between a reader and a book. An idea is to change a book as non-reactive entity into a reactive one, and to achieve interactions between a reader and a book. Here, we proposed a mechanism to feedback a reading note as response from a book to a reader's gaze information that the reader generates during reading as stimulus to the book. The technological goal of this study is to build an automatic reading notes creation system that automatically generate reading notes based on gaze information that a reader naturally generates during reading.

As the first step for that goal, we conducted experiments to measure a subject's gaze coordinate data with an eye-tracking system, and to verify whether we could identify eye movements for specific four reading patterns through the analysis of the acquired data. The results show that the characteristics of eye movements can be classified depending on the 4 reading patterns. Thus, we have confirmed it possible to identify a reader's reading patterns from the change of gaze information.

As future works, since we controlled to focus on one specific reading pattern in every experiment, we need to conduct experiments for actual cases of reading, and investigate whether we could classify reading patterns based on actual reading behaviors. That is, we need experiments to allow subjects to read freely, and analyze and classify reading patterns based on a subject's interest and intention about a book, reading his/her reading experience and custom, and so on. It means that we have to tackle the situation where not a single but multiple reading patterns are simultaneously generated.

References

1. Hamada, H., Akita, K., Fujimori, Y., Yagi, Y.: The effects of childhood reading experiences on adult consciousness, motivations and activities. Sci. Read. **58**(1), 29–39 (2016)
2. 54th Summary of Field Survey for Campus Life, National Federation of University Co-operative Associations. https://www.univcoop.or.jp/press/life/report.html
3. Matsumura, A.: Daijirin, 3rd edn. Sanseido, Tokyo (2006)
4. Kitahara, Y.: Meikyo Kokugo Dictionary, 2nd edn. Taishukan Publishing, Tokyo (2010)
5. Shimohara, K.: Artificial Life and Evolving Computer. Kogyo Chosakai Publishing, Tokyo (1998)
6. Nakajima, K., Utsumi, Y., Iwamura, M., Hirose, K.: Selection of gaze-based characteristics useful for detecting reading actions. IPSJ SIG Technical Report 2016-CVIM-202(28), pp. 1–6 (2016)
7. Kunze, K., et al.: Quantifying reading habits: counting how many words you read. In: Ubi-Comp 2015 - Proceedings of the 2015 ACM International Joint Conference on Pervasive and Ubiquitous Computing, pp. 87–96. Association for Computing Machinery, Inc., New York (2015)
8. Bulling, A., Ward, J.A., Gellersen, H., Tröster, G.: Eye movement analysis for activity recognition using electorooculoraphy. IEEE Trans. Pattern Anal. Mach. Intell. **33**(4), 741–753 (2011)
9. Yoshimura, K., Kawaichi, H., Kunze, K., Kise, K.: Relationship between document understanding and gaze information acquired by Eye-tracker. IEICE Tech. Rep. **112**(495), 261–266 (2013)
10. Kunze, K., Ustumi, Y., Shiga, Y., Kise, K., Bulling, A.: I know what you are reading: recognition of document type using mobile eye tracking. In: Proceedings of the 2013 International Symposium on Wearable Computers, pp. 113–116. Association for Computing Machinery, New York (2013)
11. Imamura, M., Cho, I.: Research on Reading Support with Eye-tracking – Attempt for Ambient Interface. Undergraduate Research of the Dept. of Intermedia Art and Science, Waseda University (2016)
12. Tobii Technology: Tobii Pro X3-120. https://www.tobiipro.com/ja/product-listing/tobii-pro-x3-120/
13. Soseki Natsume: I Am a Cat. https://www.aozora.gr.jp/cards/000148/card789.html

Notification Timing Control While Reading Text Information

Juan Zhou[1]([✉])[iD], Hao Wu[2,3], and Hideyuki Takada[4][iD]

[1] School of Environment and Society, Tokyo Institute of Technology, Tokyo, Japan
[2] Graduate School of Information Science and Engineering, Ritsumeikan University,
Kyoto, Japan
[3] Bank of Communications, Beijing, China
[4] College of Information Science and Engineering, Ritsumeikan University,
Kyoto, Japan

Abstract. The various notifications received on computers while being used interrupt users' reading experience, negatively affect their emotions, and increase the cognitive load. Many studies have focused on user behavior and user state to control the notification timing. Blinking has been applied as a type of physical activity to detect users' interests and emotions, detect driver fatigue, and design interactive robots. In this study, we focused on breakpoints while reading text information depending on the blink frequency during concentrated reading as a method to control notification timing. We constructed a system that controls notification timing based on the detected breakpoint. In the experiment, we simulated a real reading environment using the prototype of the system. We evaluated the detection times of the breakpoints and the effectiveness of the system. Although we have not proven the hypothesis that the expected breakpoint is detected based on the blink frequency, we found that the users' browsing experience was improved when they used the control system.

Keywords: Notification manager · Eye blink · Browsing experience

1 Introduction

Receiving various notifications on computers, such as e-mails, application updates, browser notifications, and SNS messages, has become increasingly common. Additionally, the use of text-reading functions, such as Safari reading mode and Firefox readability, has also increased in recent years. Users receive several notifications while reading, and it has been observed that notifications interrupt the reading experience for users, negatively affect their emotions, and increase the cognitive load if they appear at inappropriate times [3,11].

Various studies have focused on user behavior and user state to control the timing of notifications. Some studies attempted to control notifications by acquiring user actions from web camera information, mouse movement, keyboard typing

© Springer Nature Switzerland AG 2021
S. Yamamoto and H. Mori (Eds.): HCII 2021, LNCS 12765, pp. 125–139, 2021.
https://doi.org/10.1007/978-3-030-78321-1_11

information, and page or application switching timing [2, 7, 10, 12]. Furthermore, a study developed a middleware, "Attelia," that detected the breakpoints and delays in notification timing [16]. Some studies estimated user conditions using parameters such as brainwaves, heart rate, pulse rate, number of leg transitions, head movement, respiration, and cutaneous electrical activity recorded through various sensors [9, 18]. However, these studies required the users to wear several contact sensors, which interfered with their work and annoyed them.

Blinking, as a type of physical activity, has also been applied to detect the interests and emotions of users, determine driver fatigue, and design interactive robots. Therefore, in this study, we used the blinking frequency to detect the breakpoints while reading and subsequently control notifications. We developed a system that could control the notification timing based on the detected breakpoint.

2   Related Work

2.1   Effects of Notifications on Productivity

People living in the city receive more than 50 notifications every day [13]. Most systems use sounds or haptics to send notifications. However, only a few studies have considered the time when notifications should be delivered. The situation is worsened for users by the fact that some messages are unnecessary. Low-quality messages with bad timing interfere with user tasks and reduce daily productivity [4]. In particular, during the reading process, a bad notification would decrease the user's content understanding and disturb the concentration state.

By contrast, receiving notifications at the correct time (without interrupting) makes it easier for users to accept the information [11]. In some studies, the users set the notification timing according to their convenience [19], but it is difficult to set it considering all circumstances because notifications cannot be predicted. In addition, research on whether users view notifications has shown that time and place are important factors in promoting notifications. In this study, we performed notification control by focusing on the interruption timing of work, called breakpoints.

2.2   Breakpoint

A breakpoint is a concept in the field of psychology [1]. Human continuous action is a time-series set of multiple action units. There is granularity in the behavior. "Big behavior" is the largest behavioral granularity. The second largest is "medium behavior." The smallest is "small behavior." There is a breakpoint between the action units. A previous study stated that the interruption load could be reduced by providing notifications at the breakpoints [17].

It is believed that users can subjectively classify the particle size of text while reading, including the font size and space. Therefore, breakpoints occur. For example, when the action of "reading information A" changes to that of "reading information B," this information change naturally triggers the break-points. In this research, we aim to detect the breakpoints while reading character information based on the blinking frequency.

2.3 Eye Blink

Blinking is a physiological activity that is performed every day to moisten the eyes. Thus far, most studies have aimed to detect fatigue by blinking. Some studies attempted to detect driver conditions to prevent the computer vision syndrome for safety [6,17]. Blinking has also been applied to the analyse of emotional states in the learning process. Nakano et al. found that when a learning task becomes more difficult, users increase their blinking frequency. When the task becomes easier, the blinking frequency decreases [14]. Moreover, the blinking frequency is correlated with the emotional state, and a decrease in the blinking frequency indicates pleasant effects. These studies suggested that people adjust the blinking frequency according to the visual input. In addition, people unconsciously find clusters of living events and selectively blink at these breaks. Therefore, in this study, we assumed that blink-related eye movements can be used to detect the breakpoints in the reading process.

2.4 Notification System Approach

To control the notification timing, some studies have focused on user behavior or status. There are several methods for obtaining user behavior based on the information obtained from a web camera [12], mouse movement [2], or keyboard timing [7]. User behavior can be obtained from the page/application timing switch [10]. Furthermore, a middleware called "Attelia" has been used to delay the notification timing by detecting the breakpoints [16]. Other studies used various sensors, such as brainwaves, heartbeats, pulse count, head movement, breathing, and electrical activity of the skin, to obtain the user status [18]. However, these studies required users to wear contact sensors, which restrict user movements and cause them to feel monitored. Yamada et al. used peripheral cognition technology (PCT) to make the user aware of notifications automatically. PCT is a phenomenon in which the peripheral vision of users is reduced when a processing task is overloaded; the peripheral vision becomes visible when the load is reduced [21].

Some studies attempted to control notifications based on the user location or time. One study utilized multiple beacons and PCs to estimate the user's body orientation and provided information through notifications based on the direction of gaze of the user [15], considering the attenuation of the RSSI according to the human body. However, if the line of eyesight does not match the direction of the user's body or a person other than the receiver blocks the space between the receiver terminal and the beacon transmitter, the user's situation will be estimated in the wrong direction, which will cause some information to be notified on time; however, in environments where users apply personally installed beacons, the registered notification content is returned to the mobile terminal by the UUID obtained from the transmitted signal using a server [8]. Therefore, we focus on user body movements to represent controlled timing notification processing by observing eye blinking.

3   Method

3.1   Notification Timing Control for Reading Text Information

In this study, we evaluated the effect on users by providing information at appropriate times during reading. We detected the breakpoints and controlled the notifications according to the eye blinking frequency. As shown in Fig. 1, first, we acquired the eye blinking frequency within a certain period of concentration according to the user characteristics; we named it the "eye blinking frequency in concentration (BFC)." It was obtained when a user was in the middle of reading an article because it is difficult to enter the state of concentration at the beginning and end of sentences.

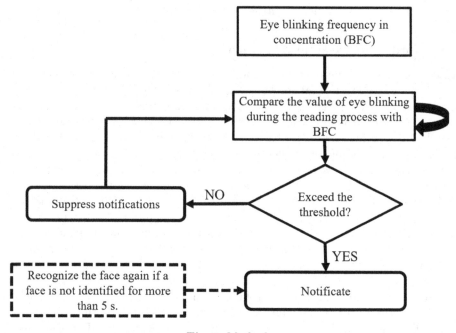

Fig. 1. Method

Next, we compared the value of the eye blinking frequency during the reading process with the BFC value detected beforehand. If the difference exceeded the threshold, we defined the user state as "off-task," and the notification was sent. Conversely, the user state was defined as "on-task," and the system suppressed the notification to prevent interruption when the difference did not exceed the threshold. If a face is not identified for more than 5 s, the state will be defined as "off-task." The system will identify it as a breakpoint and attempt to recognize the face again. To control such notifications, we developed a system to detect eye blinking frequency and control notifications.

3.2 System to Detect Eye Blinking Frequency

This system has two functions: 1) detecting the eye blinking in real time and 2) calculating the eye blinking frequency.

Face Recognition. It is necessary to implement a face recognition method to detect the blinking in real time. Nataliya et al. investigated the time complexity of computer vision algorithms for face recognition. The OpenCV library is more productive and exhibits better performance in face detection [5]. The HOG algorithm is suitable for this study because CPU processing demonstrates the best performance.

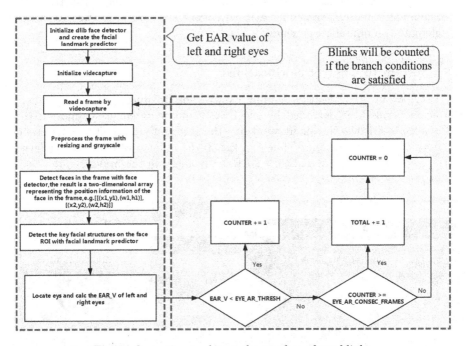

Fig. 2. Operation to detect the number of eye blinks

Detecting the Number of Eye Blinks. We used the method proposed by Tereza and Jan, which identifies blinks by comparing the value of eyes open (EAR standard value) and that in real time [20].

As shown in Fig. 2, the specific implementation method is performed in two parts. First, the EAR values of the left and right eyes are obtained, and second, the number of blinks is counted if the branch conditions are satisfied. The eye blinking conditions must satisfy the following.

1. Compared with the pre-determined standard (eye open) value of ERA and ERA in real time, if the ERA in real time is lower than the standard value of ERA, there is a tendency for blinking. However, the ERA standard value should be adjusted for each individual.
2. There are three or more frames that tend to do this.

Blinking Frequency. In order to get the BFC value, the Tobii eye tracker was originally considered for tracking the position of users' reading. However, it is difficult to obtain an accurate position of eye movements in real time. In addition, to prevent delays in line-of-sight detection, a method of manually clicking the mouse for recording where the users are reading is more suitable in this research.

The number of blinks and time elapsed between the two clicks were recorded to calculate the blinking frequency.

3.3 System to Control Notifications

As described in " Notification Timing Control for Reading Text Information," the notifications are controlled by the BFC value acquired in advance. If the difference exceeds the threshold, we define the user state as an off-task and send the notification. The screen is shown in Fig. 3; the number of blinks is displayed in the upper left, and the real-time EAR is displayed in the upper right.

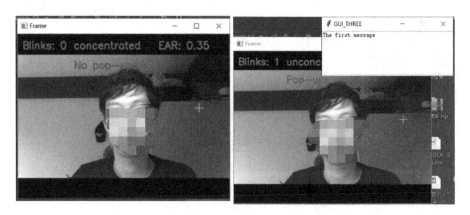

Fig. 3. Screen for demonstration and screen for notification

"Concentrated" and "Unconcentrated" are displayed in the middle of the screen. If the difference in blinking frequency in real time and BFC does not exceed the threshold value, the concentrated state will be displayed, and the notification will be delayed. If the threshold is exceeded, the user state will be identified as "Unconcentrated," and a pop-up notification will be displayed on the right side of the screen with the notification sound.

In addition, there may arise situations where the face moves away from the screen, such as going to the bathroom. Such timing can be clearly defined as a breakpoint in which the users become more receptive to notifications [17].

4   Evaluation Experiment

4.1   Purpose

We have two research questions; first, can the expected breakpoints be detected using the blinking frequency, and second, does the notification system, which is controlled by the blinking frequency, have a good effect on user experience?

In this evaluation experiment, we investigated the effect of controlling the notification timing based on the blinking frequency during the task of reading text information. The effect of notification timing control was evaluated based on the indicators of content comprehension, concentration, stress, response speed, and the number of responses.

4.2   Experiment Description

Environment. Fluorescent lamps were used in the evaluation experiment. Lenovo Y750 (Windows 10, 64-bit, Memory: 8 GB, HDD: 500 GB, 15.6-inch-wide screen) was the laptop used in this study. The distance between the users and the screen was 40–50 cm (Fig. 4).

Before conducting the evaluation experiment, we measured the blinking rate of each participant during concentrated reading. We prepared two environments: experimental environment A, which did not control the notification time and notified the users randomly, and experiment environment B, which controlled the notification time.

In experimental environment A, we used a random notification system (uncontrolled environment (NCE)). In experimental environment B, we used the developed control notification system (controlled environment (CE)). As long as our notification conditions are met, the system will notify in order. If the participants have been in the concentrated state, we will simultaneously push the three pre-prepared notifications before the end.

Reading Content and Content of Notifications. We prepared two sets of reading content on three topics. Each topic comprised three 450-word essays. As shown in Fig. 5, each article comprised a title and text. Each article was

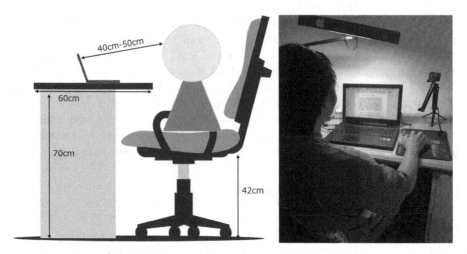

Fig. 4. Environment

separated by a line. The breakpoints we expected occur between the two topic units.

The notification was displayed in the upper right corner of the screen along with the notification sound; it can be dismissed by clicking the close button in the upper right corner. If not dismissed, the notification automatically disappeared after 5 s.

Fig. 5. Reading content

Considering that the notification content affects the response actions of participants, we set the notification content to a simple text, such as "This is the first message."

Participants and Experimental Procedure. A total of 12 informatics students participated in this experiment. We divided them into four groups of three (Table 1). We first performed an experiment on the blinking frequency of each participant, where in the degree of concentration of each participant was recorded. In this experiment, participants read an article of 796 words (20 sentences). To ensure that each participant could concentrate on reading the article, we informed them before the experiment that they were required to read the article carefully because they would be quizzed on it.

Table 1. Reading order

Group	Environment (1st round)	Reading content no. (1st round)	Environment (2nd round)	Reading content No. (2nd round)
1	A	1	B	2
2	A	2	B	1
3	B	1	A	2
4	B	2	A	1

Each participant was required to click the mouse after reading a sentence to indicate their reading position. The BFC was recorded in the middle paragraph of the article, that is, between the 3rd and 18th sentences. Subsequently, we tested and recorded the blink frequency during this time. The experimental procedure is summarized in Table 2.

Table 2. Experimental procedure

Elapsed time (min)	Contents
5	Explanation about experiment and pre-questionnaire
6	Detected blinking frequency in concentration
7	Experiment A or B (1st round)
3	Evaluation Questionnaire
7	Experiment B or A (2nd round)
3	Evaluation Questionnaire
10	Questionnaire

4.3 Evaluation Method

In this experiment, we recorded a video of participants while they were reading. After the experiment, the participants marked the position in the article where they were reading when the notification appeared according to the video recording.

We defined the breakpoints between the two articles. To evaluate whether the blinking sequence could help in detecting the defined breakpoints, we compared the relationship between the reading progress of participants and notification time and defined the breakpoints.

There is a correlation between the blinking frequency and emotional state. In the pre-questionnaire, we recorded the state of participants using a 20 grade VAS evaluation through three aspects: mental state, physical state, and eye state.

In the evaluation questionnaire, we performed a 20-grade VAS evaluation through three aspects: understanding of the articles, reading concentration, and stress level owing to the notification screen or notification sound.

4.4   Results

Blinking Frequency During Concentration. As demonstrated in Table 3, the blinking rate of each participant during the concentrated reading was different. First, participant L was the highest among the participants. In the experiment, because participant L moved the computer, we believe there was a test error in his/her data. Therefore, the data on participant L in the experimental results were not used in this study.

As shown in Fig. 6, among the remaining 11 people, the data on D, H, and K were also higher than those on others. Two of these three participants reported that their eyes were not in good condition in the pre-questionnaire. This situation can be explained by the medically proven fact that the blinking rate can increase imperceptibly during eye irritation or fatigue.

Table 3. Blinking frequency for all participants

Participant	Frequency (s)
A	0.186
B	0.175
C	0.555
D	1.230
E	0.286
F	0.293
J	0.410
H	1.245
I	0.647
J	0.307
K	1.207
L	1.924

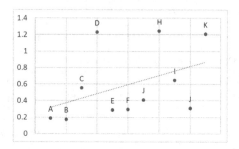

Fig. 6. Frequency of blinking during concentration (except participant L)

Understanding of Articles. In terms of content understanding, no significant difference was observed between the results for the CE (16.08) and those for the NCE(15.83).

From the results of free description, such as "There is one article I had read before" and " There is an article I read before. I think you should choose the rare articles." are obtained.

The articles used for reading in the experiment included news articles from the internet. All participants demonstrated a high understanding of the content, irrespective of the experimental environment.

4.5   Degree of Concentration During Reading

The degree of concentration in the CE was higher than that in the NCE. Significant differences were revealed by the t-test (CE: 15.75, NCE: 12, $p < 0.05$).

In the free description of the NCE, the participants wrote their opinion, such as, "The notifications received while reading affected my concentration" and "I felt that I was disturbed by the notification sounds and notifications that appeared several times in a short duration." The notification timing was set randomly without considering the user conditions. When a participant read the middle of a document, they were sometimes disturbed by the notification sound and notification.

In the free description of the CE, the participants wrote their opinion, such as, "The timing of (notification) appearance is fairly stable," "When I am not focused, notifications appear, and I can decide to keep reading or not," and "I feel like I got the notification when I was not focused on reading." Working out the concentration state of participants based on their blinking rate and controlling the notification timing enabled the participants to feel like they were reading without being interrupted by notifications.

4.6   Stress Level Owing to Notification Screen or Sound

The stress level caused by the notification screen or notifications in the CE was lower than that in the NCE. Significant differences were revealed by the t-test (CE: 4.92, NCE: 9.67, $p < 0.05$).

In the free description of the NCE, the participants wrote their opinion, such as "I was disturbed by the notification" and "I felt disturbed by multiple notification sounds and notifications." In this environment, the notification timing was set randomly, and it can be especially stressful if continuous notifications are received.

In the free description of the CE, the participants wrote their opinion, such as "The controlled environment makes it easier to concentrate and makes reading more enjoyable."

Blinking Frequency and Breakpoint. As shown in Fig. 7, participants B and E did not receive any notifications, participants F, G, and K received two notifications, and the remaining six participants received three notifications.

The marks in Fig. 7 indicate the location when they received the notification. Only one notification appeared between the two articles which is the breakpoint defined before in "Reading Contexts and Content of Notification."

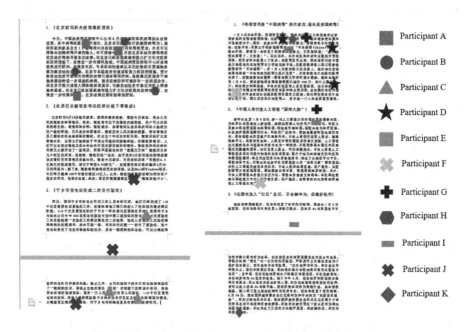

Fig. 7. Location when users received the notification

5   Discussion

5.1   Impact of Notification Timing Control on the Reading Experience

In a CE, the notifications are controlled and delayed according to the eye blinking frequency of each participant to ensure that their reading is not interrupted. In addition, by investigating the appearance position, the appearance position of notifications appeared at either the beginning or end of each article. It was considered that the participants could concentrate on the content of consecutive sentences in an article and maintain a high degree of concentration.

The results of the questionnaire about stress reveal that the CE was rated significantly lower than the NCE. One participant wrote that the CE made it easy to focus and enjoy reading. This result suggests that a system that controls the timing of notifications by detecting the blinking frequency can stabilize the notification timing to ensure that the users are less likely to be interrupted and, therefore, less stressed.

5.2   Notification and Article Contents

In response to the questionnaire, participants wrote that they would like to read more interesting articles or more novel articles. It was found that the content of articles affects the user reading experience. Additionally, a participant wrote in the questionnaire that "I do not mind the notifications (It does not matter to me whether the notifications appear or not)."

Because the notifications were too simplistic and failed to capture the attention of users, several users felt that they should continue to focus on their article without responding. Consequently, it was found that the notification content also affects the reaction of users. This aspect will be investigated in future work.

5.3   Breakpoint and Blinking Frequency Detection

Only one matched predefined breakpoint was observed. From the results of the free description, the following opinion was obtained: "Notifications appear almost at the beginning or end of an article."

In addition, the time taken by a user to move from one article to the next was fast, and the time taken by the system to decide whether to send a notification is considered to be less than a second.

Hence, considering the system delay and browsing speed, we extended the breakpoint definition section so that the first and last lines of each article were considered to be the breakpoints, and ten matched breakpoint were observed under this condition.

Even if these factors are considered, the number of sent notifications is less than half; therefore, the hypothesis "a breakpoint between the two topic units can be detected by blinking" does not hold.

6   Conclusions

In this study, we proposed notification timing control while reading text information. As a method for controlling the notification timing, we focused on the breakpoints while reading text information depending on the blinking frequency during concentrated reading. We conducted a set of evaluation experiments involving 12 people. In the experiment, we simulated a real reading environment using the system prototype. We evaluated the detection time of the breakpoints and the effectiveness of the system. Although we did not prove the hypothesis that the expected breakpoints are detected based on the blinking frequency, we found that the user browsing experience improved when they used the control system. A significant difference was observed in the items related to user concentration and stress in the system with notification timing control.

References

1. Adamczyk, P.D., Bailey, B.P.: If not now, when? the effects of interruption at different moments within task execution. In: Proceedings of the SIGCHI Conference on Human Factors in Computing Systems, pp. 271–278 (2004)

2. Avrahami, D., Hudson, S.E.: Responsiveness in instant messaging: predictive models supporting inter-personal communication. In: Proceedings of the SIGCHI Conference on Human Factors in Computing Systems, pp. 731–740 (2006)

3. Bailey, B.P., Konstan, J.A.: On the need for attention-aware systems: measuring effects of interruption on task performance, error rate, and affective state. Comput. Hum. Behav. **22**(4), 685–708 (2006). https://doi.org/10.1016/j.chb.2005.12.009

4. Bailey, B.P., Konstan, J.A.: On the need for attention-aware systems: measuring effects of interruption on task performance, error rate, and affective state. Comput. Hum. Behav. **22**(4), 685–708 (2006)

5. Boyko, N., Basystiuk, O., Shakhovska, N.: Performance evaluation and comparison of software for face recognition, based on dlib and opencv library. In: 2018 IEEE Second International Conference on Data Stream Mining & Processing (DSMP), pp. 478–482. IEEE (2018)

6. Divjak, M., Bischof, H.: Real-time video-based eye blink analysis for detection of low blink-rate during computer use. In: First International Workshop on Tracking Humans for the Evaluation of Their Motion in Image Sequences (THEMIS 2008), pp. 99–107 (2008)

7. Epp, C., Lippold, M., Mandryk, R.L.: Identifying emotional states using keystroke dynamics. In: Proceedings of the Sigchi Conference on Human Factors in Computing Systems, pp. 715–724 (2011)

8. Hashimoto, S., Yamamoto, S., Nishide, R., Takada, H.: A notification environment using user-installed beacons. In: 2017 Tenth International Conference on Mobile Computing and Ubiquitous Network (ICMU), pp. 1–4. IEEE Computer Society, Los Alamitos, CA, USA (2017). https://doi.org/10.23919/ICMU.2017.8330078

9. Hayashi, T., Haradah, F., Shimakawa, H.: Measurement of concentration based on head movement to manage progress. 72nd Nat. Convention IPSJ **9**(1), 47–50 (2010)

10. Ho, J., Intille, S.S.: Using context-aware computing to reduce the perceived burden of interruptions from mobile devices. In: Proceedings of the SIGCHI Conference on Human Factors in Computing Systems, pp. 909–918 (2005)

11. Horvitz, E.C.M.C.E.: Notification, disruption, and memory: effects of messaging interruptions on memory and performance. In: Human-Computer Interaction: INTERACT, vol. 1, p. 263 (2001)

12. Jaimes, A.: Posture and activity silhouettes for self-reporting, interruption management, and attentive interfaces. In: Proceedings of the 11th International Conference on Intelligent User Interfaces, pp. 24–31 (2006)

13. Morrison, L.G., et al.: The effect of timing and frequency of push notifications on usage of a smartphone-based stress management intervention: an exploratory trial. PloS One **12**(1), e0169162 (2017)

14. Nakano, T., Yamamoto, Y., Kitajo, K., Takahashi, T., Kitazawa, S.: Synchronization of spontaneous eyeblinks while viewing video stories. Proc. R. Soc. B: Biol. Sci. **276**(1673), 3635–3644 (2009)

15. Niwa, Y., Shiramatsu, S., Ozono, T., Shintani, T.: Realizing a direction estimation system for notification using bluetooth beacons. IPSJ Interaction B, vol. 26 (2015)

16. Okoshi, T., Ramos, J., Nozaki, H., Nakazawa, J., Dey, A.K., Tokuda, H.: Attelia: reducing user's cognitive load due to interruptive notifications on smart phones. In: 2015 IEEE International Conference on Pervasive Computing and Communications (PerCom), pp. 96–104. IEEE (2015)

17. Park, C., Lim, J., Kim, J., Lee, S.J., Lee, D.: Don't bother me. i'm socializing! a breakpoint-based smartphone notification system. In: Proceedings of the 2017 ACM Conference on Computer Supported Cooperative Work and Social Computing, pp. 541–554 (2017)
18. Sakai, T., et al.: Eda-based estimation of visual attention by observation of eye blink frequency. Int. J. Smart Sens. Intell. Syst. **10**(2), 296–307 (2017)
19. Schilit, B., Adams, N., Want, R.: Context-aware computing applications. In: 1994 First Workshop on Mobile Computing Systems and Applications, pp. 85–90. IEEE (1994)
20. Soukupova, T., Cech, J.: Eye blink detection using facial landmarks. In: 21st Computer Vision Winter Workshop, Rimske Toplice, Slovenia (2016)
21. Yamada, S., Mori, N., Kobayashi, K.: Peripheral notification that does not bother user's operations by peripheral cognition technology. Trans. Jpn Soc. Artif. Intell. **30**(2), 449–458 (2015)

Visualization and Decision-Making Support

Designing Data Visualization Dashboards to Support the Prediction of Congenital Anomalies

Tatiana Aparecida de Almeida[1,3], Ferrucio de Franco Rosa[1,2],
and Rodrigo Bonacin[1,2(✉)]

[1] UNIFACCAMP, Campo Limpo Paulista, Brazil
tatiana.almeida@ifsp.edu.br
[2] Renato Archer Information Technology Center - CTI, Campinas, SP, Brazil
{ferrucio.rosa,rodrigo.bonacin}@cti.gov.br
[3] Instituto Federal de Educação, Ciência e Tecnologia de São Paulo (IFSP Câmpus
Campinas), Campinas, SP, Brazil

Abstract. Computer based systems provides almost unlimited data storage, however, to be useful information needs to be processed and represented by suitable visualization solutions. A better visualization of information is necessary to browse and understand voluminous and complex data, such as health data. Therefore, new data visualization techniques must be investigated so that the vast amount of information makes sense for healthcare professionals, healthcare administrators and patients. More accurate and interpretative information through suitable visualization methods could, for example, contribute to mitigate problems related to congenital malformation. We investigate the design of information visualization solutions to propose new data visualization dashboards to support the prediction of congenital anomalies from patient health data. We used a national data source of Brazil. We present the design and preliminary evaluation results with experts from the health domain and related fields of research. Our proposal is intended to be useful for supporting patients, administrators and health professionals in the prediction activities.

Keywords: Information visualization · Health interfaces · Dashboards for healthcare · Healthcare decision support

1 Introduction

The evolution of computer-based systems allowed the storage of large amounts of health information. This information could be explored to provide decision support in complex situations, such as public health administration and preventive health care. However, advanced information visualization techniques must be investigated to provide an effective decision support, especially in situations when large volumes of complex data need be analyzed. According to the World Health Organization [21], in 2015 around 2.68 million newborns died in the

S. Yamamoto and H. Mori (Eds.): HCII 2021, LNCS 12765, pp. 143–162, 2021.
https://doi.org/10.1007/978-3-030-78321-1_12

neonatal period (i.e., within 4 weeks of birth), of which 330 thousand were due to congenital anomalies. In some countries, congenital anomalies are responsible for a large percentage of newborn deaths. In Brazil, for instance, around 22.8% newborn deaths are due to congenital anomalies; in the Brazilian southeast region this percentage reaches 35,9% [9]. This reinforces the need for preventive actions, as well as computational systems to assist in the prediction of congenital anomalies during prenatal care and, consequently, enabling preventive actions from the beginning.

Data science methods, such as data mining and information visualization, could be used to analyze huge amounts of data, including the factors that influence congenital anomalies cases. This analysis provides support to health care professionals, patients and health administrators in the execution of preventive actions. In this paper, we focus on the design of dashboards, which include a set of visualization techniques constructed over a real-world database.

Designing information visualization solutions is a hard task, as it should be clear and easy to understand, as well as it is necessary to take into consideration the type of visualization most appropriate for each case. The type of data to be presented, such as time data, geographical data, statistical indicators, measurements, process information, among others, also directly influences the solution to be adopted [6]. In this work, various visualization designs are integrated into dashboards to be used by health care professionals, administrators, as well as pregnant women. A dashboard is a visualization panel that presents centralized information, including indicators and metrics. Dashboards provide several views and modes of interaction. For instance, a strategic dashboard could be focused on performance indicators, whereas an analytical dashboard allows the identification of trends, and an operational dashboard usually focuses on real-time data.

Various dashboards could be used together to provide valuable visualization solutions for the health care [4]. Dowding et al. [8] present a review on the application of clinical dashboards in health care institutions. According to the authors, there is a wide variety of alternatives, as well as evidences that dashboards accessible to clinicians are associated with improvements in health care processes and treatment results. Moreover, dashboards could decrease the time spent on selecting and collecting data, and they could increase the satisfaction of health care professionals [12].

Dashboards are also used to provide health information to the population. In the coronavirus pandemic, several research institutions used dashboards as an alternative to keep the population informed about the evolution of cases of the disease. These solutions vary in terms of focus and scope. For example, the John Hopkings University dashboard[1] aims to inform the population about the cases in a global wide scale, while the Unicamp dashboard[2] focuses on cases seen at local hospitals and next their location in the city of Campinas, Brazil.

---

[1] https://gisanddata.maps.arcgis.com/apps/opsdashboard/index.html#/bda7594740fd40299423467b48e9ecf6.

[2] https://unicamp-arcgis.maps.arcgis.com/apps/opsdashboard/index.html#/3f735ecea81b419196870772a74da4a6.

In this paper, we focus on the design of dashboards to support and inform health care professionals, administrators and pregnant women about the evolutions of cases, as well as to make easy the analysis of characteristics related to congenital anomalies in all Brazilian regions, states, and towns. In our solution proposal, we use the SINASC database (Sistema de Informações de Nascidos Vivos - Live Birth Information System)[3], which integrates information on live birth in Brazil. SINASC includes over 100 characteristics of 65,873.856 births, of these 353,898 children were born with some type of anomaly.

Our dashboards include several information visualization interfaces to analyze various types of data, such as demographic, mother's health, pregnancy evolution and newborn data (including congenital anomalies). This paper also includes results from a preliminary analysis of the proposed dashboards by users.

The remainder of paper is structured as follows: Sect. 2 describes the theoretical and methodological background on information visualization and dashboards, as well as related work on dashboards for healthcare; Sect. 3 details our research method, as well as the design and development methods; Sect. 4 presents the developed dashboards and preliminary results with domain specialists; and, Sect. 5 concludes the paper and presents future work.

2   Background and Related Work

We present the theoretical and methodological foundation on which our work is based, as well as the main related works. Concepts about information visualization and concepts related to decision support dashboards used in this work are presented in Subsect. 2.1. In Subsect. 2.2 we present related works on visualization interfaces and health dashboards.

2.1   Information Visualization and Dashboards

Data visualization, a more general term than information visualization, is the representation of data in a graphic format; it aims to provide a clearer, simpler and more efficient communication of the results obtained in the data analysis process.

The growing computational power has led to the storage of large amounts of data from social networks, microscopic elements, devices, among others. As data of different formats (images, videos, texts, etc.) are stored in large quantities, advanced computing techniques became necessary to process and present them visually. In the 1990s, according to Freitas et al. [10], the work with these data was linked to the data visualization domain, which has as sub-domains, e.g., information visualization, data modeling techniques, volumetric visualization, visualization algorithms, among others.

According to Choo [7], formal decision-making is structured by rules and procedures, which characterize roles, methods and standards that will influence

---

[3] http://datasus1.saude.gov.br/sistemas-e-aplicativos/eventos-v/sinasc-sistema-de-informacoes-de-nascidos-vivos.

how organizations deal with day-to-day problems. Knowledge extraction is an important process in the context of data science; this process requires adequate information visualization methods and solutions.

Presenting data in an easy to understand way is a complex task for the designer [19]. The purpose of data visualization is to facilitate decision-making more effectively, based on the results of the graphs. The design must take into account critical aspects, such as ease of use (it should be simple for everyone), resources (powerful visualizations), support (help with visualizations and problems), and integration (integrating various data sources).

The design of visualization solutions is complex, and it's not just about adding numbers in info-graphics and waiting for people to understand; such visualizations must be clear and easy to understand, and it is necessary to take into consideration the type of visualization most appropriate for each case [16]. Various visualization solutions are used to this end, such as informational (it emphasizes a single point), comparative (it compares two or more categories), for evolution (it shows the evolution of data), for organization (it shows different ways of organizing data), and relationships (it presents correlations, distributions and relationships). The type of data to be presented, such as temporal data, topological data, statistical indicators, measurements, process information, among others, also directly influences the solution to be adopted [6].

Our proposal integrates several visualization techniques through dashboards to be used by health professionals, patients and managers. A dashboard is a panel that presents centralized information, along with indicators and metrics, to support important tasks, such as correcting failures and improving processes and strategies. Dashboards provide different views and modes of interaction. For instance, strategic dashboards focus on performance indicators, analytical dashboards allow us to identify trends, and operational dashboards usually focus on real-time data.

A health dashboard focuses on specific indicators, which directly or indirectly impact the health area [18]. As health-related data are usually obtained through a variety of systems, the quality of the information depends on the attention and precision with which it was entered into the system, as well as the level of detail that it was initially filled out. The following are some examples of the complexity and diversity of sources of health-related information: (i) information from pharmacies and drugstores on drug sales; (ii) information from rescuers and police on first aid and traffic accidents; (iii) information from hospitals and ICUs about deaths; (iv) information from health plans on private consultations (doctors, dentists, physiotherapists, etc.); and (v) information from health centers regarding vaccines.

2.2   Related Work

We present the main works related to our proposal. The works arise from an exploratory search, which begins with a historical perspective and uses previous bibliographic reviews on related subjects. Presenting a systematic review is outside the scope of this Subsection; the aforementioned reviews provide an overview of the subject.

According to Carroll et al. [5], in the last 20 years there has been an increase in the need for computerization regarding analytical visualization in public health, as a result of increased investments in information systems. Thus, these investments have generated new tools for public health activities, including systems developed at the federal, state and local levels, as well as organizational research. Advances in electronics and interoperability, computational technology, biotechnology, and other methods have led the public health area to develop itself and put these advances into practice [14]. Khan et al. [13] point out that these advances were particularly high to support the management of infectious diseases in pandemics or potential events of bio-terrorism.

Infectious diseases are a burden on the population and it could be analyzed in terms of geographical distribution, clinical risk factors, demographic data, or sources of exposure [11]. According to Andre et al. [2], although many public health reports include epidemic curves and maps, novel forms of visualization, such as graphs of social networks and phylogenetic trees, have been increasingly used to characterize disease outbreaks. According to AvRuskin et al. [3], these types of complex data allow public health professionals and researchers to integrate, synthesize and visualize information related to the surveillance, prevention and control of diseases. Thus, tracking the distribution of diseases with geographic location tools helps in detecting the clustering of diseases and in analyzing their spread.

A literature review on the use of dashboards for electronic health record information visualization is presented by Khairat et al. [12], which aimed to find proposals related to the intensive care area. 17 papers were identified and analyzed. According to the authors, the use of dashboards provided a reduction in the time of data collection, difficulty in the data collection process, cognitive load, time to complete the task, and errors, in addition to improving situational awareness, compliance with evidence-based security guidelines, usability, and navigation.

A literature review on visualization techniques in the medical field of research was conducted by Dowding et al. [8]; 4 papers were selected and are summarized and discussed, in the view of the objectives of our paper, in the following paragraphs.

Ahern et al. (2012) [1] studied patients with hypertension who used dashboards to monitor their blood pressure. The information is used for Internet communication with the primary care team. A dashboard with lights shows patients who are out of reach or in need of assistance.

McMenamin et al. [15] use first aid practices as scenarios to present a dashboard of patients linked to an electronic medical record system. The dashboard uses colors as indicators to highlight patients' health conditions; for example, informing that a patient is a smoker. Each indicator works as a tool to manage the patient's health, in order to verify other smoking-related interventions, such as alcoholism and breast cancer.

Zaydfudim et al. [22] propose a control dashboard, in the form of a screen saver, with indicators for each patient. The indicators are grouped in 3 colors: green (in compliance), red (out of compliance), and yellow (alert). Data were collected by nurses and medical staff. In addition, managers received daily reports on these compliance levels.

Pablate (2009) [20] proposes a screen saver containing a control panel that provides feedback to anesthetists in the preoperative and administration phases.

An overview of works presenting visualization and dashboard solutions for health was presented. However, no studies were found in the literature that integrate visualization methods, including geographically referenced information, in order to support the prediction of congenital malformations. This aspect, as well as the focus on user-centered design and the extensive audience, which includes health professionals, managers and pregnant women, makes the work presented in this article a unique contribution.

3   Design and Development Method

We highlight how the visualization interfaces and dashboards were designed, implemented and evaluated. To achieve our objective of supporting the prediction of congenital anomalies, our proposal is a 5-steps approach, as follows:

- Step 1: *Review of solutions and scientific literature about health care information visualization.* In this step, we aim to substantiate the choice of visualization methods, as well as the development of design alternatives.
- Step 2: *Analyze the chosen visualization techniques.* In this step, we use real data and results from the application of Data Mining techniques on congenital malformation.
- Step 3: *Propose dashboard designs.* In this step, we aggregate selected techniques in an integrated visualization solution for congenital malformation data analysis. The proposal of this step involved the participation of patients (e.g., a pregnant woman looking for information on incidence of malformation) and domain experts (e.g., health and demographic professionals) in a user-centered view, prioritizing usability aspects [17].
- Step 4: *Iterative prototype implementation.* In this step, we implement the design proposed in step 3 in a cyclic iterative process with the participation of patients and health experts. Visualization and usability aspects are evaluated in functional prototypes of the proposed dashboards.
- Step 5: *Investigation and analysis of the results.* In this step, we analyze the results along with domain experts and patients.

In the next paragraphs, we highlight the implementation activities (Step 4). The proposed approach begins with the selection of the database. We used the Live Birth Information System (SINASC, in Portuguese) database, from Brazil; it provides information about live births. The data set used comprises the years 2014 to 2017, and it has more than 100 attributes.

After obtaining the data set, we excluded information that is not relevant for predicting congenital malformation, e.g., the system version. We created the rules, separating the data from the mother, the pregnancy and the newborn, and grouped by the states to understand what to predict for states and the country. Later, we created the tests, tested the design, developed statics calculations and recreated the rules. Finally, we present the dashboards with the results of the analysis by states.

The results obtained and the rules created are sent to the dashboard; it must allow navigation in the data sources, as well as the recalculation of the data when excluding or including any attribute. We incorporate usability principles and techniques to make the system more intuitive, such as a simple interface, fields with simplified and representative names (so that demonstrate what they refer to), and colors appropriate to the type of information presented (e.g., to highlight something, to attract attention, among others).

The solution allows a comprehensive analysis of the data from mothers, pregnancy and newborn, focusing on the analysis of congenital malformations (congenital anomalies). Our solution is divided into chapters, groups, category and subcategory by state, as well as data from the mother, pregnancy, and newborns who did not have the congenital malformation. For better understanding, we created maps with the location where the children are concentrated, separated by chapters. Data can be filtered by state and year of analysis.

Functionalities were prioritized according to feedback from a demographic expert. Thus, the initial (current) version of the prototype has dashboards that allow:

1. Select the year to be analyzed.
2. Select the type of birth: with or without congenital malformation.
3. The General Data Dashboard allows us to view and select: (i) absolute birth values: Yes, No and Ignored; (ii) ranking by city; (iii) number of cities to be shown; (iv) classification in ascending or descending order.
4. The Newborn Analysis Dashboard allows us to view and select: (i) year to be analyzed; (ii) type of birth to be analyzed; (iii) state on the map to be analyzed; (iv) gender; (v) race; (vi) father's age; (vii) weight; (viii) type of childbirth; (ix) place of birth; (x) Apgar1[4] ; (xi) Apgar5 ; and (xii) type of anomaly.
5. The Mother Analysis Dashboard allows us to view and select: (i) Mother's age (average, annual variation, and ranking of the average per city); (ii) race; (iii) educational level; (iv) marital status; (v) number of previous pregnancies (average, annual variation, and ranking of the average per city); (vi) mother's occupation; (vii) ages of mother and father; (viii) number of live children (average, annual variation, and ranking of the average per city); and (ix) number of dead children (average, annual variation, and ranking of the average per city).

---

[4] The Apgar score is a method to quickly summarize the health of newborn children, the Apgar1 refers to health at the first minute and Apgar5 in the fifth minute.

6. The Gestation Data Dashboard allows us to view and select: (i) gestation time; (ii) type of pregnancy; (iii) type of birth; (iv) place of birth; and (v) quantity of prenatal consultations.

In the implementation the following tools were used: (i) Tableau[5], for data visualization, which is an analysis platform that allows us to build dashboards with advanced functionalities; (ii) a single file data source, with HYPER[6] extension, and (iii) QGIS[7] to obtain a result focused on the visualization of geographic information, for the analysis of georeferenced data. QGIS is a free software that allows us to create maps by using geographic data. To develop the map focused on the location of the newborn child with congenital malformation, we classified the malformations (anomalies) and municipality (geo code) by category, and we used the SINASC data for 2017 in our first prototype.

4   Prototype and Preliminary Study

We present a description of the current version of the prototype (Subsect. 4.1), as well as a preliminary study carried out along with a domain expert (Subsect. 4.2). We named our prototype as Congenital Anomalies Prediction Dashboards (CAPDashboards),

4.1   CAPDashboards Prototype Description

First, we highlight some examples of data visualization that can be obtained, starting with more generic results and then filtering for more detailed data, until reaching the proposed dashboards and georeferenced views. The set of all databases represents 65,873,856 births, of these 353,898 children were born with some type of anomaly. In Fig. 1, we show the evolution of births, with and without anomalies.

Based on the initial analyzes (e.g., total number of births and annual variation for live births and stillbirths), the general data dashboard was created (Fig. 2). This dashboard includes total births, number of births per type (filtered by year), number of births assigned as "Yes" (with malformation), "No" (without malformation), and "Ignored" (unknown). It also includes the annual variation, comparison of the total births of the year selected in the filter (pink) compared to the previous year (gray) opened in the month. The ranking by city shows the total number of births in cities (filtered by year).

The dashboard for analyzing mothers data allows us to filter by year, by state, or if the mother had a child with malformation; for instance, in Fig. 3 we present the result of the filter for the year 2017, state of SP, and a newborn with malformation.

---

[5] https://www.tableau.com/.

[6] Hyper is a high-volume data processing technology that offers fast analytics performance considering big data.

[7] https://www.qgis.org/.

In Fig. 4, we present a dashboard prototype focusing on analysis of pregnancies. It allows us to analyze and define the profiles of pregnant women from all over the country, selecting by state, by year (from 2014 to 2016), whether they had a newborn with malformation or not, length of pregnancy that started prenatal care, type of labor, place of birth, time of the cesarean section, induced labor, type of presentation for the state of RN, and the averages by city.

In Fig. 5, we present the dashboard with data on newborns, including data on the existence of a malformation, weight, race, gender, whether the labor was premature, distribution on the APGAR1 scale, and distribution on the scale APGAR5.

The newborn dashboard also includes the Anomalies Table (Fig. 6) to count the number of births with congenital malformations (filtered by year and type of birth) by chapter, group, category and subcategory. The arrow in Fig. 6 indicates the hide-columns functionality to increase the granularity of the information. With this dashboard the users can analyze, by state, the number of newborn children with congenital malformation, as well as view the numbers of the entire national territory by chapter, group, category and subcategory.

In addition to the aforementioned dashboards, a set of georeferenced interfaces was created. The number of cases of newborn children with congenital malformation in 2017 was weighted by the population of the municipalities, according to the estimation of the IBGE (Brazilian Institute of Geography and Statistics) for 2017, and standardized in the number of cases per 100 thousand inhabitants, being connected to the shapefile of the Brazilian municipalities[8] through a "georeferenced-code".

In Fig. 7(a) we present a map based representation of the Brazilian's 2010 Municipalities Human Development Index (MHDI - IDHM in Portuguese)[9]. In Fig. 7(b), the quantity of all newborn children with congenital malformation, from Brazil (during 2017) is presented. Despite the distance of terms (2010 and 2017), the users can identify, for example, an eventual relation between MHDI and congenital malformation.

In Fig. 8, we present other examples of maps, with quantity of cases (per 100,000 inhabitants) of newborn children with congenital malformations (various types of anomaly). Categories are described according to the IDC-10[10].

4.2   Preliminary Study

We aim to evaluate the feasibility of the CAPDashboards prototype, to identify interface problems and to elicit improvement suggestions for the next version.

The study was carried out during January 2020. 25 people with different profiles participated in this study, namely: (i) 4 demography experts (2 professors and 2 researchers); (ii) 5 heath care professionals (3 nurses, 1 dentist, and 1

---

[8] https://portaldemapas.ibge.gov.br/.

[9] https://www.br.undp.org/content/brazil/pt/home/idh0/rankings/idhm-municipios-2010.html.

[10] https://icd.who.int/browse10/2019/en.

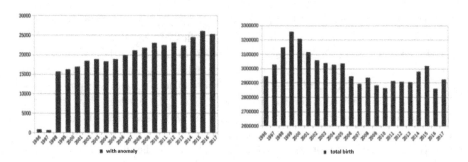

Fig. 1. Evolution of the number of births with anomaly and the total number of births in Brazil (1996–2017).

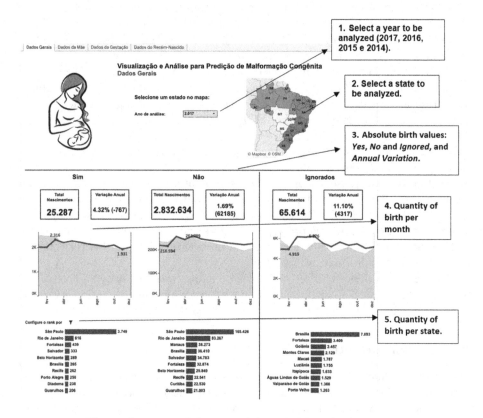

Fig. 2. Overview of the general data dashboard prototype.

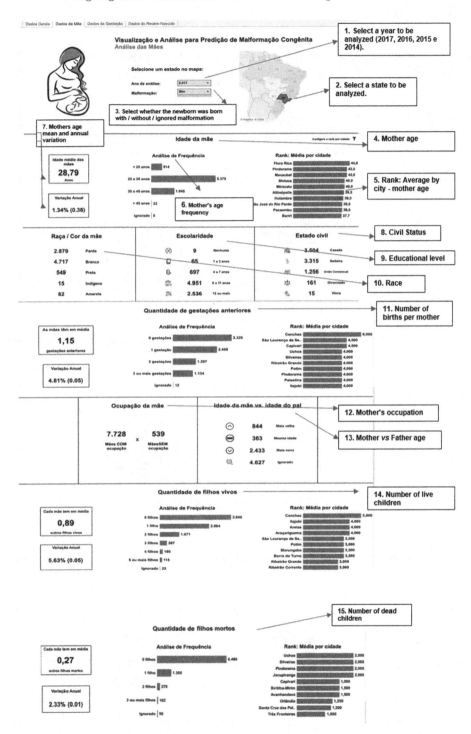

Fig. 3. Overview of the mother dashboard prototype.

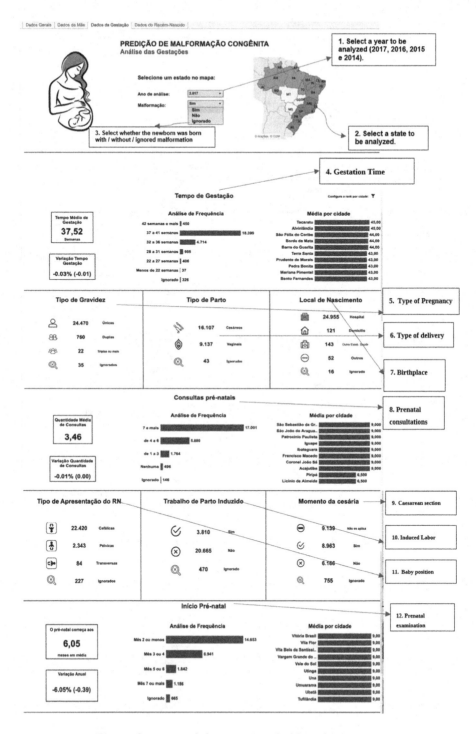

Fig. 4. Overview of the gestation dashboard prototype.

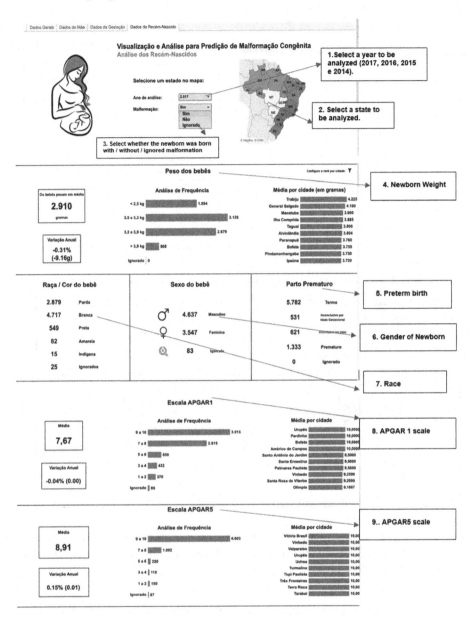

Fig. 5. Overview of the newborn dashboard prototype.

veterinary); (iii) 2 computer scientists; and (iv) 14 people from other specialties and professions on patient role (e.g., 3 teachers, 2 call center operators, 2 undergraduate students, etc.). 17 participants are female and 8 male; and 12 single, 9 married and 4 divorced. 12 participants know a child with (any) congenital malformation.

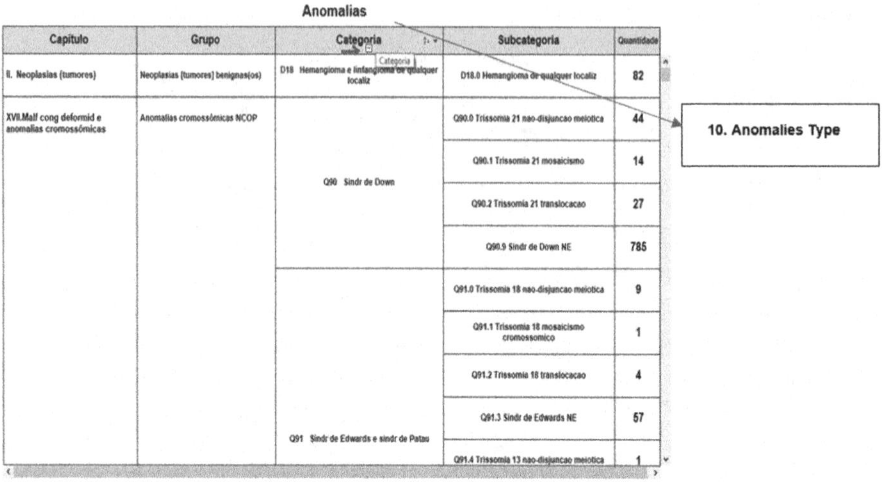

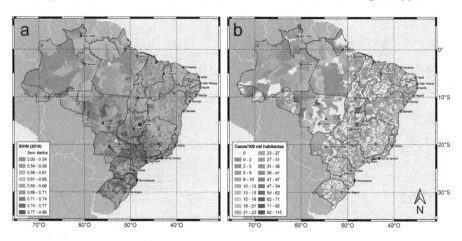

Fig. 6. Overview of anomalies table of the newborn dashboard prototype.

Fig. 7. (a) Map representing the MHDI (2010) of Brazilian municipalities and (b) a Map with incidence of newborn with congenital malformation of year 2017, cases/100 thousand inhabitants.

Firstly, all participants signed a consent form to the online evaluation. In sequence, the participants interacted with the CAPDashboards, for as long as they feel necessary to explore the interface, and then answered questions in the Google Forms. This form included the following questions:

– Positive and negative Likert Scale questions[11] were used in order to verify opinions. In Total we had 14 positive and 4 negative statements, as follows (the negative ones are identified with *(negative) marking*):

---

[11] 1. Strongly disagree, 2. Disagree, 3. Neutral, 4. Agree, 5. Strongly agree.

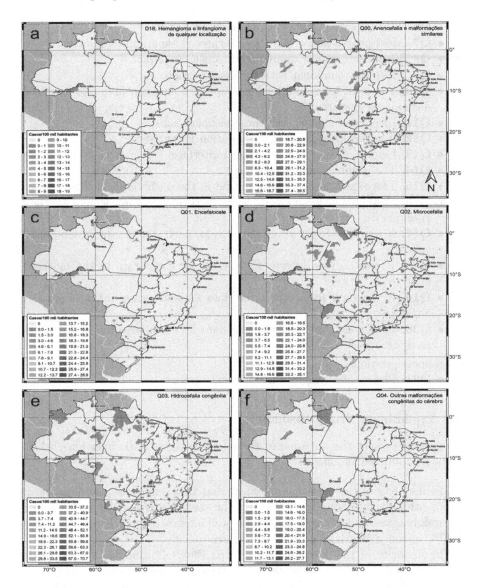

Fig. 8. Maps of the incidence (cases/100 thousand inhabitants) of congenital malformation according to the type of anomaly.

- **Q1.** I would like to use these dashboards.
- **Q2.** I found the visualizations unnecessarily complex *(negative)*.
- **Q3.** I found the visualization dashboards easy to use.
- **Q4.** A support of an expert is necessary to be able to use and complete the analyzes *(negative)*.
- **Q5.** The visualizations were well integrated.

- **Q6.** The data visualization is inconsistent *(negative)*.
- **Q7.** Most people would learn to interpret these graphics quickly.
- **Q8.** Visualizations are complicated to use *(negative)*.
- **Q9.** I feel very confident with the visualization.
- **Q10.** You must learn to interpret graphics before continuing to use these dashboards *(negative)*.
- **Q11.** I feel comfortable with these visualizations.
- **Q12.** It was easy to find the information I wanted.
- **Q13.** I enjoyed using the Dashboard interface.
- **Q14.** The graphics interface is nice.
- **Q15.** The organization of information in the dashboards is clear.
- **Q16.** I had difficulties in interpreting the state values in the map interface *(negative)*.
- **Q17.** It was easy to interpret the data for each state in the general analysis.
- **Q18.** It was easy to interpret the data for each state in the mother's analysis.
- **Q19.** It was easy to interpret the data for each state in the pregnancy analysis.
- **Q20.** It was easy to interpret the data for each state in the newborn analysis.

– An open question:
  - **Q21.** Please, give suggestions for improvements.

4.3 Preliminary Study Results

In Fig. 9 we present the outcome to the positive Likert Scale questions, which can be summarized as follows:

– The majority of the participants affirm that they would use the dashboards (8% strongly agree, and 52% agree); the dashboards are easy to use (8% strongly agree, and 44% agree); and it is well integrated (68% agree). However, 40% agreed with "most people would learn quickly".
– The majority of the participants declared they felt confident (8% strongly agree, and 44% agree); comfortable (8% strongly agree, and 56% agree); they found information easily (4% strongly agree, and 52% agree); and they considered the organization of information clear (20% strongly agree, and 52% agree). The majority also enjoyed using the CAPDashboards (4% strongly agree and 48% agree), and considered its interface as nice (4% strongly agree, and 64% agree).
– Regarding map-based interfaces, the majority considered easy to interpret in the four perspectives: (i) General (4% strongly agree, and 68% agree); (ii) Mother (4% strongly agree, and 56% agree); (iii) Pregnancy (4% strongly agree, and 72% agree); and (iv) Newborn (4% strongly agree, and 76% agree).

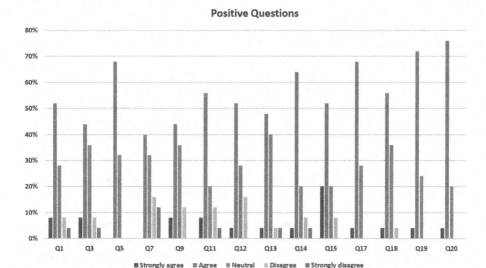

Fig. 9. Overview of results for positive likert scale questions.

Figure 10 presents the outcomes to the negative likert scale questions. The majority disagrees that the visualizations are unnecessarily complex (16% strongly disagree, and 40% disagree); that the support of an expert is necessary (4% strongly disagree, and 48% disagree); that visualizations are inconsistent (12% strongly disagree, and 46% disagree); that visualizations are complicated to use (56% disagree), and that is necessary to learn interpreting graphics (20% strongly disagree, and 48% disagree). The majority also disagrees about having difficulties in interpreting the state values in map-based interfaces (16% strongly disagree, and 40% disagree).

Regarding Q21, we obtained seven suggestions, which can be summarized as follows:

1. "I have suggestions for better visualizations, but I don't know if the Tableau software tool allows it."
2. "I think it would be good to include a search filter by city, or to allow choosing a city x to analyze its information."
3. "The information is great and very well organized. I would only suggest adding web accessibility requirements to the dashboard."
4. "There are lots of information on the same page. I believe it should be more synthetic."
5. "Well presented. As a suggestion, use more distinct colors to facilitate visualization, which would be interesting, especially when it is necessary to open several tabs to cross data."
6. "I believe that because I view it on my mobile, it is not the best option to analyze. It would be better to view the data printed on the paper."
7. "It is necessary to filter by target audience. A more direct vocabulary is needed."

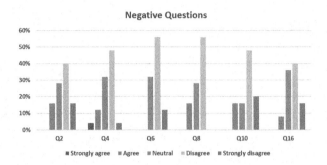

Fig. 10. Overview of results for negative likert scale questions.

In general, the suggestions indicate the need for improving visualizations and providing new filters. The participants also expressed the need for accessibility and more minimalists (synthetic) solutions. The use of colors and adapted vocabularies were proposed as alternatives to improve the interface. One participant also pointed out difficulties in viewing the dashboards' information on mobile devices.

5   Conclusion

In the public health realm, computer systems are used, for example, in the prediction of epidemics, identification of disease epicenters, as well as in the analysis and prediction of congenital malformations. The proper design of visualization solutions is essential for the large amount of information stored in computer systems to be correctly interpreted by healthcare professionals, managers and patients.

Various techniques can be used to visualize health information, e.g., heat maps and graphs (bubbles, bars, dispersion, among others). However, these forms of visualization in general present the information in a static way, not allowing dynamic navigation, where data can be filtered according to the users' needs.

We use information visualization techniques and Human-Computer Interaction principles, as well as geo-referenced information, to propose intuitive and flexible dashboard designs, with several filters and alternative views. We are aiming to allow users with different levels of computer knowledge to be able to use health information for decision making.

As future work, we expect to build solutions for visualizing statistical analysis results, as well as visualizing the results from applying machine learning techniques. We also plan long-term evaluation by health professionals, managers and patients, with the database constantly updated.

References

1. Ahern, D.K., Stinson, L.J., Uebelacker, L.A., Wroblewski, J.P., McMurray, J.H., Eaton, C.B.: E-health blood pressure control program. J. Med. Pract. Manage. **28**(2), 91–100 (2012)
2. Andre, M., et al.: Transmission network analysis to complement routine tuberculosis contact investigations. Am. J. Publ. Health **97**(3), 470–477 (2007). https://doi.org/10.2105/AJPH.2005.071936, 17018825[pmid]
3. AvRuskin, G.A., Jacquez, G.M., Meliker, J.R., Slotnick, M.J., Kaufmann, A.M., Nriagu, J.O.: Visualization and exploratory analysis of epidemiologic data using a novel space time information system. Int. J. Health Geograph. **3**(1), 26 (2004). https://doi.org/10.1186/1476-072X-3-26
4. Buttigieg, S.C., Pace, A., Rathert, C.: Hospital performance dashboards: a literature review. J. Health Organ. Manag. **31**(3), 385–406 (2017)
5. Carroll, L.N., Au, A.P., Detwiler, L.T., Fu, T.C., Painter, I.S., Abernethy, N.F.: Visualization and analytics tools for infectious disease epidemiology: a systematic review. J. Biomed. Inform. **51**, 287–298 (2014)
6. Chen, M., et al.: Data, information, and knowledge in visualization. IEEE Comput. Graph. Appl. **29**(1), 12–19 (2009). https://doi.org/10.1109/MCG.2009.6
7. Choo, C.W.: The knowing organization: how organizations use information to construct meaning, create knowledge and make decisions. Int. J. Inform. Manag. **16**(5), 329–340 (1996)
8. Dowding, D., et al.: Dashboards for improving patient care: review of the literature. Int. J. Med. Inform. **84**(2), 87–100 (2015). https://doi.org/10.1016/j.ijmedinf.2014.10.001
9. Fontoura, F.C., Cardoso, M.V.L.M.L., Rodrigues, S.E., Almeida, P.C.D., Carvalho, L.B.: Ansiedade de mães de recém-nascidos com malformações congênitas nos períodos pré e pós-natal. Rev. Latino-Am. Enfermagem **26**, e3080 (2019). https://doi.org/10.1590/1518-8345.2482.3080
10. Freitas, C.M.D.S., Chubachi, O.M., Luzzardi, P.R.G., Cava, R.A.: Introdução à visualização de informações. Rev. Inform. Teórica aplicada. Porto Alegre. **8**(2), 143–158 (2001)
11. Hay, S.I., et al.: Global mapping of infectious disease. Philosop. Trans. R. Soc. London. Ser. B Biol. Sci. **368**(1614), 20120250–20120250 (2013). https://doi.org/10.1098/rstb.2012.0250, 23382431[pmid]
12. Khairat, S.S., Dukkipati, A., Lauria, H.A., Bice, T., Travers, D., Carson, S.S.: The impact of visualization dashboards on quality of care and clinician satisfaction: integrative literature review. JMIR Hum. Factors **5**(2), e22–e22 (2018). https://doi.org/10.2196/humanfactors.9328, 29853440[pmid]
13. Khan, A.S., Fleischauer, A., Casani, J., Groseclose, S.L.: The next public health revolution: public health information fusion and social networks. Am. J. Publ. Health **100**(7), 1237–1242 (2010). https://doi.org/10.2105/AJPH.2009.180489. 20530760[pmid]
14. Lopez, D., Blobel, B.: Semantic interoperability between clinical and public health information systems for improving public health services. Stud. Health Technol. Inform. **127**, 256–67 (2007)
15. McMenamin, J., Nicholson, R., Leech, K.: Patient Dashboard: the use of a colour-coded computerised clinical reminder in Whanganui regional general practices. J. Prim. Health Care **3**(4), 307–310 (2011)

16. Nediger, M.: How to choose the best types of charts for your data (2019). https://venngage.com/blog/how-to-choose-the-best-charts-for-your-infographic/
17. Nielsen, J.: Usability Engineering. Morgan Kaufmann, Burlington (1994)
18. Nogueira, P., Martins, J., Rita, F., Fatela, L.: Dashboard da saúde: passado, presente e futuro. uma perspectiva da evolução em portugal. Int. J. Inform. Manag. **1**(2), 1–12 (2017)
19. Ouellete, C.: Five best survey data visualization tools (2019). https://optinmonster.com/best-survey-data-visualization-tools/
20. Pablate, J.: The effect of electronic feedback on anesthesia providers' timely preoperative antibiotic adminstration. Ph.D. thesis, University of North Florida (2009)
21. WHO: Congenital anomalies (2016). https://www.who.int/news-room/fact-sheets/detail/congenital-anomalies
22. Zaydfudim, V., et al.: Implementation of a real-time compliance dashboard to help reduce SICU ventilator-associated pneumonia with the ventilator bundle. Arch. Surg. **144**(7), 656–662 (2009)

Improving User Experience Through Recommendation Message Design: A Systematic Literature Review of Extant Literature on Recommender Systems and Message Design

Antoine Falconnet[1], Wietske Van Osch[1], Joerg Beringer[2], Pierre-Majorique Léger[1], and Constantinos K. Coursaris[1(✉)]

[1] HEC Montréal, Montréal, QC H3T 2A7, Canada
constantinos.coursaris@hec.ca
[2] Blue Yonder, Coppell, TX 75019, USA

Abstract. This paper reports the findings of a Systematic Literature Review of extant literature on Recommender Systems (RS) and message design. By identifying, analyzing and synthesizing relevant studies, we aim to generate a contemporary mapping of studies related to user-RS interaction, extend the body of knowledge regarding effective recommendation messages, inform practitioners about the effect of recommendation message design choices on the user's experience, and motivate researchers to conduct related future research on new RS message factors identified in the literature. To conduct this SLR, 132 papers were collected and analyzed; after assessing their relevance and quality, 41 papers were selected, classified, interpreted and synthesized under a strict methodology producing the results reported in this paper, and concluding with a concept matrix outlining opportunities for future research on how to optimize the design of RS in support of a managerial decision-making context.

Keywords: User experience · Recommender Systems · Message design · Systematic Literature Review

1 Introduction

To help businesses, employees, and customers improve their decision-making quality while reducing their decision-making effort, Recommender Systems (RSs) have been used in many domains since the mid-1990s (e.g., e-commerce, management, education, healthcare, government) [1–4]. For example, Amazon.com, a well-known e-commerce vendor, recommends products to its customers through an RS. Another example, Netflix uses an RS to offer its users movies according to their preferences.

Due to this growing use and importance of RSs, researchers have studied their performance [5], algorithmic accuracy [6–9], elicitation recommendation methods [9–12] and explanations [13–16] and demonstrated the system design's effects on user experience. Studies have shown that the recommendation's nature (i.e. the message's content

S. Yamamoto and H. Mori (Eds.): HCII 2021, LNCS 12765, pp. 163–181, 2021.
https://doi.org/10.1007/978-3-030-78321-1_13

and format) will impact users' beliefs, attitudes and behavioral intentions, particularly in the context of managerial decision-making [17, 18]; however, these past studies were mostly focused on system-level (e.g. interface) design implications [19]. Also, given that recommendation explanations can improve user experience [20, 21], there is a need for future research to understand how the recommendation message format and content could be improved to further improve users' decision-making [22–24].

Whereas knowledge on the impact of message design choices on the reader's perceptions and behaviors are more developed in other areas [25–30], limited research exists on what would optimize the design of recommendation messages [31]. Building on the work by [32], this paper aims to further explore the antecedents to effective recommender system design by developing the theoretical grounding of related concepts.

The aim of this study is to carry out a systematic literature review of relevant studies in order to: 1) Generate a contemporary mapping of studies related to user-RS interactions; 2) Extend the body of knowledge by synthesizing the retrieved studies' findings into a framework describing the design factors that impact the effectiveness of recommendation messages and inform practitioners including system designers about the effects of recommendation message design choices on the user's experience; 3) Propose new RS message factors to study and motivate researchers to conduct related future research to deepen our understanding. Hence, this study is guided by the following research questions:

RQ1 What comprises the current knowledge base of the antecedents to effective RS message design?

RQ2 What statistically significant results from past research can inform current scholars and practitioners of optimal RS message design practices?

RQ3 What are the opportunities for future research subsequently potentially revealing guidelines on how to optimize RS message design in a managerial decision-making context?

This article is presented as follows. First, we describe the research methodology by presenting the protocol we followed to conduct this Systematic Literature Review. Next, we present the analysis and results of our review by answering our three research questions. Finally, we conclude the paper by a discussion of our work, what it brings to the scientific and practical fields, and opportunities for future research.

2   Research Methodology

This paper is a Systematic Literature Review of extant literature concerning RS and message design [33]. A review protocol was defined in order to conduct the literature review and consists of the four stages below: i) searching for literature in scientific databases, ii) reviewing and assessing the search results, iii) analyzing and synthesizing the results, and iv) reporting the review findings.

To find the appropriate studies, queries with relevant search terms were carried out in international databases of authoritative academic resources and publishers, international journals and selected conference proceedings. Details on the search queries are presented in Table 1.

Table 1. List of search keywords and scientific databases used to identify papers.

Search keywords	Scientific databases
• "Recommender System" • "Recommendation System" • "Recommender System message" • "Recommendation System message" • "Recommender System message design" • "Recommendation System message design" • "Recommender System user acceptance" • "Recommendation System user acceptance" • "Recommendation message design" • "Message design" • "Message design acceptance" • "Message design guidelines" • "Warning message design" • "Persuasive message design" • "Information Systems message design" • "Trust in Recommender System" • "Trust in Recommendation System" • "Explanation in Recommender System" • "Explanation in Recommendation System"	• Google Scholar • ACM Digital Library • ABI/INFORM collection (Proquest) • IEEE Xplore • SpringerLink • ScienceDirect • Journal of the Association for Information Systems • Information Systems Journal • Information Systems Research • Journal of Information technology • Association for Information Systems Transactions on Human-Computer Interaction • Management Information Systems Quarterly • Journal of Management Information Systems

The timeframe of the research was specified to the last decade (2010–2020), during which researchers conducted numerous user-centric studies and the importance of recommendation presentation emerged.

When all papers had been identified, we subsequently proceeded to assess their applicability to this literature review applying two filtering stages, focused on relevance and quality respectively. In the first filtering stage, we applied a set of inclusion and exclusion criteria to assess the relevance of the article to the purposes of this literature review, as defined in Table 2.

Applying these inclusion and exclusion criteria, 132 papers corresponding to the search terms and criteria were collected, which were subsequently subjected to a quality review.

After identifying and collecting the papers, they were analyzed in order to assess the quality of their content according to the following criteria: 1) Citations (numerous and varied); 2) Clear and detailed presentation of the results and their implication and contribution to the field; 3) Brings new knowledge and/or proposes relevant future research to be carried out.

Applying these quality criteria to the remaining 132 papers, 91 papers were removed. Then, the remaining 41 papers were classified according to the research discipline and

Table 2. Inclusion and exclusion criteria used to filter the 132 identified papers.

Include papers about or published in…	Exclude papers…
• RS user acceptance • RSs user-centric studies • Message design • Peer-reviewed conferences, workshops, and journals • English • Between from 2010 to 2020	• Not addressing RS or message design • Papers addressing RSs but centered on methods and techniques (algorithm, elicitation recommendation, RS types, data mining etc.) • Without empirical evidence

results. Finally, non-statistical methods were used to evaluate and interpret findings of the collected papers and conduct the synthesis of this review.

3   Analysis and Results

3.1   RQ1 What Comprises the Current Knowledge Base of the Antecedents to Effective RS Message Design?

In the following, the current knowledge base of the antecedents to effective RS message design are summarized. Given the similarity of identified topics, we grouped existing findings into four (4) major groups of knowledge: (1) users' perceptions and behaviors, (2) information design, (3) recommendation message format and content and (4) explanations in recommendation messages.

User's Perceptions and Behaviors
Based on [17, 18] developed a user-centric evaluation framework for recommender systems named **ResQue** in order to determine the essential qualities of an effective and satisfying RS and the key determinants motivating users to adopt a recommender technology. The model shows that RS transparency and perceived usefulness influence the user's trust and confidence in the RS, which are linked to the intention to use and/or purchase. In their turn, control, perceived usefulness, and perceived ease of use of the RS affect the overall satisfaction of the user, which is also linked to users' use intentions of the RS. The authors also explain that to create a transparent system, the recommendation must be accurate and contain explanations. The user feels control over the RS if the interaction is adequate. Also, the RS is perceived as easy to use due to an adequate interface and perceived as useful if the recommendations are accurate, novel, diverse and contain sufficient information.

Information Design
Information design is a concept whose main goal is the clarity of communication. So, its principles are beneficial to any context where the aim is to provide information to someone, e.g. via a recommendation system message. [25, 34, 35] have studied information design and gathered theories and guidelines for message designers. Based on the observation that several authors have pointed out that the message's form follows

function, authors argue that the message's content is more important than its actual execution [34, 35]. Thus, the information in each message will have to be structured and adapted to the needs of the intended receivers. Also, because the intended receivers must have easy access to facts and information when they need it, information should be conveyed in a clear manner from the sender (i.e., recommender system) to the receiver (i.e., recommender system's user). Thus, complicated language which will impair the understanding of the recommendation system message should be avoided.

Recommendation Message Format and Content

The procedure of the recommendation's presentation depends on the implemented algorithms. Depending on the algorithm chosen by the designers, recommendations will vary. However, no matter the algorithm chosen, the recommendation message itself, and more specifically its design, will have a significant influence on the user experience, their perceptions and behavior.

Effectively, information sufficiency, defined by [36] as "the content presented to the user that should be enough for him/her to understand the situation and to act on it while saving time and effort if he or she do not have this information or in a different form", is a characteristic of an effective recommendation message; similarly important characteristics include transparency, flexibility, and accessibility because "the recommendation should allow the shopper to easily navigate through" [36]. Moreover, [37] recommend to RS designers to pay attention to the display format of the recommendations by using navigational efficacy, design familiarity, and attractiveness. Also, these same authors identified in the literature that the user's perceived credibility of the system influences the user's attitudes and behaviors.

Like [17, 38] consider that the accuracy of the recommendations is an important factor in the decision-making process preceding the uptake of the recommendations, but their potential value for the user are also important. On their side, [39] argue that the degree of trust users put in the system plays an important role in the acceptance of a recommendation. In addition to transparency and perceived usefulness, explanations contribute to user trust in RS [40–42]. Furthermore, these factors can affect each other. For example, the accuracy and the diversity of recommendations positively affect user trust and lead to an increased adoption rate of recommendations [43].

Thus, there are numerous factors that impact user's evaluations of RSs and recommendation messages, and these factors are influenced by the RS message characteristics [37]. Indeed, the content and the format of recommendations have significant and varied impact on users' evaluations of recommender systems, thereby influencing the user's decision-making process [37, 44].

While text explanations were perceived as more persuasive than every visual format [45], the navigation and layout of recommendation presentation interfaces are also important, because they significantly influence users' satisfaction and their overall rating of the systems; also, interface design and display format influenced RS users' behaviors.

Explanations in Recommendation Messages

Another important factor influencing the effectiveness of an RS message identified in the literature pertains to information, and more specifically explanations.

Explanation is a component of the explanation interface that consists of three elements: explanation, presentation and interaction, and is itself composed of three elements: (i) the content (ii) the provisioning mechanism and (iii) the presentation format [46].

There are two types of explanations, i.e. explaining the way the recommendation engine works and explaining why the user may or may not accept the recommendation [21]. Explanations are used to give more information to the users on the recommendation in order to help them to better understand the systems' outcomes and be sure they make the best choice, which leads to satisfaction, transparency, confidence, perceived ease of use and perceived usefulness [22]. Although they cannot compensate for poor recommendations, they can increase user acceptance of RSs, help users make decisions more quickly, convince them to accept the recommendation and develop users' trust in the system as a whole [22, 47]. So, a recommendation should use an explanation model that would help users understand the recommendation reasoning process [22] and positively influence transparency, scrutability, trust, effectiveness, efficiency, persuasiveness, and satisfaction [48].

However, users respond to explanations differently according to their context and intent, so there is a need to jointly optimize both recommendation (i.e., the solution) and explanation selections [49]. Different explanations have been tested over the last decade. In a taxonomy for generating explanations in RS, [48] argue that structural characteristics such as length, writing style or the confidence that is conveyed in explanations can be used as additional dimensions. Moreover, the authors encourage RS designers to exclude from an explanation all information and knowledge that are not relevant for answering a request. Also, users have a preference for knowledgeable explanations and the recommender may be formulated along the line of "You might (not) like Item A because…" to create an effective explanation [21]. Thereby, the richness of explanations plays a pivotal role in trust-building processes and recommendation should incorporate explanatory components that imitate more closely the way humans exchange information [50]. To enact it, [50] recommend combining different explanation styles in the recommendation, which lead to explanations with a higher perceived value and trust in the recommendation. Furthermore, arguments should contain only pertinent and cogent information, while titles are preferred to be presented in a natural and conversational language. For the recommendation of low-risk products, users were found to prefer short sentences. However, for high-risk products, users were found to prefer long and detailed sentences. Long and strongly confident explanations can be more effective in the acceptance of interval forecasts [46]. As context and intent are important in the decision-making process, personalized explanations are often linked to improved transparency, persuasiveness and satisfaction, when compared to non-personalized explanations [51]. In addition, generating familiar recommendations with detailed information and explanations regarding the underlying logic of how the recommendation was generated increases the users' perceived credibility of the system [37]. Successful recommendations also need to take into account user perceptions of recommendation properties such as diversity and serendipity, user short-term information needs, user context, and mood, i.e. what users are thinking [52].

Thus, in answering the above-stated RQ1 ("What comprises the current knowledge base of the antecedents to effective RS message design?"), a concept matrix of the current knowledge base of the antecedents to effective recommender system (RS) message design is presented in Table 3.

Table 3. Concept matrix: current knowledge base of the antecedents toeffective recommender system (RS) message design.

	Users perceptions and behaviors	Information message	Recommendation message content and format	Explanations in recommendation message
Coursaris et al. (2020)	x		x	
Kunkel et al. (2019)				x
McIenerney et al. (2018)				x
Nunes and Jannach (2017)				x
Paniello et al. (2016)			x	
Guanawardana and Shani (2015)			x	
Jameson et al. (2015)			x	
Al-Taie and Kadry (2014)				x
Pettersson (2014)		x		
Pettersson (2012)		x		
Tintarev and Masthoff (2012)				x
Yoo and Gretzel (2012)			x	x
Friedrich and Zanker (2011)				x
Gedikli and Jannach (2011)				x
Mandl et al. (2011)			x	
Pu et al. (2011)	x			
Tintarev and Masthoff (2011)				
Ozok et al. (2010)			x	
Pettersson (2010)		x		
Total	2	3	7	8

Having presented the current knowledge base of the antecedents to effective RS message design, we now present statistically significant results from prior research that may inform scholars and practitioners' RS message design practices.

3.2   RQ2 What Statistically Significant Results from Past Research Can Inform Current Scholars and Practitioners of Optimal RS Message Design Practices?

In what follows, we begin by focusing on relevant results from two studies. These two studies provide an ideal starting point as they offer an overarching classification of message design elements. Then, we turn to studies that extend the two previous studies by studying fewer factors in depth.

[17] proposed the *ResQue* model after conducting a user-centric study. This study has led to a better comprehension of the factors that impact the acceptance of recommendations and the use intentions of RSs. Indeed, all of the hypotheses presented in the *ResQue* model presented in the previous section have been statistically validated. Based on this work, [32] proposed 36 hypotheses to study the effects of message design on the user acceptance of Recommender Systems and the system-generated recommendations. Among the 36 hypotheses, 23 were statistically validated. The most interesting results for the RS message design practices found by [32] are that information specificity of the recommendation impacted information sufficiency and information transparency. In turn, information sufficiency (i.e., both problem and solution) influences perceived usefulness, and information transparency positively impacts confidence in recommendation and perceived ease of use of the RS. This study also showed that a higher information specificity of both the problem and the solution increase the user's recommendation acceptance and reduce the decision-making time (when controlled for message length), while information sequence such that information is presented from problem to solution reduces the decision-making time.

[20] studied the influence of knowledgeable explanations on users' perception of a recommender system. The author conducted an online experiment on a real-world platform and found that knowledgeable explanations significantly increase the perceived usefulness of a recommender system. In another study on presenting explanations in RS [53], the authors developed an innovative study design for measuring the persuasion potential of different explanation styles (sentences, facts or argument styles) by comparing participants' robustness of preferences in face of additional explanations. Results indicate that fact-based explanations (i.e. only facts with keywords without sentences) have a stronger impact on participants' preference stability than sentence-based explanations. Further, argumentative facts and argumentative sentences impact stronger the users' preference than the solely facts. Thus, fact-based explanations and argumentative explanation style are preferred by the users than full sentences explanations.

[54] presents a system design featuring interactive explanations for mobile shopping recommender systems in the domain of fashion. Based on a framework for explanation generation [55] and a previously developed mobile recommender system [56] they developed a model of mobile recommender systems and generated explanations to increase transparency. Based on the fact that explanations must be concise and include variations in wording, they defined positive and negative argument aspects. They found that positive arguments convince the user of the relevance of the recommendation, whereas negative arguments increase the user's perceived honesty of the system regarding the recommendation.

In addition, [31] were interested in the transparency of a Recommender Assistant (RA). Their study reveals that transparent RA requiring low cognitive effort increases the user's perceived sense of control. Furthermore, results showed that participants' perceived RA credibility, decision quality, and satisfaction were positively affected by a transparent RA and were not impacted by the cognitive effort needed to access and understand the explanations of the RA.

Also, RS interface design and trust have previously been studied. [40] studied users' trust factors in music recommender systems by comparing different recommendation presentation factors (i.e. presentation, explanation, and priority). They found that the type of explanation to be used in an RS depends on the desired effect on the user. More precisely, persuasive explanation is suited to support the competence facet of the RS, while displaying a rating (there, IMDb score) will promote the honesty and objectivity of the RS. Then, [57] conducted a user study on the influence of interface design choices along two axes (i.e. information scent and information access cost) on feedback quality and quantity. The authors found that people have a preference for descriptions with a higher level of detail. Indeed, people preferred the interfaces that provided a strong information scent over the ones with weak information scent.

[58] developed a UTAUT2-based framework and tested it in a quantitative study with 266 participants on social recommender systems. The results of the survey showed that the integration of user's social information could improve the intention to use a recommender system. Thus, users prefer a system that provides recommendations based on a combination of user's social networking information, profile information and reading behavior, especially when the recommendations are from well-known and trustworthy people.

Thus, in answering the above-stated RQ2, a concept matrix of the statistically significant results from past research informing current scholars and practitioners of optimal RS message design practices is presented in Table 4.

Table 4. Concept matrix: statistically significant results from past research informing current scholars and practitioners of optimal RS message design practices.

Article	Study	Results
Pu et al. (2011)	The *ResQue* model tested several factors for their impact on the acceptance of recommendations and the use intentions of the RS	All hypotheses have been statistically validated
Coursaris et al. (2020)	A research model composed of 36 hypotheses to study the effects of message design on the user acceptance of Recommender Systems and the system-generated recommendations	23 hypotheses on 36 are validated

(continued)

Table 4. (*continued*)

Article	Study	Results
Zanker (2012)	An online experiment on a real-world platform on the impact of knowledgeable explanations on the RS users' perceptions	Knowledgeable explanations significantly increase the perceived usefulness of a recommender system
Zanker and Schoberegger (2014)	An innovative study design for measuring the persuasion potential of different explanation styles (sentences, facts or argument styles)	Fact-based explanations and argumentative explanation style are preferred by the users than full sentences explanations
Lamche et al. (2014)	A system design featuring interactive explanations for mobile shopping recommender systems in the domain of fashion to increase transparency and honesty in RS	Positive arguments convince the user of the relevance of the recommendation, whereas negative arguments increase the user's perceived honesty of the system regarding the recommendation
Bigras et al. (2019)	A within-subject laboratory experiment conducted with twenty subjects investigating how assortment planners', perceptions, behavior, and decision quality are influenced by the way recommendations of an artificial intelligence (AI)-based recommendation agent (RA) are presented	Transparent RA requiring low cognitive effort increases the user's perceived sense of control. RA credibility, decision quality, and satisfaction were positively affected by a transparent RA and were not impacted by the cognitive effort needed to access and understand the explanations of the RA
Holliday et al. (2016)	A study on users' trust factors in music recommender systems by comparing different recommendation presentation factors (i.e. presentation, explanation, and priority)	The type of explanation to be used in an RS depends on the desired effect on the user
Schnabel et al. (2018)	A user study on the influence of interface design choices along two axes (i.e. information scent and information access cost) on feedback quality and quantity	People have a preference for descriptions with a higher level of detail. They prefer the interfaces that provided a strong information scent over the ones with weak information scent
Oechslein et al. (2014)	A UTAUT2-based framework and tested it in a quantitative study with 266 participants on social recommender systems	The integration of user's social information could improve the intention to use a recommender system

3.3   RQ3 What Are Opportunities for Future Research Subsequently Potentially Revealing Guidelines on How to Optimize RS Message Design in a Managerial Decision-Making Context?

In order to unify the presentation of the above results from prior research, an iterative research process was applied. First, we describe in the following three identified domains of message design that provide insights and inspiration for the domain of recommendation message design, i.e., narrative messages, software update warning messages and information messages. Then, Table 5 presents relevant dimensions and characteristics for opportunities in RS message design.

Table 5.   Concept matrix: opportunities for future research on how to optimize RS message design in a managerial decision-making context.

Concept	Article	Opportunities for future research
Information messages	Coursaris et al. (2020)	What is the effective combination of problem information specificity, solution information specificity and information sequence in a recommendation message?
	Makkan et al. (2020)	What is the effect of affective language in recommendation messages on user behaviors and perceptions?
	Matsui and Yamada (2019)	How subjective and objective language influence the user's perception and intent for acceptance and future usage?
	Schreiner et al. (2019)	Are short and precise recommendations more efficient than long recommendations?
	Zhang et al. (2019)	What is the effect of typographical on user's perceptions and behaviors?
	Li et al. (2017)	Do apologies, reward and praise in recommendations create positive affective responses and influence users' decisions?
	Nunes and Janach (2017)	Are long explanations more persuasive than short explanations?
	Al-Taie and Kadry (2014)	Are long explanations more persuasive than short explanations?
	Harbach et al. (2013)	Are long explanations more persuasive than short explanations?
	Bravo-Lillo et al. (2011)	Are long explanations more persuasive than short explanations?

(continued)

Table 5. (*continued*)

Concept	Article	Opportunities for future research
Software update warning messages	Fagan et al (2015)	Which designs help alleviate annoyance and confusion while increasing importance and noticeability of the recommendation message? Which place and design of buttons help to reduce the level of confusion in a recommendation message?
Narrative messages	Barbour et al. (2015)	Do recommendation messages using narrative's style are more explicit, more understandable and involve less information overload?
	Niederdeppe et al. (2014)	Does the narrative's style in recommendation messages increase the persuasion of the recommendation?
	Weber & Wirth (2014)	Does the narrative's style in recommendation messages encourage a greater belief in the recommendation?
	Jensen et al. (2013)	Does the narrative's style in recommendation messages reduce feelings of being overloaded?
	Moyé-Gusé et al. (2011)	Do recommendation messages using narrative's style are more enjoyable, produce involvement, identification, and parasocial interaction than non-narrative?
	Moyé-Gusé and Nabi (2011)	Do recommendation messages using narrative's style are more enjoyable, produce involvement, identification, and parasocial interaction than non-narratives.
	Niederdeppe et al. (2011)	Do narratives mask the persuasive intent of the recommendation and increase its persuasion?
	Appel & Richter (2010)	Do recommendation messages using narrative's style reduce or circumvent negative reactions to persuasion and counterarguing and evoke emotional and cognitive responses?

Insights from Narrative Message Design Implications.
[59] and [60] conducted studies on the influence of narrative messages on readers. They compared narrative messages vs. non-narrative message and found that narratives are more enjoyable, produce involvement, identification, and parasocial interaction with

characters than non-narratives. [61] and [62] also studied the perceptions of narrative messages and observed that narrative messages reduce or circumvent negative reactions to persuasion and counterarguing and evoke emotional and cognitive responses. Also, when the legitimacy burden is greater, narratives are more persuasive because they mask the persuasive intent of the message [61, 63]. Furthermore, they encourage a greater belief in the realism of claims or the authenticity of the narrative world through a suspension of disbelief [64]. From a design perspective, narratives convey information in ways that may reduce feelings of being overloaded [65], have a more explicit story structure, are more understandable and involve less information overload [66]. Despite these benefits, the narrative message's style has never been used in recommendation messages. Thus, it would be interesting to conduct studies with narratives in recommender messages and observe their influence on the users' perceptions and behaviors.

Insights from the Design of Warning Update Software Message.
[67] studied warning message updates to identify design features that may significantly influence the level of confusion, annoyance, noticeability, and perceived importance experienced by users once a software update message is delivered. In their study, the authors showed 14 warning messages with different designs to the participants. The participants were asked to evaluate the warning message on various criteria and give feedback regarding their feelings. The study demonstrated that better designs could help alleviate annoyance and confusion while increasing importance and noticeability of updates. To this end, the update message must clearly mention what software is being updated, the risks, use simple language, explain the reason(s) behind the update, and explain the benefits clearly. Also, the placement and design of buttons underscore the importance of well-designed buttons on usability, which also help to reduce the level of confusion. Moreover, scare tactics, used to increase noticeability and importance in messages, correlates with higher annoyance, which may affect the message effectiveness in the long run.

Insights from the Information Message Design.
[68] studied message content and the format of recommendation messages and suggested that recommendation messages could elicit positive affective responses from users by utilizing affective language. Also, apologies, reward, or praise could be used to create positive affective responses [69]. Thus, affective language could increase (i.e. more positive) users' perceptions of the recommendation, recommendation acceptance, and use intentions.

[30] used typographical cues in a scientific Q&A forum context to observe users' perceptions. The results were not statistically significant, but the authors remain confident about the effect of typographical cues in messages on users and recommend testing them in a different context. [32] tested one typographical cue proposal (i.e. bold keyword vs. plain text) in a decision-making context but it did not impact the user's perceptions and behaviors. Other identified typographical cues that can be tested include italicizing, underlying, and using color among others.

Also, an extended study of [32] could be performed by focusing on the three factors (i.e., problem information specificity, solution information specificity and information sequence) that were found to be statistically significant in affecting recommendation and system-level outcomes, in order to find the best combination(s) of design choices regarding the presentation of the recommendation message.

In [70], the authors compared subjective versus objective language for the recommendation of a virtual agent and suggested testing the subjective language to measure the user's perception of the system and the recommendation, as well as their intent to accept the recommendation and use the system in the future.

Like [46, 51] suggest that long explanations in recommendation messages are more persuasive than short explanations. However, other studies recommend creating short and precise recommendations [71–73]. Thus, length and details or precision of the recommendation message should be tested in a managerial decision-making context in order to determine the effective length of a recommendation message, and how detailed the recommendation should be.

Thus, in answering the above-stated RQ3, a concept matrix of the opportunities for future research on how to optimize RS message design in a managerial decision-making context is presented in Table 5.

4 Discussion and Concluding Comments

This paper provides a valuable contribution and step forward in our understanding of the antecedents to effective recommendation message design in a managerial decision-making context. It does so by giving an overview of the current knowledge base, citing prior research, and summarizing opportunities for future research with a concept matrix.

The purpose of this literature review was to provide an overview of the current knowledge base of the antecedents to effective RS message design, identify statistically significant results from past research so as to inform current scholars and practitioners of optimal RS message design practices, and detect opportunities of future research subsequently potentially revealing guidelines on how to optimize RS message design in a managerial decision-making context. This work has been conducted on the basis of the systematic literature review protocol described in Sect. 3. We retrieved 132 papers between 2010 and 2020, which were filtered for relevance and quality before being included for the final literature review analysis. After this filtering, 41 papers were preserved. Among this papers, nineteen (19) papers were used to inform the current knowledge of the antecedents to effective RS message design, eight (8) studies with statistically significant results were identified to inform current scholars and practitioners of optimal RS message design practices, and nineteen (19) papers were used to identify opportunities of future research subsequently potentially revealing guidelines on how to optimize RS message design in a managerial decision-making context and allocated in three (3) sections (i.e. narrative messages, warning software message updates and message information design). The opportunities are presented in Table 2.

Our literature review shows that relevant studies on this topic exist but are far from comprehensive and knowledge is fragmented between several studies that have never been unified. Indeed, most studies in this research space concern themselves with recommender systems and recommendations more broadly, but only a few have investigated the factors of relevance in optimizing the design of recommendation messages. Despite the limited prior research, these studies demonstrated the importance of recommendation message design and its impact on user experience.

Still, large research gaps in this domain persist. Two user-centric studies [17, 32] presented more holistically the qualities of effective and satisfactory recommendation messages, and in turn recommender systems. Those two studies have so far been extended only by a few studies that are primarily centered on explanations provided in recommendations; an even smaller number of studies has focused on presentation details such as the recommendation's message length, information detail, information sequence, message format or typographical cues.

Future research is needed to enrich our understanding of the antecedents to effective recommendation messages, with three prime areas for opportunities for future research: narrative messages, software update warning messages, and information messages.

Narrative message inquiries could focus on the effect of narratives on the persuasion of the recommendation and information overload.

Studies related to software update warning messages could explore the effectiveness of a better design on the annoyance and confusion experienced by users, while increasing the importance and noticeability of the recommendation message and the placement and design of buttons on the level of confusion evoked by a recommendation message.

Our understanding of information messages could be extended by (i) exploring the effect of the explanation's length in recommendation messages on persuasion, (ii) the impact of affective, subjective and objective language on user behaviors and perceptions, and (iii) by defining the effective combination of problem information specificity, solution information specificity, and information sequence in a recommendation message.

While the contribution of this work is significant, limitations are also inherent. Due to the number of selected keywords used in the search, the volume and diversity of retrieved papers was limited.

Beside this limitation, we anticipate a significant impact on future research on the antecedents to effective RS message design in a managerial decision-making context. This paper provides the first systematic review, synthesis, and overview regarding knowledge, guidelines and future research opportunities for RS message design in a managerial decision-making context. Furthermore, we identified three research gaps that could be addressed by researchers through future research.

Acknowledgement. This study was financially supported by Blue Yonder, NSERC and Prompt (Grant number IRCPJ/514835-16).

References

1. Candillier, L., Jack, K., Fessant, F., Meyer, F.: State-of-the-art recommender systems. In: Collaborative and Social Information Retrieval and Access: Techniques for Improved User Modeling, pp. 1–22. IGI Global (2009)
2. Lops, P., de Gemmis, M., Semeraro, G.: Content-based recommender systems: state of the art and trends. In: Ricci, F., Rokach, L., Shapira, B., Kantor, P.B. (eds.) Recommender Systems Handbook, pp. 73–105. Springer, Boston, MA (2011). https://doi.org/10.1007/978-0-387-858 20-3_3
3. Lu, J., Wu, D., Mao, M., Wang, W., Zhang, G.: Recommender system application developments: a survey. Decis. Support Syst. **74**, 12–32 (2015)

4. Wiesner, M., Pfeifer, D.: Health recommender systems: concepts, requirements, technical basics and challenges. Int. J. Environ. Res. Public Health **11**(3), 2580–2607 (2014)
5. Sarwar, B., Karypis, G., Konstan, J., Riedl, J.: Application of dimensionality reduction in recommender system-a case study. Minnesota Univ Minneapolis Dept. of Computer Science (2000)
6. Hayes, C., Cunningham, P.: An on-line evaluation framework for recommender systems. Trinity College Dublin, Department of Computer Science (2002)
7. Torres, R., McNee, S.M., Abel, M., Konstan, J.A., Riedl, J.: Enhancing digital libraries with TechLens+. In: Proceedings of the 4th ACM/IEEE-CS Joint Conference on Digital Libraries, pp. 228–236, June 2004
8. McNee, S.M., Riedl, J., Konstan, J.A.: Being accurate is not enough: how accuracy metrics have hurt recommender systems. In: CHI'06 Extended Abstracts on Human Factors in Computing Systems, pp. 1097–1101, April 2006
9. Armentano, M.G., Abalde, R., Schiaffino, S., Amandi, A.: User acceptance of recommender systems: influence of the preference elicitation algorithm. In: 2014 9th International Workshop on Semantic and Social Media Adaptation and Personalization, pp. 72–76. IEEE, November 2014
10. Gena, C., Brogi, R., Cena, F., Vernero, F.: The impact of rating scales on user's rating behavior. In: Konstan, J.A., Conejo, R., Marzo, J.L., Oliver, N. (eds.) UMAP 2011. LNCS, vol. 6787, pp. 123–134. Springer, Heidelberg (2011). https://doi.org/10.1007/978-3-642-22362-4_11
11. Dooms, S., De Pessemier, T., Martens, L.: An online evaluation of explicit feedback mechanisms for recommender systems. In: 7th International Conference on Web Information Systems and Technologies (WEBIST-2011), pp. 391–394. Ghent University, Department of Information Technology (2011)
12. Lee, Y.E., Benbasat, I.: Research note—the influence of trade-off difficulty caused by preference elicitation methods on user acceptance of recommendation agents across loss and gain conditions. Inf. Syst. Res. **22**(4), 867–884 (2011)
13. Herlocker, J.L., Konstan, J.A., Riedl, J.: Explaining collaborative filtering recommendations. In: Proceedings of the 2000 ACM Conference on Computer Supported Cooperative Work, pp. 241–250, December 2000
14. Symeonidis, P., Nanopoulos, A., Manolopoulos, Y.: Providing justifications in recommender systems. IEEE Trans. Syst. Man Cybern. Part A Syst. Hum. **38**(6), 1262–1272 (2008)
15. Gedikli, F., Jannach, D., Ge, M.: How should I explain? A comparison of different explanation types for recommender systems. Int. J. Hum.-Comput. Stud. **72**(4), 367–382 (2014)
16. Bilgic, M., Mooney, R.J.: Explaining recommendations: Satisfaction vs. promotion. In: IUI Workshop: Beyond Personalization, San Diego, CA (2005)
17. Pu, P., Chen, L., Hu, R.: A user-centric evaluation framework for recommender systems. In: Proceedings of the Fifth ACM Conference on Recommender Systems, pp. 157–164, October 2011
18. Xiao, B., Benbasat, I.: E-commerce product recommendation agents: use, characteristics, and impact. MIS Q. **31**(1), 137–209 (2007)
19. Ricci, F., Rokach, L., Shapira, B.: Recommender systems: introduction and challenges. In: Ricci, F., Rokach, L., Shapira, B. (eds.) Recommender Systems Handbook, pp. 1–34. Springer, Boston, MA (2015). https://doi.org/10.1007/978-1-4899-7637-6_1
20. Zanker, M.: The influence of knowledgeable explanations on users' perception of a recommender system. In: Proceedings of the Sixth ACM Conference on Recommender Systems, pp. 269–272, September 2012
21. Tintarev, N., Masthoff, J.: Designing and evaluating explanations for recommender systems. In: Ricci, F., Rokach, L., Shapira, B., Kantor, P.B. (eds.) Recommender Systems Handbook, pp. 479–510. Springer, Boston, MA (2011). https://doi.org/10.1007/978-0-387-85820-3_15

22. Tintarev, N., Masthoff, J.: Evaluating the effectiveness of explanations for recommender systems. User Mod. User-Adap. Inter. **22**(4–5), 399–439 (2012)
23. Konstan, J., Riedl, J.: Recommender systems: from algorithms to user experience. User Mod. User-Adap. Interact. **22**(1), 101–123 (2012)
24. Knijnenburg, B.P., Willemsen, M.C.: Evaluating recommender systems with user experiments. In: Ricci, F., Rokach, L., Shapira, B. (eds.) Recommender Systems Handbook, pp. 309–352. Springer, Boston, MA (2015). https://doi.org/10.1007/978-1-4899-7637-6_9
25. Pettersson, R.: Introduction to message design. J. Vis. Lit. **31**(2), 93–104 (2012)
26. Flowerdew, L.: A combined corpus and systemic-functional analysis of the problem-solution pattern in a student and professional corpus of technical writing. Tesol Q. **37**(3), 489–511 (2003)
27. Keller, P.A.: Regulatory focus and efficacy of health messages. J. Consum. Res. **33**(1), 109–114 (2006)
28. Bennet, A., Bennet, D.: Organizational Survival in the New World. Routledge, Milton Park (2004)
29. Bennet, A., Bennet, D.: The decision-making process in a complex situation. In: Burstein, F., Holsapple, C.W. (eds.) Handbook on Decision Support Systems 1. International Handbooks Information System, pp. 3–20. Springer, Heidelberg (2008). https://doi.org/10.1007/978-3-540-48713-5_1
30. Zhang, Y., Lu, T., Phang, C.W., Zhang, C.: Scientific knowledge communication in online Q&A communities: linguistic devices as a tool to increase the popularity and perceived professionalism of knowledge contribution. J. Assoc. Inf. Syst. **20**(8), 3 (2019)
31. Bigras, É., Léger, P.M., Sénécal, S.: Recommendation agent adoption: how recommendation presentation influences employees' perceptions, behaviors, and decision quality. Appl. Sci. **9**(20), 4244 (2019)
32. Coursaris, C.K., Falconnet, A., Berger, C.: Improving user acceptance of recommender systems and their recommendations: effect of message design. In: SIGHCI Pre-ICIS Workshop 2020 (2020)
33. Paré, G., Trudel, M.C., Jaana, M., Kitsiou, S.: Synthesizing information systems knowledge: a typology of literature reviews. Inf. Manage. **52**(2), 183–199 (2015)
34. Pettersson, R.: Information design–principles and guidelines. J. Vis. Lit. **29**(2), 167–182 (2010)
35. Pettersson, R.: Information design theories. J. Vis. Lit. **33**(1), 1–96 (2014)
36. Ozok, A.A., Fan, Q., Norcio, A.F.: Design guidelines for effective recommender system interfaces based on a usability criteria conceptual model: results from a college student population. Behav. Inf. Technol. **29**(1), 57–83 (2010)
37. Yoo, K.H., Gretzel, U., Zanker, M.: Persuasive Recommender Systems: Conceptual Background and Implications. Springer, New York (2012). https://doi.org/10.1007/978-1-4614-4702-3
38. Gunawardana, A., Shani, G.: Evaluating recommender systems. In: Ricci, F., Rokach, L., Shapira, B. (eds.) Recommender Systems Handbook, pp. 265–308. Springer, Boston, MA (2015). https://doi.org/10.1007/978-1-4899-7637-6_8
39. Jameson, A., et al.: Human decision making and recommender systems. In: Ricci, F., Rokach, L., Shapira, B. (eds.) Recommender Systems Handbook, pp. 611–648. Springer, Boston, MA (2015). https://doi.org/10.1007/978-1-4899-7637-6_18
40. Holliday, D., Wilson, S., Stumpf, S.: User trust in intelligent systems: a journey over time. In: Proceedings of the 21st International Conference on Intelligent User Interfaces, pp. 164–168, March 2016
41. Berkovsky, S., Taib, R., Conway, D.: How to recommend? User trust factors in movie recommender systems. In: Proceedings of the 22nd International Conference on Intelligent User Interfaces, pp. 287–300, March 2017

42. Sharma, A., Cosley, D.: Do social explanations work? Studying and modeling the effects of social explanations in recommender systems. In: Proceedings of the 22nd International Conference on World Wide Web, pp. 1133–1144. May 2013

43. Panniello, U., Gorgoglione, M., Tuzhilin, A.: Research note—in CARSs we trust: how context-aware recommendations affect customers' trust and other business performance measures of recommender systems. Inf. Syst. Res. **27**(1), 182–196 (2016)

44. Mandl, M., Felfernig, A., Teppan, E., Schubert, M.: Consumer decision making in knowledge-based recommendation. J. Intell. Inf. Syst. **37**(1), 1–22 (2011)

45. Kouki, P., Schaffer, J., Pujara, J., O'Donovan, J., Getoor, L.: Personalized explanations for hybrid recommender systems. In: Proceedings of the 24th International Conference on Intelligent User Interfaces, pp. 379–390, March 2019

46. Al-Taie, M.Z., Kadry, S.: Visualization of explanations in recommender systems. J. Adv. Manage. Sci. **2**(2), 140–144 (2014)

47. Gedikli, F., Ge, M., Jannach, D.: Understanding recommendations by reading the clouds. In: Huemer, C., Setzer, T. (eds.) EC-Web 2011. LNBIP, vol. 85, pp. 196–208. Springer, Heidelberg (2011). https://doi.org/10.1007/978-3-642-23014-1_17

48. Friedrich, G., Zanker, M.: A taxonomy for generating explanations in recommender systems. AI Mag. **32**(3), 90–98 (2011)

49. McInerney, J., et al.: Explore, exploit, and explain: personalizing explainable recommendations with bandits. In: Proceedings of the 12th ACM Conference on Recommender Systems, pp. 31–39, September 2018

50. Kunkel, J., Donkers, T., Michael, L., Barbu, C.M., Ziegler, J.: Let me explain: impact of personal and impersonal explanations on trust in recommender systems. In: Proceedings of the 2019 CHI Conference on Human Factors in Computing Systems, pp. 1–12, May 2019

51. Nunes, I., Jannach, D.: A systematic review and taxonomy of explanations in decision support and recommender systems. User Mod. User-Adap. Interact. **27**(3), 393–444 (2017)

52. Zhao, Q., Chang, S., Harper, F.M., Konstan, J.A.: Gaze prediction for recommender systems. In: Proceedings of the 10th ACM Conference on Recommender Systems, pp. 131–138, September 2016

53. Zanker, M., Schoberegger, M.: An empirical study on the persuasiveness of fact-based explanations for recommender systems. In: Joint Workshop on Interfaces and Human Decision Making in Recommender Systems, vol. 1253, pp. 33–36, September 2014

54. Lamche, B., Adıgüzel, U., Wörndl, W.: Interactive explanations in mobile shopping recommender systems. In: Joint Workshop on Interfaces and Human Decision Making in Recommender Systems, vol. 14, September 2014

55. Carenini, G., Moore, J.D.: Generating and evaluating evaluative arguments. Artif. Intell. **170**(11), 925–952 (2006)

56. Lamche, B., Trottmann, U., Wörndl, W.: Active learning strategies for exploratory mobile recommender systems. In: Proceedings of the 4th Workshop on Context-Awareness in Retrieval and Recommendation, pp. 10–17, April 2014

57. Schnabel, T., Bennett, P.N., Joachims, T.: Improving recommender systems beyond the algorithm. arXiv preprint arXiv:1802.07578 (2018)

58. Oechslein, O., Fleischmann, M., Hess, T.: An application of UTAUT2 on social recommender systems: Incorporating social information for performance expectancy. In: 2014 47th Hawaii International Conference on System Sciences, pp. 3297–3306. IEEE, January 2014

59. Moyer-Gusé, E., Chung, A.H., Jain, P.: Identification with characters and discussion of taboo topics after exposure to an entertainment narrative about sexual health. J. Commun. **61**, 387–406 (2011)

60. Moyer-Gusé, E., Nabi, R.L.: Comparing the effects of entertainment and educational television programming on risky sexual behavior. Health Commun. **26**, 416–426 (2011)

61. Niederdeppe, J., Shapiro, M.A., Porticella, N.: Attributions of responsibility for obesity: Narrative communication reduces reactive counterarguing among liberals. Hum. Commun. Res. **37**, 295–323 (2011)
62. Appel, M., Richter, T.: Transportation and need for affect in narrative persuasion: a mediated moderation model. Media Psychol. **13**, 101–135 (2010)
63. Niederdeppe, J., Shapiro, M.A., Kim, H.K.: Narrative persuasion, causality, complex integration, and support for obesity policy. Health Commun. **29**, 431–444 (2014)
64. Weber, P., Wirth, W.: When and how narratives persuade: the role of suspension of disbelief in didactic versus hedonic processing of a candidate film. J. Commun. **64**, 125–144 (2014)
65. Jensen, J.D., King, A.J., Carcioppolo, N., Krakow, M., Samadder, N.J., Morgan, S.: Comparing tailored and narrative worksite interventions at increasing colonoscopy adherence in adults 50-75. Soc. Sci. Med. **104**, 31–40 (2013)
66. Barbour, J.B., Doshi, M.J., Hernández, L.H.: Telling global public health stories: narrative message design for issues management. Commun. Res. **43**(6), 810–843 (2016)
67. Fagan, M., Khan, M.M.H., Nguyen, N.: How does this message make you feel? A study of user perspectives on software update/warning message design. Human-Centric Comput. Inf. Sci. **5**(1), 36 (2015)
68. Makkan, N., Brosens, J., Kruger, R.: Designing for positive emotional responses in users of interactive digital technologies: a systematic literature review. In: Hattingh, M., Matthee, M., Smuts, H., Pappas, I., Dwivedi, Y.K., Mäntymäki, M. (eds.) I3E 2020. LNCS, vol. 12067, pp. 441–451. Springer, Cham (2020). https://doi.org/10.1007/978-3-030-45002-1_38
69. Li, H., Chatterjee, S., Turetken, O.: Information technology enabled persuasion: an experimental investigation of the role of communication channel, strategy and affect. AIS Trans. Hum.-Comput. Interact. **9**(4), 281–300 (2017)
70. Matsui, T., Yamada, S.: The effect of subjective speech on product recommendation virtual agent. In: Proceedings of the 24th International Conference on Intelligent User Interfaces: Companion, Los Angeles, California, USA, 16–20 March, pp. 109–110 (2019)
71. Schreiner, M., Fischer, T., Riedl, R.: Impact of content characteristics and emotion on behavioral engagement in social media: literature review and research agenda. Electron. Commer. Res., 1–17 (2019)
72. Harbach, M., Fahl, S., Yakovleva, P., Smith, M.: Sorry, I don't get it: an analysis of warning message texts. In: Adams, A.A., Brenner, M., Smith, M. (eds.) FC 2013. LNCS, vol. 7862, pp. 94–111. Springer, Heidelberg (2013). https://doi.org/10.1007/978-3-642-41320-9_7
73. Bravo-Lillo, C., Cranor, L.F., Downs, J., Komanduri, S., Sleeper, M.: Improving computer security dialogs. In: Campos, P., Graham, N., Jorge, J., Nunes, N., Palanque, P., Winckler, M. (eds.) INTERACT 2011. LNCS, vol. 6949, pp. 18–35. Springer, Heidelberg (2011). https://doi.org/10.1007/978-3-642-23768-3_2

Research on Innovative Application Mode of Human-Computer Interaction Design in Data Journalism

Rui Fang[1], Qiang Lu[1], and Feng Liu[2(✉)]

[1] School of Publishing, University of Shanghai for Science and Technology,
Shanghai, People's Republic of China
[2] School of Journalism and Communication, Shanghai University,
Shanghai, People's Republic of China

Abstract. In the context of intelligent communication, with the advancement of artificial intelligence technology, technologies such as voice interaction, image recognition, gesture recognition, and brain-computer interface have been further developed and used, and interactive data news is also considered to be the development of data news trends and directions. However, the current human-computer interaction technology used in data journalism is relatively simple. And it still has some problems. In the future, interactive visualization needs to improve its interactive operation and immersive experience. At the same time, it needs to update data in real time and achieve advanced interaction, encourage crowdsourcing production, and combine VR technology to give users a stronger sense of "presence".

Keywords: Human-computer interaction design · Data news · Application model

1 Introduction

Intelligent technology is changing the form of communication, and intelligent communication with human-computer interaction as the core is beginning to emerge. In the context of intelligent communication, as computer technology is increasingly integrated into all areas of human life, human-computer interaction is with us anywhere, anytime. Human-computer interaction design has gradually expanded from primitive graphical interface interaction to voice interaction, gesture recognition, and brain-computer interface. At the same time, with the advent of the era of big data, data journalism based on massive amounts of data has become a new force and entered the journalism industry in the third wave of revolution. Contemporary data journalism is part of the digital journalism industry. One of its characteristics is the use of interactive technology, which gave birth to interactive narratives. Data journalism on the new media platform also innovates the expression of news narrative with the help of interactive narrative. Data shows that two-thirds of the "Best Visualization" works in the Data Journalism Award

© Springer Nature Switzerland AG 2021
S. Yamamoto and H. Mori (Eds.): HCII 2021, LNCS 12765, pp. 182–191, 2021.
https://doi.org/10.1007/978-3-030-78321-1_14

use interactive visualization to varying degrees [1]. This has transformed the form of data news from a single static information chart into an interactive data visualization work. Interactive data journalism is also considered the trend and direction of data journalism development. However, there are still a lot of problems in the design of human-computer interaction in current data journalism. Human-computer interaction technology has not been developed simultaneously because of the achievements of information technology. The current human-computer interaction method used in data journalism is still based on the traditional one-handed one-eye mode [2]. This article will first briefly introduce the design of human-computer interaction. Then sort out the current status and short-comings of human-computer interaction design in data journalism under the background of intelligent communication. In the last part, this article will explore new models of innovative applications of human-computer interaction design in data journalism.

2   Advantages of Human-Computer Interaction Design

Human-computer interaction is a technology that studies people, computers and their mutual influence. Human-computer interaction is the communication between humans and computer systems. It is a two-way information exchange of various symbols and actions between humans and computer systems. Here "interaction" is defined as a kind of communication, that is, information exchange, and it is a two-way information exchange. The user can input information to the computer, and the computer can feed back information to the user. This form of information exchange can appear in many ways. For example, the keystrokes on the keyboard, the movement of the mouse, the symbols and graphics on the display screen, etc. It can also exchange information with sounds, gestures or body movements. First of all, human-computer interaction design has a wealth of interactive means. It can be human body parts, such as human body, hands, feet, head, eyeballs, expressions, postures, etc. Through these means, users can perform various interactive behaviors such as voice interaction, motion recognition interaction, and eye tracking. In addition, interactive devices can also be used as a means of human-computer interaction. Such as common interactive devices such as mouse, keyboard, interactive handle, etc. The user can select a certain point or area in the image through the mouse or keyboard, and complete operations such as zooming and dragging the virtual object at that point or area. This type of method is simple and easy to operate, but requires the support of external input devices.

Secondly, the human-computer interaction design has a direct and efficient operation mode. Machines are an extension of human functions. If humans can control machines as easily as their own bodies, the gap between humans and machines can be completely eliminated. Human-computer interaction design can enable humans and computers to achieve as efficient and natural interaction as between humans. Artificial intelligence technologies such as speech recognition, image analysis, gesture recognition, semantic understanding, and big data analysis can help computers better perceive human intentions. This makes the user's operation more direct and efficient. At the same time, the humanization of human-computer interaction is also improved. It also eliminates tedious commands or menu operations. Humans can control machines more easily, and machines have become an extension of humans.

Finally, the human-computer interaction design enables users to have a more realistic experience. One of the main characteristics of virtual reality technology in human-computer interaction technology is "immersion." Virtual reality technology presents users with more realistic visual images by constructing three-dimensional virtual scenes. At the same time, it provides functions such as experience interaction, cultural guidance and interactive communication, allowing users to interact with the elements of the three-dimensional scene through the participation of multiple dimensions such as sight, hearing, touch, and smell. So that the user can be immersed in the virtual scene. It guides users to a sense of immersion from both physical and psychological aspects, and gives users a more realistic experience.

3   Current Status and Shortcomings of Human-Computer Interaction Design in Data Journalism

Thomas Rollins believes: "News organizations must realize that news is pushed through computers, and the essence of computers is interaction. The notion that news organizations rely on static and non-interactive ways to attract audiences is outdated [3]". Therefore, the use of human-computer interaction technology in data journalism can provide a large amount of valuable data to users in a personalized way to achieve accurate information dissemination. In addition, a good interactive experience can attract more users, guide more users to participate, and achieve deeper communication effects.

Current data journalism uses multiple dimensions and highly expressive interactive visualization works to present news events. At the same time, it is supplemented by text explanations, showing a new reporting mode. Through interactive visualization works, users can obtain news content by clicking on the mouse, moving interactive elements and so on. The use of human-computer interaction design has changed the linear reading mode of traditional news, and greatly stimulated users' desire to explore. And to some extent, it also gives users greater autonomy. At the same time, the use of interactive data tools has gradually become a trend. When users use interactive data tools while reading, the background will collect the audience's behavior and the data that users generated in real time according to the interactive instructions that have been set. Then the background will analyze the collected data. In the end, the background will push personalized information to the audience. We often see the emergence of data tools in interactive data news, such as calculators for calculating numbers, geographic data tools for positioning, etc. The "New York Times" data news section once posted an interactive data news called "Is It Better to Rent or Buy?". Through this work, users can calculate the cost of renting a house and buying a house. It can help users to judge whether renting a house or buying a house is cost-effective. It is precisely because of the use of these interactive data tools that data journalism is more functional.

Although the application of human-computer interaction design in data journalism has made great progress compared with before. However, the advantages of human-computer interaction design are not perfectly reflected in the current data news creation.

First of all, the current data journalism using human-computer interaction technology is relatively small, and the interaction method is relatively single. Although voice interaction, gesture recognition, and somatosensory technology have made great progress at the application level, they are still less used in data journalism. Most of the interactive technology in data journalism is graphical interface interaction. However, most graphical interface interactions are done through mouse clicks, keyboard input and touch controls. On the surface, touch seems to liberate the mouse and keyboard. But in essence, it is no different from a mouse click. This single repetitive operation will bring fatigue to the user [4]. In addition, this type of interaction that requires clicks or touches has certain operating difficulties for some disabled people. What's more, when users are immersed in virtual environments such as AR or VR, frequent mouse clicks or touches will reduce the user's experience.

The next, there is a lack of two-way interaction settings in interaction design. At present, the two-way interaction is not strong in most data news works. Most of the data journalism works are simply listings of data and charts. And lack of two-way interaction with the audience. However, the interaction of data news cannot be limited to one-way content delivery, but to supplement and update the data through interaction with users to enrich the content of the news. In the socialized communication pattern of the mobile Internet and media, news is not only for people to "see", but also for people to "use". Technology has turned traditional audiences into users and changed their media behavior. Users become a "combination" of consumers and producers of news content, which is not in the traditional media era.

Then, some interactive data journalism products lack post-maintenance and fail to update the old data and replace the correct data in time. The most important point that interactive data journalism is different from static information schema data journalism and traditional journalism is that it changes the argument that journalism is "fragile". Interactive data news is no longer a one-time display of information. It surpasses the news content itself to a certain extent and becomes an application product with a tool nature. If this kind of non-one-off display is to be realized, the editing team needs to maintain the database continuously, add the latest data provided by the reader, and correct errors in the data. In addition, openness and sharing are not only the basic concepts of data journalism, but also the communication characteristics of the Mobile Internet era. In order to maintain the vitality of news reports, news reports need to continuously update and maintain their core data and news stories. In December 2015, The Washington Post produced a data journalism titled "Investigation: People Shot and Killed by Police This Year". However, after entering 2016, there have been more shooting incidents that caused casualties in the United States. The Washington Post continued to pay attention to this, and on July 27, the journalism was updated. Until 2020, the data is still being updated. This approach effectively extends the communication value of data news products. However, only a small number of interactive data journalism will periodically update the data and perform post-maintenance of the data, which also greatly reduces the effectiveness of some data journalism [5].

4 The Emergence of New Technologies Promotes Continuous Upgrading of Human-Computer Interaction Technology

The development of human-computer interaction is a process from people adapting to machines to machines adapting to people. Summarizing the development history of human-computer interaction, it can be roughly divided into the following stages. The first stage: the manual operation stage, represented by punched paper. The second stage: the interactive command language stage, where the user operates the computer through a programming language; The third stage: the graphical user interface stage, the Windows operating system is the representative of this stage. The fourth stage: the emergence of intelligent human-computer interaction such as voice interaction and virtual reality [6]. We are currently in the fourth stage of the development of human-computer interaction technology. The development of new technologies has continuously upgraded human-computer interaction methods. The next part will briefly introduce voice interaction technology, virtual reality technology and intelligent interaction technology.

First of all, the voice interaction. Voice interaction mainly includes the following steps. Step one is the voice recognition. The sound wave signal is extracted through a microphone. Next, the vibration signal of the sound wave is converted into an electric signal, and then the electric signal is processed. The signal characteristics after analysis and processing are matched with the text information in the database. Step two is the semantic recognition. Let the computer "understand" the meaning of the sentence through the recognized text information. Step three is the speech synthesis. It can be divided into two parts: online speech synthesis and offline speech synthesis. Online speech synthesis synthesizes a synthetic voice close to human voice through a cloud database, but the sound quality and synthesis rate will be affected by the network environment. Offline speech synthesis requires the user to download the local speech package in advance, and the computer directly calls the local speech package for synthesis after receiving the sentence to be synthesized. Its timbre is slightly inferior to online speech synthesis. But the advantage is that it is not restricted by the network environment, and the synthesis speed is faster [7].

The earliest voice interaction technology is an interactive voice response system. The user interacts by dialing, but the system cannot answer the user's questions. The system can only broadcast pre-recorded sounds, such as voice mail, dial prompts, etc. Moreover, this interactive voice response system has a narrow application range, low interaction efficiency and a rigid interaction mode. It cannot solve practical problems in users' lives. Due to these drawbacks of the interactive voice response system, the system cannot solve many practical problems of users. So smart products such as Siri, Google, Mi AI, Amazon Echo, DUER and other voice interactions were born. These voice interaction products combined with AI technology have been widely praised since they came out. The birth and success of the products have accelerated the development of voice interaction technology. With the improvement of AI technology, speech recognition and semantic understanding technologies have gradually matured and perfected, and the humanization of voice interaction has become possible. The form of interaction has also advanced from a mechanical dialogue to a more fluent multi-round dialogue, and can even recognize multiple languages and regional dialects, which makes voice interaction a qualitative leap in flexibility and experience.

The second one is virtual reality technology. Virtual reality technology is actually a kind of simulation technology. It integrates the cutting-edge technology of many disciplines such as computer graphics, human-machine interface, multimedia fusion technology, sensor technology, etc. Virtual reality technology mainly includes the following aspects. The first one is to simulate the environment. It refers to the computer-generated dynamic three-dimensional images, and requires a high degree of fidelity. Next, it is perception. The ideal virtual reality technology should have all the perception capabilities that humans have. In addition to the visual perception capabilities corresponding to computer vision, it should also have the capabilities of hearing, touch, motion perception, pressure perception and even smell. The last is natural skills. It means that virtual reality can detect and track various human movements, such as head, eyeballs, gestures, limbs, etc. And convert the sensed action signals into data for analysis and real time feedback, where the feedback objects are the user's facial features and skin and other sensory organs. It is mainly realized through interactive devices. Such as eye tracking technology. The eye tracking system can record the user's focus and eye movement. The system uses infrared light to illuminate the human eye. The camera sensor is used to record the human eye response, and the software is used to determine the pupil position and eye direction. Humans mainly perceive the surrounding environment through vision, so the eyes can show what the user is thinking. The device can distinguish the focus of the user's attention and their reaction through the eyes. The advantages of eye tracking technology can be reflected in the experience of games, VR, and AR. Although virtual reality technology is constantly upgrading, it still faces many challenges. Excessive use of virtual reality and technology may affect physical health. For example, the "VR dizziness" symptoms that people often discuss at present refer to the continuous use of VR products for a period of time, which will produce dizziness-like effects similar to motion sickness. This symptom is caused by the inconsistency between the video seen by the vision and the condition perceived inside the body. In addition, the limited sensor accuracy and network transmission speed affect the user experience of virtual reality technology to varying degrees. However, with the application of 5G communication networks, the hardware and software requirements of virtual reality technology will be met [8].

The third one is intelligent interaction. Intelligent interaction is the process of interaction between humans and intelligent systems designed to improve the performance of intelligent systems. It is an effective means to improve robot performance in the field of intelligent robots. Intelligent interaction is also an important part of machine adaptive learning. Adaptive learning means that in the process of human interaction with the machine, the machine accumulates user feedback or actively asks the user for the key information needed for model learning and improvement. And based on this information, the machine will appropriately modify the model and its parameters to improve the accuracy of model prediction. These series of operations will make the model continue to learn and modify, and then improve and evolve itself [9]. Intelligent interaction is currently mainly used in personalized services of service robots. For example, for a home service robot, it needs to adapt to different home environments and provide personalized services for each family member. However, at the beginning, the robot cannot fully establish the cognition of all users. In other words, the robot cannot understand each family member and provide corresponding personalized services. This requires robots to actively learn the information and preferences of different users through intelligent interaction. Then gradually establish relevant cognition and enhance the understanding of each member, so as to better serve different family members. Therefore, the use of intelligent interaction technology can more accurately initiate functions and responses that are closer to the user's intention. It enables users to obtain a more natural and personalized experience.

5 Innovative Application Models of Human-Computer Interaction Design in Data Journalism

The innovative application of human-computer interaction design in data journalism is fundamentally a deepening of the audience's reading experience and the enhancement of user participation. Charts and data in traditional news only play a supporting role. In data journalism, pictures and data are the narrative language of news. It has the advantages of intuition, image and logical rigor, which enhance the audience's reading experience [10]. When the audience is reading data journalism, they need to actively participate in the work and strengthen the interaction with the work. Therefore, strengthening the application of human-computer interaction technology in data journalism is not only the need for the development of data journalism, but also the need for improving the audience's reading experience in news dissemination.

First of all, it is necessary to adapt to the user's usage habits, while optimizing interactive operations. With the advent of the mobile internet era, the mobile trend of news dissemination is getting stronger and stronger. Obtaining new information through mobile devices has become one of the main ways for users to obtain information. However, most of the interactive visualization data news works at this stage can only be displayed and interacted on the PC, but not on the mobile terminal. Moreover, some visual interactive operations cannot perfectly adapt to the operating habits of mobile client users. Therefore, the interaction designer should start from the user experience. The human-computer interaction design of data journalism should be combined with the mobile screen size, interaction method, and fragmented reading habits of mobile users. As mentioned above, the screen touch interaction method has operational difficulties for some people with disabilities. It will also reduce the user experience. So, mainstream media should pay close attention to the development of intelligent voice information. They can carry out in-depth cooperation with scientific research institutions, technology companies, and terminal manufacturers. Jointly innovate and develop targeted and highly matched intelligent voice interactive data journalism products. When the user interacts with the terminal through voice, it not only eliminates the operational difficulties of some disabled persons, but also optimizes the user experience. Such as the "Financial Times" and Google's "Hidden City" voice interaction project launched at the end of 2018. The first issue focused on Berlin, Germany. It launched a map of Berlin with "hidden information". Readers use voice interaction to unlock hidden content. The president of the "Financial Times" personally introduces the new Berlin airport and tells users which places are worth visiting. At the same time, we need to optimize interactive operations and be good at applying advanced interactions and game interaction design. At present, the visual experience in most interactive visualization reports is too simple and the interactive means are too single. In addition, the interactive picture is too simple to provide users with beautiful reading and immersive experience. Therefore, we should use dynamic interactive views as much as possible in visualized data journalism. We can try to hide part of the data in the dynamic interactive chart, and guide the audience to click on the corresponding icon to view the relevant data. When the reader manipulates the mouse to interact, it can stimulate the audience's interest in exploring, so that the audience can independently read the data a second time. In addition, the audience can also actively discover and solve problems through the game. The reward and punishment mechanism set up around the problem will also trigger the audience's interest in participation and increase the audience's understanding of news reports. The New York Times specially opened a "You Draw It" section. It is committed to allowing users to deepen their understanding of various social issues and enhance the fun of interaction by drawing diagrams. Advanced interaction and game interaction will become the future development trend of data journalism interactivity. At the same time, it can also bring new reading experience and feelings to the audience.

Secondly, leveraging 5G technology and combining virtual reality technology to give audiences a stronger "presence" experience. 5G technology has the characteristics of high speed, large capacity, low latency, and low energy consumption. It can provide fully mobile and fully connected data support for human-computer interaction technology in data journalism. In addition, sensors and the Internet of Things can generate a huge amount of data information, providing a data foundation for in-depth data reporting. This also provides the possibility for predictive reporting of event development and trends based on authentic data, which is also the value and advantage of data journalism.

Moreover, the use of virtual reality technology can enable users to achieve a high degree of "presence" and "immersion." It allows readers to get close to the scene infinitely and get a "live" experience. Wall Street Journal once in "Nasdaq Coming to Another Stock Market Bubble? " had tried to use VR data visualization. This work uses the combination of VR technology and line graphs to let the audience experience the ups and downs of the Nasdaq Composite Index from 1994 to 2016 in the form of a roller coaster ride. It vividly embarked on an expedition to the American stock market [11]. Moreover, with the continuous upgrading of virtual reality technology, the audience's sense of experience will continue to strengthen. In the future, force feedback technology and eye tracking technology may also be used. It alleviates the decrease in experience when the user is in an immersive virtual environment and must use the virtual keyboard-based text input behavior. Through the use of advanced technical means, to provide users with a new human-computer interaction experience.

Finally, data needs to be updated and maintained in a timely manner. At the same time, users need to be encouraged and guided to crowdsource production. The biggest difference between interactive data journalism and traditional news is that interactive data journalism is no longer a simple one-time data display. It surpasses news content to some extent and becomes an application product with a tool nature. The simple one-time data display in the past has been transformed into a tool-like application product. If we want to realize this kind of non-disposable display, we need the editing team to continuously update and maintain the data, add the latest data provided by readers, and correct and update the data that has changed.

For example, the data news visualization works of Bloomberg and The Guardian pay much attention to the update and maintenance of data. Its interactive works will continue to update and maintain data within a certain period of time, which is not a simple one-time data display in the past. Among them, The Guardian "Australian Covid-19 Map" will update the works in real time based on data from each state. Bloomberg "Covid-10 Vaccine Tracker" will be updated from time to time based on the latest information. "Comparison of Oil Prices in Various Countries" is updated every quarter. "U.S. State Economic Health Data" and "Bloomberg Best" are updated weekly. "Bloomberg Billionaires Index" uses daily real time data. The data involved in financial reports often changes at any time according to the latest developments in the situation, market and policy changes. Regular updates of data visualization works in the time dimension can not only reduce information misleading caused by expired data to readers, but also show data changes in different time periods through the superposition of new and old data. Provide users with a new perspective on reading and exploration.

Moreover, some data news production is currently affected by traditional news production thinking. That is, producers control the rights of news production, and users cannot participate in news production. As mentioned above, the interaction of data journalism cannot be limited to one-way content delivery, but to supplement data through interaction with users to enrich the news content. For example, the Guardian launched the "Word of the year poll" project in December 2020. It invites the public to participate in the 2020 annual vocabulary selection through public opinion polls. And in a short time, 6185 pieces of data uploaded by users have been sorted out. Then the data visualization team of The Guardian made these data into visual data journalism, and announced the results to the public. The Guardian guides the public to upload data actively. It interprets the data in a visual way, gives the data new vitality, and makes the content of the article more vivid.

6 Conclusion

Negroponte in Being Digital had predicted that "The challenge of the next decade will go far beyond providing people with larger screens, better music, and easier-to-use graphical input devices. The challenge will be: Let the computer know you, understand your needs, understand your words, expressions and body language [12]". Today, the era of intelligent communication has arrived, and human-computer interaction design has also developed rapidly. Interactive data journalism is also considered the trend and direction of data journalism development. Interactive data journalism is a great innovation to traditional journalism forms under the background of the development of human-computer interaction design. As mentioned above, strengthening the application of human-computer interaction design in data journalism is not only the need for the development of data journalism, but also the need for improving the audience's reading experience in news dissemination. In the future, the interactive visualization of data journalism will be bound closely with the innovation of human-computer interaction design. As a result, a new form of news expression with a higher level of intelligence, a stronger user experience, and a tool nature can be produced. But at the same time, the application of human-computer interaction design in interactive data news also needs to be wary of formalism and technological supremacy. Human-computer interaction design should be human-centered and dedicated to bringing users a better reading and using experience.

Acknowledgements. This paper is supported by Shanghai social science foundation project "Research on the reproduction of newsroom space under the background of 5G" (2019BXW004) and "Creative research on children's books from the perspective of traditional culture education – Based on the perspective of situational learning" (2020bwy010).

References

1. Fang, J.: Introduction to Data Journalism. China Renmin University Press, Beijing (2015)
2. Fan, J.J., Tian, F., et al.: Some thoughts on human-computer interaction in intelligent age. Sci. China **48**, 361–375 (2018)
3. Alastair Reid. Newsgames: Future media or a trivial pursuit? https://www.journalism.co.uk/news/newsgames-future-media-or-a-trivial-pursuit-/s2/a554350/
4. Wang, W.L.: Analysis on the future development trend of human-computer interaction in intelligent age. Broadcast. Realm **5**, 57–60 (2020)
5. Jiang, R.X., Peng, L.: From static presentation to deep data exploration-interactive graphic application in bloomberg. Journal. Mass Commun. Mon. **21**, 65–69 (2014)
6. Wang, S.M.: Current situation and future prospect of human-computer interaction technology. Public Commun. Sci. Technol. **3**, 142–144 (2019)
7. Yu, L.L.: Summary of speech recognition technology and applications. Mod. Electron. Tech. **13**, 43–45 (2013)
8. Liao, S.Y.: Characteristics and application of virtual reality technology. Public Commun. Sci. Technol. **21**, 127–128, 135 (2018)
9. Li, J.L., Sun, X.H., Guo, W.W.: Active response interactive design of automobile based on intelligent interaction. J. Graph. **4**, 668–674 (2018)
10. Ding, Y.J., Kang, Z.: Innovation of data news: upgrading of interaction and visualization. Youth Journal. **5**, 45–46 (2018)
11. Zhu, T.Z.: Practice and challenges of VR data news. New Media Res. **1**, 110–112 (2019)
12. Negroponte: Being Digital. Hainan Publishing House, Hainan

Evaluating the Impact of Algorithm Confidence Ratings on Human Decision Making in Visual Search

Aaron P. Jones, Michael C. Trumbo, Laura E. Matzen, Mallory C. Stites, Breannan C. Howell, Kristin M. Divis, and Zoe N. Gastelum[✉]

Sandia National Laboratories, Albuquerque, USA
{ajones3,mctrumb,lematze,mcstite,bchowel,kmdivis,
zgastel}@sandia.gov

Abstract. As the ability to collect and store data grows, so does the need to efficiently analyze that data. As human-machine teams that use machine learning (ML) algorithms as a way to inform human decision-making grow in popularity it becomes increasingly critical to understand the optimal methods of implementing algorithm assisted search. In order to better understand how algorithm confidence values associated with object identification can influence participant accuracy and response times during a visual search task, we compared models that provided appropriate confidence, random confidence, and no confidence, as well as a model biased toward over confidence and a model biased toward under confidence. Results indicate that randomized confidence is likely harmful to performance while non-random confidence values are likely better than no confidence value for maintaining accuracy over time. Providing participants with appropriate confidence values did not seem to benefit performance any more than providing participants with under or over confident models.

Keywords: Visual search · Human-machine teaming · Cognition

1 Introduction

The human ability to effectively analyze visual data is failing to keep pace with rapid growth in the ability to collect and store data (Keim et al. 2008). Machine learning (ML) algorithms have been proposed as a method of providing aid to human decision makers faced with an overabundance of data in domains such as automated vehicle operation, detection of system failures, medical diagnostics, and threat detection (Du et al. 2020; Goddard et al. 2014; Körber et al. 2018). ML algorithms have demonstrated performance equal to or exceeding that of humans on simple classification tasks involving sorting of datapoints into pre-determined categories. Examples include classifying the topics of research abstracts (Goh et al. 2020), recognizing traffic signs (Ciresan et al. 2011), geolocating photographs (Weyand et al. 2016), and navigating highspeed video games (Fuchs et al. 2020). Humans, however, continue to outperform automatic image analysis

© National Technology & Engineering Solutions of Sandia, LLC 2021
S. Yamamoto and H. Mori (Eds.): HCII 2021, LNCS 12765, pp. 192–205, 2021.
https://doi.org/10.1007/978-3-030-78321-1_15

when the task requires background knowledge for proper interpretation, as may be the case in the domains of satellite or medical image analysis (Kneusel and Mozer 2017).

It has therefore been suggested that automatic classifiers may be used to improve, rather than replace, human analysis of images by, for example, highlighting regions of an image that the classifier deems likely to contain a target (Gastelum et al. 2020). In this fashion, algorithms perform an object detection and segmentation function that goes beyond classification by providing an indicator of the location of a target (Druzhkov and Kustikova 2016). This allows these human-machine teams to function such that the ML algorithm provides information that helps guide the human who ultimately makes the decision.

Whether or not algorithm guidance is adopted by various domains will depend on a number of interrelated factors such as user compliance (to what extent human users accept the algorithm recommendations), trust in the system, perceived and actual accuracy of the algorithm, and performance of the human-machine team (Khasawneh et al. 2003; Merritt and Ilgen 2008; Yin et al. 2019), as well as domain-specific considerations such as the tolerance for particular rates and types of errors (Gastelum et al. 2020).

Previous research indicates that performance synergy between human experts and automatic classification algorithms is enhanced when the method of drawing attention to probable targets takes the form of a graded highlighting approach that accounts for classifier confidence (Kneusel and Mozer 2017). It is possible, however, that classifier confidence may not accurately reflect target probability in all cases. In order to evaluate the cognitive impact of model bias, in the current study we used simulated ML algorithm output over a visual search task along with manipulated confidence values that varied from no confidence, random confidence, over confidence, under confidence, and appropriate confidence. The results will inform optimal presentation of object detection algorithms.

2   Method

2.1   Participants

This study was reviewed and approved by the Sandia National Laboratories Human Studies Board (HSB). 180 participants were recruited via Amazon Mechanical Turk, which has been found to have a population that is at least as representative of the United States population as more traditional subject pools and allows for rapid collection of data from a large number of participants (Paolacci et al. 2010). All participants were at least 18 years old and completed an online informed consent procedure prior to study participation. All participants were operating as Mechanical Turk Masters Qualified workers at the time of participation, a status that reflects a high level of performance reported by a large number of requesters across a wide variety of tasks. Participants were paid $2.00 for their time. Five participants were excluded due to low overall experimental accuracy, which indicated that they did not follow the task instructions. This left a total of 175 participants who completed the task across five conditions (34 for the random condition, 36 for the over confidence condition, 35 for the under confidence condition, 34 for the appropriate confidence condition, and 36 for the no confidence condition).

2.2 Stimuli

Stimuli consisted of the simulated output of a ML algorithm over a visual search task (similar to that reported in Biggs et al. 2013) in which participants are asked to determine whether or not a "perfect" capital T with a centered crossbar is present in each image. Distractors were Ts with an offset crossbar which may make the shape look more like a capital "L." The targets and distractors appeared in grey against a mottled grey background and were oriented in any of four rotations (0°, 90°, 180°, or 270°). Each image contained a total of 10 letters, half of which were colored light grey and half of which were colored dark grey. Participants were additionally presented with information they were told was produced by a ML algorithm in the form of a blue bounding box placed around a letter indicating the algorithm had determined that letter to be a perfect T (see Fig. 1 for an example). Additionally, a confidence score appeared above each bounding box in each condition except the no confidence condition, indicating how "sure" the machine learning algorithm was of the selected T. These confidence scores ranged from 0.01–0.99, where 0.01 is the lowest possible confidence of a target, and 0.99 is the highest. Confidence scores were binned into perceived levels of confidence (as determined by the research team), such as 0.03 (low confidence), 0.51 (moderate confidence), or 0.97 (high confidence). In images designed to indicate that the algorithm did not find a perfect T there was not a bounding box on the image. A simulated algorithm was used in order to allow for experimental manipulation of data to balance conditions and allow for valid statistical comparisons between groups of participants.

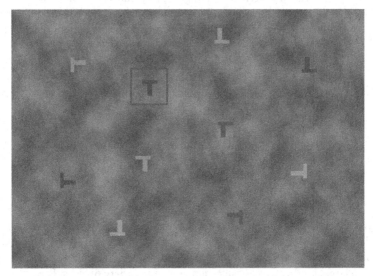

Fig. 1. An example of a "Target Present" image that contains a perfect T and a correctly placed bounding box indicator, with a high confidence score of 0.93

There were 120 experimental trials, with 10 additional "catch" trials, in which participants were explicitly instructed via text on screen to respond target absent or target present. These trials were used to ensure that participants were paying attention to the task. All participants included in analysis responded with at least 80% accuracy to the catch trials. Each trial contained either zero or one perfect T, with five possible image types, as follows (note that the distributions below refer to the responses of the simulated algorithm):

1. *Correct Rejection (CR).* There was no target present in the image and the algorithm did not indicate the presence of a target. 36 of the 120 experimental trials were of this type (30%).
2. *False Alarm (FA).* There was no target in the image and the algorithm incorrectly indicated a distractor "L" as a target. 12 of the 120 experimental trials were of this type (10%).
3. *False Alarm + Miss (FA + Miss).* There was a target in the image and the algorithm incorrectly indicated a distractor "L" as a target while failing to identify the target "T" that was present. 12 of the 120 experimental trials were of this type (10%).
4. *Hit.* There was a target "T" present in the image and the algorithm correctly identified it. 48 of the 120 experimental trials were of this type (40%).
5. *Miss.* There was a target "T" present in the image and the algorithm failed to identify it. 12 of the 120 experimental trials were of this type (10%).

Additionally, we systematically varied the confidence type between subjects, while ground truth model accuracy was held constant at 70%. The four confidence types were random confidence, over confidence, under confidence, and appropriate confidence. A condition in which confidence was never displayed was also included (the no confidence condition). The confidence values were generated by the research team and value ranges were determined based on discussion with machine learning subject matter experts (see Table 1).

Table 1. Confidence conditions and associated value ranges displayed to participants. Note that for the "appropriate confidence" condition high confidence values were reported for relatively easy hit trials while low confidence values were reported for the more difficult false alarm and false alarm + miss trials.

Confidence condition	Range of values
Random confidence	0.01–0.99 (randomly distributed)
Over confidence	0.85–0.99 (all trial types)
Under confidence	0.01–0.15 (all trial types)
Appropriate confidence	0.85–0.99 (Hit trials); 0.01–0.15 (FA/FA + Miss trials)

2.3 Procedure

Participants completed the experiment via Amazon Mechanical Turk. After completing the consent form, the participants read the instructions for the task. They were told that their task was to determine, as quickly as and as accurately as possible, whether or not each image contained a perfect T with a centered crossbar. Then they were told "As you complete this task, you will also see information produced by a machine learning algorithm. The algorithm places a blue box around the shape that it thinks is a perfect T. Above each box is a confidence score, ranging from 0–1, indicating how 'sure' the machine learning algorithm is of the selected T, where 0 is the lowest possible confidence of a target, and 1 is the highest. You will see confidence scores represented as two-digit decimals, such as 0.03 (low confidence), 0.51 (moderate confidence), or 0.97 (high confidence). If the algorithm thinks there is NOT a perfect T in the image, it does not place a box on the image. The indicator from the machine learning algorithm may not be accurate 100% of the time, but it is intended to help you search the images faster".

Participants were then shown three example image types, as follows:

1. An image containing a perfect T along with a correctly placed bounding box indicator and a confidence score (a "hit" example).
2. An image that did not contain a perfect T and did not have a bounding box (a "correct rejection" example).
3. An image containing a perfect T in which the model placed the bounding box and associated confidence score incorrectly over a distractor (a "false alarm + miss" example).

Participants were then informed of the potential for the other image types (false alarms, in which the algorithm indicates the presence of a target but a target is not present in the image, and misses, in which a target is present but the algorithm fails to identify it).

Participants were then reminded that the indicators would not be accurate 100% of the time, and that their job was to determine whether or not there was a T in the image even if the indicator was incorrect. In all of the experimental blocks, the images were scaled to be 550 pixels wide × 436 pixels tall. Participants clicked on one of two radio buttons below the image, labeled "Target Present" and "Target Absent", to select their response. When the participants clicked on the buttons, a black rectangle matching the size of the stimuli was presented for 200 ms and then the next trial appeared. The order of the stimuli and the catch trials was randomized. On average, it took the participants approximately 8 min to complete the task.

3   Results

For the reaction time analyses, trials were removed if they were ± 3SD from the individual's mean, although trials with RTs under 15 s were kept if the 3SD cutoff was 15 s or less. This resulted in removing a total of 111 trials from the RT analysis (out of a total of 21000 trials).

3.1   Effect of Confidence on Accuracy for Target Present vs. Target Absent Trials

Our first question was whether or not confidence condition had an effect on performance. One-Way ANOVAs were conducted for overall mean accuracy and overall mean RTs across all confidence conditions, with confidence condition as a between subjects factor. Results suggest no effect of confidence condition on overall mean accuracy ($F(4,174174) = 0.5151, p = 00.72$) or response times ($F(4,174) = 1.63, p = 0.17$).

Our next question was whether or not the confidence values had an impact on accuracy for different types of trials. The results are shown in Fig. 2. A 5×5 mixed-design ANOVA was conducted with image type (hits, correct rejections, misses, false alarms, and false alarms + misses) as a within-subjects factor and confidence condition (no confidence, appropriate confidence, over confidence, under confidence, and random confidence) as a between-subjects factor. The ANOVA showed that there was a significant effect of image type on participants' accuracy ($F(4,680) = 128.93, p < 0.001$), but there was not a significant effect of confidence condition ($F(4,680) = 0.65, p = 0.63$) or an interaction between confidence condition and image type ($F(16,680) = 1.12, p = 0.33$).

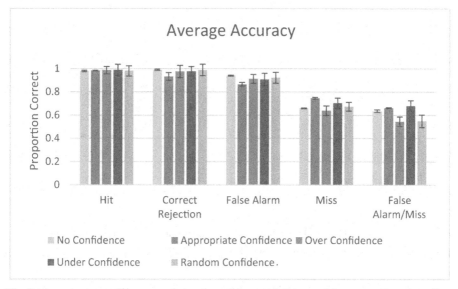

Fig. 2. Average proportion correct for each confidence condition and item type. Error bars show the standard error of the mean

Pairwise comparisons using Bonferroni correction showed that the participants' performance was significantly higher for hits and CRs than for any of the image types in which the bounding box (or lack thereof) was erroneous (all $ts > 2.98$, all $ps < 0.04$). For the three error types, participants had the highest accuracy for FAs, where the bounding box was placed on a distractor and no targets were present in the image. Their next highest accuracy was for misses, where a target was present, but there was no bounding box on the image. Finally, participants had the lowest performance for the FA + miss condition, where there was a target in the image, but the bounding box was instead placed around a distractor. The participants' accuracy for the FAs was significantly higher than their accuracy for the misses and FA + misses (both $ts > 10.52$, both $ps < 0.001$), and their accuracy for the misses was significantly higher than their accuracy for the FA + misses ($t(174) = 3.26, p < 0.02$).

Next, we tested whether the confidence conditions impacted the participants' RTs for different image types. Only correct trials were included in this analysis. These results are shown in Figs. 3 (correct model outputs) and 4 (incorrect model outputs). Once again, a 5×5 mixed-design ANOVA was conducted with image type as a within-subjects factor and confidence condition as a between-subjects factor. There was a significant effect of confidence condition ($F(4,656) = 2.60$, $p < 0.04$) and a significant effect of image type ($F(4,656) = 126.65$, $p < 0.001$), but there was not a significant interaction between the two ($F(16,656) = 1.30$, $p = 0.18$).

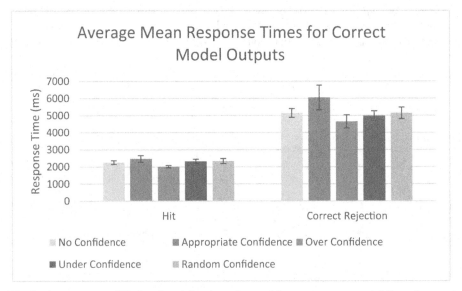

Fig. 3. Average mean RTs for the trials where the model output was correct (hits and correct rejections). The error bars represent the standard error of the mean.

Pairwise comparisons across the image types using Bonferroni correction showed that participants' responses were fastest for the hits. In the hit trials, participants could simply look at the bounding box and confirm that the letter inside was a T. The next fastest conditions were the miss and FA + miss conditions, which were both significantly slower than the hit trials (both $ts > 14.26$, both $ps < 0.001$) but did not differ significantly from one another ($t(167) = 2.37$, $p = 0.17$). In both of these conditions, a target was present in the image, but the bounding box was missing or placed incorrectly. The erroneous bounding boxes slowed the participants down, but they were able to end their visual search upon finding the target. The two slowest image types, the FAs and CRs, were the two that did not contain a target, meaning that participants had to look at every letter in the image to confirm that there was no T present. The average RTs for these two conditions did not differ significantly from one another ($t(167) = 2.73$, $p = 0.07$), and they were significantly slower than any other condition (all $ts > 3.34$, $p < 0.01$) (Fig. 4).

According to our planned comparisons, two-tailed t-tests were used to assess the impact of confidence condition on the participants' RTs for each image type. For the

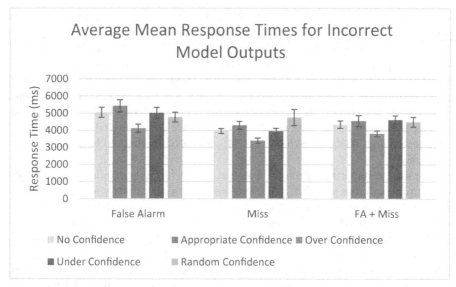

Fig. 4. Average mean RTs for the trials where the model output was incorrect (false alarms, misses, and false alarm + misses). The error bars represent the standard error of the mean.

hits, the t-tests showed that the participants in the over confidence condition had significantly faster RTs than participants in the appropriate ($t(40) = 2.18$, $p < 0.04$), under ($t(50) = 2.12$, $p < 0.04$), and random confidence ($t(46) = 2.07$, $p < 0.05$) conditions. The difference between the over confidence condition and the no confidence condition approached significance ($t(57) = 1.93$, $p = 0.06$). There were no significant differences among the appropriate, under, random, and no confidence conditions (all $ts < 0.94$, all $ps > 0.35$). For the other image types where the mock model output was correct, the correct rejections, there were no significant differences in RTs across conditions (all $ts < 1.71$ all $ps > 0.09$).

For the FAs, we observed once again that participants in the over confidence condition had faster RTs than participants in the other conditions. This difference was significant when comparing the over confidence condition to the no confidence ($t(65) = 2.39$, $p < 0.02$), appropriate confidence ($t(54) = 2.94$, $p < 0.01$), and under confidence ($t(62) = 2.15$, $p < 0.04$) conditions. There was not a significant difference in RTs between the over confidence and random confidence conditions ($t(61) = 1.71$, $p = 0.09$). There were also no significant differences when comparing the appropriate, under, random, and no confidence conditions to one another (all $ts < 1.39$, all $ps > 0.17$). Recall that only trials where the participants answered correctly were included in the RT analysis. A handful of participants responded incorrectly to all of the FA trials (one in the no confidence condition and two each in the appropriate, over, and random confidence conditions), so those participants did not contribute any data to these comparisons.

A similar pattern emerged for the misses and FA + misses. For the misses, there were no significant differences in mean RT when comparing the appropriate, under, random, and no confidence conditions (all $ts < 1.56$, all $ps > 0.12$). However, there were significant differences in mean RT when comparing the over confidence condition to each of the other conditions (all $ts > 2.36$, all $ps < 0.03$).

For the FA + misses, there were no significant differences in mean RT when comparing the appropriate, under, random, and no confidence conditions (all $ts < 0.84$, all $ps > 0.41$). When comparing the over confidence condition to the others, there was a significant difference in mean RT between the over and under confidence conditions ($t(62) = 2.70$, $p < 0.01$). The difference between the over confidence condition and the no confidence ($t(65) = 1.90$, $p = 0.06$), appropriate confidence ($t(48) = 1.98$, $p = 0.05$), and random confidence ($t(49) = 1.94$, $p = 0.06$) approached but did not reach significance.

Note that an important difference between the FA + miss condition and the other model error conditions was that participants could respond correctly in the FA + miss condition "by accident." In the FA condition, there was a bounding box but no target in the images. To produce the correct response of "target absent," participants had to identify the fact that the bounding box was incorrect. In the miss condition, there was a target in the image, but no bounding box. To produce the correct response of "target present," participants had to search the image to find the target. A number of participants did not do this, since overall accuracy rates were low in the miss condition. Furthermore, there were two participants in every confidence group who responded incorrectly to *all* of the miss trials. In the FA + miss condition, there was a target and a bounding box in the image, but the bounding box was placed incorrectly. Participants could reach the correct answer of "target present" in two ways. First, they could check the item inside of the bounding box, confirm that it was not a target, and then search the rest of the image until finding the target. Second, they could trust the bounding box and respond "target present" without checking the image. There is evidence that at least some participants used the second approach, because there were 10 participants who responded correctly to all or almost all of the FA + miss items, but incorrectly to all or almost all of the miss items and FA items. This indicates that there was a subset of participants who responded based on the presence or absence of the bounding boxes without checking to verify that they were correct.

There were seven participants who responded incorrectly to all of the FA + miss trials (three in the random confidence condition, two in the appropriate confidence condition, and one each in the over and no confidence conditions). When looking at the other model error conditions, this group of participants performed very well on the FA items and poorly on the miss items. This pattern indicates that these participants checked the letter inside of the bounding box and identified it as a non-target but failed to search the rest of the image. This strategy served them well for the FA items, but not for the FA + miss items. This group of participants also failed to search the miss items thoroughly, leading to poor performance in that condition as well.

3.2 Effect of Confidence Values on Responses Over Time

Our next question was whether the presence of confidence values changed how participants responded over time. A series of logistic regressions were run to investigate how participants' performance changed over time. The outcome was accuracy (binary – accurate, inaccurate) and the continuous predictor was trial number. Regressions were run first for target present and target absent trial types for each confidence condition (no confidence, random confidence, over confidence, under confidence, appropriate confidence). Further regressions were run for each trial type (hit, false alarm, miss, correct rejection, false alarm + miss), within confidence condition separately. Variables were inspected for normality. All pairwise comparisons were run with a Bonferroni alpha correction.

For these binary logistic regression analyses, results suggest no effect within the target absent condition ($\beta = 0.00104$, $z = 0.72$, $p = 0.47$), but a significant effect within the target present condition ($\beta = -0.002144$, $z = -3.12$, $p < 0.01$). Given this effect, separate regressions were run for only target present trials for each confidence condition separately. In random ($\beta = -0.00242$, $z = -1.74$, $p = 0.08$), over ($\beta = -0.00124$, $z = -0.083$, $p = 0.04$), under ($\beta = -0.00157$, $z = -0.092$, $p = 0.404$), and appropriate confidence ($\beta = -0.00098$, $z = -0.61$, $p = 0.539$), trial order was not a significant predictor of accuracy. However, it was significant in the no confidence condition ($\beta = -0.00419$, $z = -2.71$, $p < 0.0101$), where trial order was negatively related to the probability of a correct response. This effect was further investigated by looking at individual trial types within the no confidence condition. No effect was observed for CRs ($\beta = 0.00953$, $z = 1.22$, $p = 0.22$), FAs ($\beta = -0.00425$, $z = -0.78$, $p = 0.4444$), or hits ($\beta = 0.00553$, $z = 1.24$, $p = 0.2222$). However, significant effects were found for both FAs + misses ($\beta = -0.00650$, $z = -2.40$, $p = 0.0202$), and misses ($\beta = -0.00752$, $z = -2.82$, $p < 0.101$).

3.3 Effect of Confidence Values on Perception of Model Accuracy

Finally, we asked whether the confidence conditions affected the participants' perceptions of the mock model's accuracy. At the end of the experiment, participants were asked how accurate they thought the model was. Their response choices ranged from 0% to 100% correct with 10% intervals. The results are shown in Fig. 5. The saturated green bars show the number of participants who chose the correct response (70%) in each of the confidence conditions. Note that the pattern of responses is very similar for the no confidence and over confidence conditions, with more participants selecting the correct response than any other response. In the under and random confidence conditions, the participants tended to underestimate the accuracy of the model, with nearly two-thirds of the participants saying that the model's accuracy was 60% or less. In the appropriate confidence condition, 8 of the 34 participants correctly guessed the model's accuracy. The appropriate confidence condition had both the lowest number of participants who underestimated the model's accuracy (15) and the highest number of participants who overestimated the model's accuracy (11).

4 Discussion

The objective of this study was to evaluate the impact of manipulating confidence information included with simulated ML object detection in a visual search task as measured by participant accuracy and RTs. Image types included correct rejection (CR; no target present and the algorithm did not indicate a target as present), false alarm (FA; no target

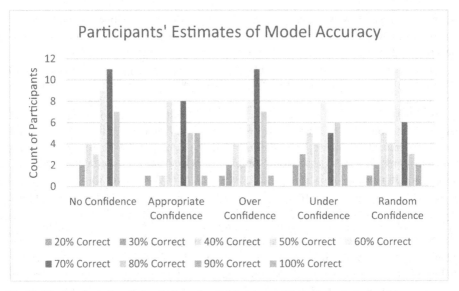

Fig. 5. The number of participants in each confidence condition who selected each response on the question about their perception of the model's accuracy.

present, but the algorithm indicated a target as present), false alarm + miss (FA + miss; the algorithm indicated a distractor as a target while a target was present elsewhere in the image), hit (the algorithm correctly identified a target as present), and miss (the algorithm failed to identify a present target).

Confidence values were manipulated in a between-subjects fashion such that participants were provided with a model that displayed appropriate confidence (high confidence for hit images, low confidence for FA and FA + miss), a model that was under confident (low confidence for hit, FA, and FA + miss images), a model that was over confident (high confidence for hit, FA, and FA + miss images), a model that displayed random confidence (a confidence value ranging from 0.01 to 0.99 was assigned to each trial in which a confidence value was displayed), and a no confidence condition in which bounding boxes were displayed as in the other conditions, but without providing a confidence value. Note that for all conditions aside from the no confidence condition (in which confidence values were never given), only for trials on which the simulated ML algorithm indicated the presence of a target via a bounding box was a confidence value included (hit, FA, FA + miss).

Though participants performed better when the model output was correct vs. when the model output was incorrect, our hypothesis that the participants in the appropriate confidence condition would perform best in terms of accuracy and RTs was not supported by the data. Participants in the appropriate confidence condition performed no better than those in the no confidence, under confident, or random confidence conditions. In terms of RT, participants provided with an over confident model consistently responded faster than participants in the other conditions. While participants exposed to an over confident model responded faster, suggesting they believed the model was accurate, they did not rate the model as any more accurate than did participants in the no confidence condition, so exposure to an overconfident model did not result in overestimation of model accuracy. As a general finding, an analysis of RTs corresponds with prior work in the domain of search termination – that is, participants responded faster to trials with a target present, likely because as soon as participants found the target, they concluded their search (Wolfe 2012).

Taken together, these results show that the appropriate model did little to distinguish itself from models with consistent bias. It is worth noting that participants in the appropriate condition had both the lowest number of participants who underestimated the model's accuracy (15) and the highest number of participants who overestimated the model's accuracy (11). This might suggest that the appropriate confidence indicators made the model seem more accurate or reliable, while the under and random confidence indicators made the model seem less reliable (see Fig. 5).

A regression analysis demonstrated no difference in accuracy over time aside from within the no confidence condition in which accuracy for miss and FA + miss trials decreased over time. This suggests that having confidence values helps keep participants engaged in the task and reduces errors over time for the most difficult trial types. It is possible that because of the simple nature of the visual search task, participants adjusted to the bias of the under and over confident conditions early enough in the task that there was no change in accuracy over time.

There are a number of notable limitations to this work that future efforts should endeavor to consider. The current study used simulated machine learning outputs rather than output from an actual ML algorithm. This allowed us to manipulate whether the outputs were correct or incorrect at the stimulus level, and to control the exact proportion of correct and incorrect responses at the experiment level. The drawback of using outputs from an actual ML model is that the stimuli for which the model made an incorrect decision could be systematically different from those it answered correctly, producing confounds in the stimuli. The method employed of creating simulated model outputs allowed us to isolate the impacts of ML confidence values on human performance outside of the bounds of any single ML model implementation.

There are a large number of factors that could influence human interaction with ML algorithms, including but not limited to perceived understanding of the model (Cramer et al. 2008), error type and error frequency (Gastelum et al. 2020; Rice 2009), the method used to draw user attention to an object (Gastelum et al. 2020), code structure, algorithm factors, and affordances for learning (Lyons 2019), predictability, and explainability (Toreini et al. 2020). We did not systematically vary these additional factors, and in some sense the description to participants that they were being assisted by a ML algorithm was arbitrary. We did not describe how the algorithm would have arrived at its conclusions, what machine learning is, or question participants about what the term machine learning meant to them.

Real world domains that may make use of ML algorithms are therefore quite complex, and following an understanding of isolated, controlled adjustments it will be critical to examine how these factors might interact in order to have an understanding of how to optimally implement ML algorithm assistance in complex contexts.

5 Conclusion

ML algorithms have been demonstrated to improve the performance of a system by enhancing human decision making in a number of domains. In order to understand the optimal implementation of algorithmic assistance, it is necessary to isolate particular facets of that implementation and examine the resultant impact on human performance of manipulating that facet. In this study, we evaluated different models of confidence values to participants, and found that randomized confidence is likely harmful while non-random confidence values are likely better than no confidence value for maintaining accuracy over time. Surprisingly, providing participants with appropriate confidence values did not seem to benefit performance any more than providing participants with under or over confident models.

Acknowledgements. Sandia National Laboratories is a multimission laboratory managed and operated by National Technology and Engineering Solutions of Sandia, LLC, a wholly owned subsidiary of Honeywell International Inc., for the U.S. Department of Energy's National Nuclear Security Administration under contract DE-NA0003525. SAND2021-2142 C. This work was funded by Sandia National Laboratories' Laboratory-Directed Research and Development program, through the Computational and Information Sciences investment area.

References

Biggs, A.T., Cain, M.S., Clark, K., Darling, E.F., Mitroff, S.R.: Assessing visual search performance differences between Transportation Security Administration Officers and nonprofessional visual searchers. Vis. Cognit. **21**(3), 330–352 (2013). https://doi.org/10.1080/13506285.2013.790329

Cireşan, D., Meier, U., Masci, J., Schmidhuber, J.: A committee of neural networks for traffic sign classification. In: The 2011 International Joint Conference on Neural Networks, pp. 1918–1921. IEEE, July 2011

Cramer, H., et al.: The effects of transparency on trust in and acceptance of a content-based art recommender. User Mod. User-Adapt. Interact. **18**(5), 455–496 (2008). https://doi.org/10.1007/s11257-008-9051-3

Druzhkov, P.N., Kustikova, V.D.: A survey of deep learning methods and software tools for image classification and object detection. Pattern Recognit. Image Anal. **26**(1), 9–15 (2016). https://doi.org/10.1134/S1054661816010065

Du, N., Huang, K.Y., Yang, X.J.: Not all information is equal: effects of disclosing different types of likelihood information on trust, compliance and reliance, and task performance in human-automation teaming. Hum. Factors **62**(6), 987–1001 (2020). https://doi.org/10.1177/0018720819862916

Fuchs, F., Song, Y., Kaufmann, E., Scaramuzza, D., Duerr, P.: Super-human performance in gran turismo sport using deep reinforcement learning. arXiv preprint arXiv:2008.07971 (2020)

Gastelum, Z.N., et al.: Evaluating the cognitive impacts of errors from analytical tools in the international nuclear safeguards domain. In: Proceedings of the Institute of Nuclear Materials Management Annual Meeting, July 2020

Goddard, K., Roudsari, A., Wyatt, J.C.: Automation bias: empirical results assessing influencing factors. Int. J. Med. Inf. **83**(5), 368–375 (2014)

Goh, Y.C., Cai, X.Q., Theseira, W., Ko, G., Khor, K.A.: Evaluating human versus machine learning performance in classifying research abstracts. Scientometrics **125**(2), 1197–1212 (2020). https://doi.org/10.1007/s11192-020-03614-2

Khasawneh, M.T., Bowling, S.R., Jiang, X., Gramopadhye, A.K., Melloy, B.J.: A model for predicting human trust in automated systems. Origins, 5 (2003)

Keim, D.A., Mansmann, F., Schneidewind, J., Thomas, J., Ziegler, H.: Visual analytics: scope and challenges. In: Simoff, S.J., Böhlen, M.H., Mazeika, A. (eds.) Visual Data Mining. LNCS, vol. 4404, pp. 76–90. Springer, Heidelberg (2008). https://doi.org/10.1007/978-3-540-71080-6_6

Kneusel, R.T., Mozer, M.C.: Improving human-machine cooperative visual search with soft highlighting. ACM Trans. Appl. Percept. (TAP) **15**(1), 1–21 (2017). https://doi.org/10.1145/312 9669

Körber, M., Baseler, E., Bengler, K.: Introduction matters: manipulating trust in automation and reliance in automated driving. Appl. Ergon. **66**, 18–31 (2018). https://doi.org/10.1016/j.apergo. 2017.07.006

Lyons, J.B., Wynne, K.T., Mahoney, S., Roebke, M.A.: Trust and human-machine teaming: a qualitative study. In: Artificial Intelligence for the Internet of Everything, pp. 101–116. Academic Press (2019). https://doi.org/10.1016/b978-0-12-817636-8.00006-5

Minitab 19 Statistical Software: [Computer software]. State College, PA: Minitab, Inc. (2020). www.minitab.com

Merritt, S.M., Ilgen, D.R.: Not all trust is created equal: dispositional and history-based trust in human-automation interactions. Hum. Factors **50**(2), 194–210 (2008). https://doi.org/10.1518/ 001872008X288574

Paolacci, G., Chandler, J., Ipeirotis, P.G.: Running experiments on amazon mechanical turk. Judgm. Decis. Mak. **5**(5), 411–419 (2010). https://ssrn.com/abstract=1626226

Rice, S.: Examining single-and multiple-process theories of trust in automation. J. Gen. Psychol. **136**(3), 303–322 (2009). https://doi.org/10.3200/GENP.136.3.303-322

Toreini, E., Aitken, M., Coopamootoo, K., Elliott, K., Zelaya, C.G., van Moorsel, A.: The relationship between trust in AI and trustworthy machine learning technologies. In: Proceedings of the 2020 Conference on Fairness, Accountability, and Transparency, pp. 272–283, January 2020. https://doi.org/10.1145/3351095.3372834

Wolfe, J.: When do I quit? The search termination problem in visual search. In: Dodd, M., Flowers, J. (eds.) The Influence of Attention, Learning, and Motivation on Visual Search. Nebraska Symposium on Motivation, pp. 183–208. Springer, New York (2012). https://doi.org/10.1007/ 978-1-4614-4794-8_8

Weyand, T., Kostrikov, I., Philbin, J.: PlaNet - photo geolocation with convolutional neural networks. In: Leibe, B., Matas, J., Sebe, N., Welling, M. (eds.) ECCV 2016. LNCS, vol. 9912, pp. 37–55. Springer, Cham (2016). https://doi.org/10.1007/978-3-319-46484-8_3

Yin, M., Wortman Vaughan, J., Wallach, H.: Understanding the effect of accuracy on trust in machine learning models. In: Proceedings of the 2019 Chi Conference on Human Factors in Computing Systems, pp. 1–12, May 2019. https://doi.org/10.1145/3290605.3300509

NEARME: Dynamic Exploration
of Geographical Areas

Noemi Mauro[✉], Liliana Ardissono, Federico Torrielli, Gianmarco Izzi,
Claudio Mattutino, Maurizio Lucenteforte, and Marino Segnan

Computer Science Department, University of Torino, Torino, Italy
{noemi.mauro,liliana.ardissono,claudio.mattutino,maurizio.lucenteforte,
marino.segnan}@unito.it, {federico.torriell,gianmarco.izzi}@edu.unito.it

Abstract. Web GIS offer precious data to explore geographic areas
but they might overload the user with large amounts of information if
(s)he is unable to specify efficient search queries. Services such as Open-
StreetMap and Google Maps support focused information search, which
requires people to exactly define what they are looking for. However,
what can be searched within a specific area mainly depends on what
is located there. Thus, the question is how to provide the user with
an overview of the available data (s)he can look for, instead of forcing
her/him to search for information in a blind way.

This paper attempts to address this issue by introducing the NEARME
exploration model. NEARME offers a search lens which, positioned on a
geographic map, enables the user to discover the categories of Points of
Interest that are available in the selected area (e.g., services and Cultural
Heritage items) and to choose the types of information to be displayed,
based on a faceted-exploration search model. NEARME is based on a
semantic representation of geo-data and it is integrated in the OnToMap
Participatory GIS, which supports geographic information sharing. We
carried out a preliminary user study with 25 participants to assess the
User Experience with our model. The results show that NEARME is perceived as easy to use, understandable, attractive and that it efficiently
supports exploratory search using geographic maps.

Keywords: GIS · Geographic information search · Semantic filtering
lenses

1 Introduction

Geographic information search may challenge users in different ways:

- Its exploratory nature makes it hard for people to define efficient search
  queries. If users are not familiar with a region, they simply do not know which
  Points of Interest (PoIs) they can find there. Thus, they should be guided in
  discovering the available options. This is in line with Marchiorini's discussion
  [21] that, in exploratory search, users typically have ill-formed goals because

© Springer Nature Switzerland AG 2021
S. Yamamoto and H. Mori (Eds.): HCII 2021, LNCS 12765, pp. 206–217, 2021.
https://doi.org/10.1007/978-3-030-78321-1_16

they are not familiar with what they are looking for. Therefore, they are unable to specify efficient search queries aimed at satisfying their information needs. As a consequence, they can be overloaded by possibly large amounts of irrelevant data.

– Users might be interested in exploring in detail the PoIs they can find near themselves, or in a very specific geographic area such as the neighborhood of a place, rather than viewing complex maps that cover large regions.

These issues can be addressed by changing the way geographic information systems interact with users. For instance, suppose that somebody has dinner in a restaurant and looks for nearby places to spend the rest of the night. Different options might be considered, such as visiting a pub, a club, or going to the bowling. Our idea is that, instead of asking the user to specify focused search queries that might lead to zero solutions, or visualizing whatever is located nearby, the system might show an overview of the categories of PoIs that are available in the selected area (e.g., clubs and pubs might be present, but not bowlings) and let her/him focus the search accordingly.

Starting from the concepts developed in the research about faceted search support [14,19], we propose to guide the user by means of context-dependent search criteria that take the solution space into account to enhance information exploration. Specifically, we propose NearMe, an interactive model that uses an augmented lens metaphor to assist map-based information filtering and data visualization in restricted geographic areas. By placing the NearMe lens on a map, the user receives a list of data categories corresponding to the types of PoIs located in the area under the lens. Then, (s)he can select the categories to be visualized in an informed way, being guaranteed that (s)he will not receive an empty set of results [14,28]. Moreover, the user can interactively revise the content of the lens to customize the view on the available PoIs, and (s)he can bookmark PoIs to make them permanent in the map. Combined with the possibility of sharing the map with other users in a persistent way, this model enables people to build Personal Information Spaces [5] for the organization of individual and group activities.

We carried out a preliminary user study to test the functions offered by NearMe. The experiment involved 25 participants and obtained very positive User Experience (UX) results. NearMe was perceived as easy to use, understandable and attractive. Moreover, participants felt that it efficiently supports exploratory search. These results encourage a general adoption of this model in geographic information exploration.

The remainder of this paper is organized as follows: Sect. 2 provides some background and positions our work in the related one. Section 3 describes the NearMe model. Sections 4 and 5 present the validation methodology we applied and the results of our user study. Section 6 concludes the paper.

2   Background and Related Work

2.1   Geographic Information Search Support

Web GIS constrain the information they present on maps in different ways. For instance, both OpenStreetmap [24] and Google Maps [13] steer the visualization of data categories to the zoom level of a map in order to progressively display more details while the user focuses on more restricted areas. This information filtering strategy does not enable people to select the types of PoIs they want to view. Thus, it fails to support the visualization of custom maps reflecting individual information needs.

Google Maps also helps the user explore the region around a place. However, it assumes that (s)he knows what (s)he is looking for, and that (s)he submits a search query satisfying her/his information goals in a rather precise way. As previously discussed, this type of interaction is challenging in exploratory search. Moreover, it limits the discovery of places that could interest the user but which do not belong to the data categories (s)he specifies. Furthermore, Google Maps assumes that the user focuses on a specific type of information within each search task. As each search query focuses on a single data category, results cannot be jointly visualized in the map. Therefore, the system exposes the user to a fragmented view of results, instead of providing her/him with a unified presentation of the area of interest.

Drive navigation systems, such as TomTom [29] and Google Maps, show nearby Points of Interest depending on the user's geographic location. However, they focus on routing and on the visualization of broadly inspected data related to driving, like speed cameras, fuel dispensers, and so forth. They also show other types of PoIs, if mapped, but they do not enable the user to select the categories (s)he is interested in (e.g., historical buildings versus hotels) and they do not provide in-depth information about PoIs. Waze [30] community driven GPS navigation app allows its users to report any travel-related information, which is then used to optimize route planning and to provide real-time traffic updates. In addition to specific places relevant to driving, it displays some categories of PoIs using distinct search icons. In this way, specific sites around the user can be viewed on the map by choosing one category out of parking, gas stations, food, drive-thru, cafes, and similar. Although the user can select a specific place near her/him to get detailed information about it, only sites belonging to a single information category at a time can be visualized on the map. In summary, we conclude that the above listed services are not sufficient to support the user while visiting a geographical area.

Personalized mobile guides, such as [2,7,10,22], help people find PoIs relevant to their interests. They also present detailed information about places. However, they are based on the management of long-term user profiles that have to be bootstrapped before providing effective suggestions. Thus, they can hardy react to short-term search goals.

Fig. 1. NEARME: selection of information categories to be visualized within a lens.

NEARME differs from all the above listed models because:

- It provides the user with a synthesis of the categories of PoIs (e.g., schools, bus stations, libraries, cafes) that are available in the selected area and it enables her/him to manage a custom map by choosing the relevant data to be visualized.
- It supports the inspection of details about the items displayed in the map. Moreover, it enables the user to bookmark PoIs in order to create persistent maps that describe the areas of interest.
- It enables a flexible exploration of a geographic area by moving the lens on the map, thus exploring small areas in a continuous way.

2.2 Map-Based Information Visualization

As discussed by Kraak et al., maps are commonly used as visual thinking and analysis tools [1,16]. Moreover, in participatory decision-making [8,12], they are used as shared representations to collect information in long-lasting distributed projects. However, the presentation of large amounts of data may overload the user, also depending on her/his visual spatial abilities [9]. Various models have thus been designed to enable a selective visualization of information, either based on spatial or on time multiplexing.

Lobo et al. [20] found that Translucent Overlay [11,18] and Blending Lens [6, 26] are the best performing visualization models for interactive map comparison. However, these models work on two layers only. Differently, we are interested in

Fig. 2. Data visualization within the NEARME lens. (Color figure online)

enabling a joint search on multiple types of related information, such as schools, pharmacies and bus stops. In order to address this limitation, we propose a lens-based visualization model integrated with a faceted-search one [23] that supports interactive information filtering and semantic data visualization.

3   NEARME

NEARME exploits the OnToMap Web collaborative GIS [3,4] as a semantic data container, information sharing service and map viewer.[1] OnToMap defines the data categories in an OWL [25] ontology that is mapped to the domain representations of external data sources in order to retrieve information from them. The ontology also specifies graphical details for map visualization. For instance, it stores the colors and icons to be displayed in the map when visualizing the items belonging to each category.

Figure 1 shows a step of interaction with NEARME. In order to explore a geographical area, the user draws a circle on the map. In turn, the system shows a list of checkboxes corresponding to the categories of the items located in the bounding box. For each category c, the bounding area includes at least one PoI belonging to c. The user can choose the relevant categories by clicking on the checkboxes. For instance, in the sample interaction of Fig. 1, the user has

---

[1] OnToMap is used as data container in "co3project: co-create, co-produce, co-manage", https://www.projectco3.eu/it/.

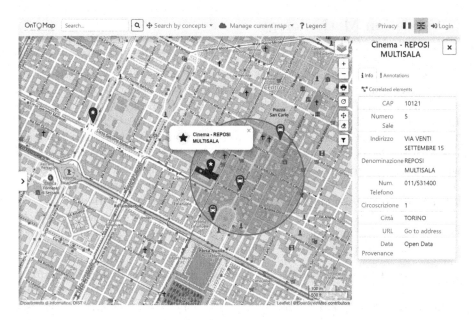

Fig. 3. Details about a Point of Interest.

selected categories Cinemas, Museums and Public Transportation Stops. In turn, the system visualizes in the map the related items; see Fig. 2.

In the map, PoIs are depicted using representative icons, or geometries, if they are available. Moreover, color-coding [15] is used to identify their categories. For instance, in Fig. 2 there are a cinema (REPOSI MULTISALA), represented as a violet geometry, a museum (yellow pointer and yellow area) and three public transportation stops (magenta pointers). The user can bookmark the visualized PoIs to make them persistent in the map. Bookmarked items are identified by pointers with yellow stars. In the example, the user has bookmarked cinema REPOSI MULTISALA, but (s)he previously bookmarked other three PoIs: a church (brown pointer), a parking lot (blue marker), and a shop (green marker).

The user can also interact with the PoIs visualized in the map to view their details. As shown in Fig. 3, in that case, OnToMap shows a table that reports specific information about them. This makes it possible to use the map as catalog presenting geographic data at different levels of detail. Notice that the user can drag the lens to other areas of the map, or revise the selected filters, to retrieve other Point of Interests. Overall, the lens complements the basic search support offered by OnToMap by enabling an incremental and interactive exploration of geo-data, focusing on small areas to reduce information overload.

4   Validation Methodology

To test the user experience with NEARME, we carried out an experiment involving 25 participants. People joined in the user study on a voluntary basis, without any compensation, and they signed a consent to the treatment of personal data.

The user study took place live, in video calls with shared screen due to COVID-19 pandemic. One person at a time performed the study which lasted about 20 min. Initially, the participant watched a video showing how the lens works. After that, (s)he interacted with OnToMap and with the lens on a sample map to get acquainted with the system. Then, we asked her/him to answer a pre-test questionnaire designed to assess demographic information, cultural background, as well as familiarity with map-based online applications.

During the study, we asked people to solve two map learning tasks in which they had to identify and possibly bookmark certain PoIs. The tasks (translated from the Italian language) are the following ones:

1. *Suppose that you arrive at the Porta Nuova train station to spend the afternoon in the city. We ask you to use NEARME to find information related to museums, cinemas, urban parks and bike sharing stations. Moreover, we ask you to bookmark three places that you would like to visit.*
2. *Suppose that you are going to live in a place nearby Torino Porta Susa train station. We ask you to use NEARME to find information about schools, parking lots, post offices, bike sharing stations and drugstores that you will find near your place.*

Participants carried out the two tasks by interacting with the NEARME lens. We did not put any time restrictions on the execution of the tasks.

After having completed the tasks, each participant answered a questionnaire to assess her/his User Experience (UX) with the system. Moreover, (s)he could provide free-text feedback by answering an open question. In order to measure UX, we used the Italian version of the UEQ questionnaire [17]. UEQ supports a quick assessment of a comprehensive impression of user experience covering perceived ergonomic quality, hedonic quality, and attractiveness of a software product. Questions are proposed as bipolar items, e.g., [annoying 1 2 3 4 5 6 7 enjoyable]. Moreover, in order to check user attention, half of the items start with the positive term (e.g., "good" versus "bad") while the other ones start with the negative term (e.g., "annoying" versus "enjoyable") in randomized order. The range of values of the questions is mapped to the $[-3, +3]$ interval for the UX evaluation, where -3 is very negative. Each question corresponds to an individual UX aspect and it belongs to one of six *UEQ factors* that describe broader user experience dimensions: "Attractiveness", "Perspicuity", "Novelty", "Stimulation", "Dependability", and "Efficiency". Table 1 shows the bipolar items of our questionnaire.

5 Results

5.1 Data About Participants

For the user study we recruited 25 participants (52% women; 48% men; 0% not declared). Their age is between 21 and 60 years, with a mean value of 29.36. 68% of participants are students of different Schools of the University of Torino, 16% are full-time employees and 16% are teachers. Overall, 44% have a Bachelor or Master degree, 52% have a Secondary School diploma and 4% have a middle school certificate. Participants' background can be split as follows: 64% scientific, 16% technical, 16% humanities, and 4% linguistic.

Regarding familiarity with technology, 56% of participants declared to have a middle level, 20% beginners, 20% experts and 4% totally unfamiliar with it. Moreover, 48% of people stated that they use online maps such as Google Maps every week, 32% monthly, 12% daily and 8% a couple of times per year.

Table 1. Detailed results of the UEQ [27] questionnaire. The Aspect column shows the bipolar item that is the object of the question posed to the user. The Factor column shows the UX Factor to which the aspect belongs. All results are positive, as denoted by the ↑ symbol preceding the mean value for each question.

Question	Mean	Variance	St. Dev.	Aspect	Factor
1	↑ 2.0	1.5	1.2	Annoying/enjoyable	**Attractiveness**
2	↑ 2.2	0.8	0.9	Not understandable/understandable	**Perspicuity**
3	↑ 2.0	1.2	1.1	Creative/dull	**Novelty**
4	↑ 2.4	0.7	0.8	Easy to learn/difficult to learn	**Perspicuity**
5	↑ 2.1	0.6	0.8	Valuable/inferior	Stimulation
6	↑ 1.7	0.8	0.9	Boring/exciting	Stimulation
7	↑ 2.5	0.3	0.6	Not interesting/interesting	Stimulation
8	↑ 1.2	1.4	1.2	Unpredictable/predictable	Dependability
9	↑ 1.6	2.8	1.7	Fast/slow	**Efficiency**
10	↑ 1.8	1.6	1.3	Inventive/conventional	**Novelty**
11	↑ 2.4	0.4	0.6	Obstructive/supportive	Dependability
12	↑ 2.4	0.4	0.6	Good/bad	**Attractiveness**
13	↑ 2.2	1.2	1.1	Complicated/easy	**Perspicuity**
14	↑ 2.0	1.0	1.0	Unlikable/pleasing	**Attractiveness**
15	↑ 2.0	1.4	1.2	Usual/leading edge	**Novelty**
16	↑ 2.4	0.4	0.6	Unpleasant/pleasant	**Attractiveness**
17	↑ 2.4	0.5	0.7	Secure/not secure	Dependability
18	↑ 2.0	0.8	0.9	Motivating/demotivating	Stimulation
19	↑ 2.0	1.5	1.2	Meets expectations/does not meet expectations	Dependability
20	↑ 2.4	1.2	1.1	Inefficient/efficient	**Efficiency**
21	↑ 2.3	1.7	1.3	Clear/confusing	**Perspicuity**
22	↑ 2.2	0.8	0.9	Impractical/practical	**Efficiency**
23	↑ 2.6	0.3	0.6	Organized/cluttered	**Efficiency**
24	↑ 2.2	0.6	0.8	Attractive/unattractive	**Attractiveness**
25	↑ 2.2	0.8	0.9	Friendly/unfriendly	**Attractiveness**
26	↑ 2.0	1.1	1.1	Conservative/innovative	**Novelty**

Table 2. User Experience results grouped by UX factor. Values are obtained by averaging the ratings given by participants to individual UX aspects of UEQ.

	NEARME
Attractiveness	↑ 2.187
Perspicuity	↑ 2.300
Efficiency	↑ 2.220
Dependability	↑ 1.990
Stimulation	↑ 2.060
Novelty	↑ 1.950

5.2   User Experience Results

Answers to Individual UEQ Aspects. Table 1 shows the mean values of participants' answers to the questions of UEQ. As the scores given by participants are mapped to the $[-3, 3]$ interval, values ≥ 0.8 denote positive evaluations. Moreover, values in $(-0.8, 0.8)$ are neutral and lower values than -0.8 are negative. It can be noticed that:

- all the mean scores received by the questions are ≥ 1.2;
- 22 mean values out of 26 are ≥ 2.0;
- 14 mean values out of 26 are ≥ 2.2.

Moreover, standard deviation values are low. They range between 0.6 and 1.3, with the only exception of the efficiency value, which is slightly higher, and equal to 1.7. This means that, overall, participants appreciated our model quite a lot.

The questions that received the lowest scores concern the predictability (question 8, meanvalue = 1.2) and speed (question 9, meanvalue = 1.6) of the system. However, these values are counterbalanced by the other aspects belonging to the same UX factors. Specifically, users evaluated NEARME as supportive (question 11), secure (question 17) and meeting their expectations (question 19). Moreover, even though participants perceived the system as a bit slow, they considered it efficient, practical and organized.

Results by UX Factor. Table 2 provides the reader with a more compact description of results. The individual questions of UEQ are grouped by UX factor and the mean values of each question are further averaged to obtain a single, representative value of the factor, such as Attractiveness, Perspicuity, and so forth. This gives a more general view of users' perceptions, which abstracts from the possible fluctuations occurred in individual questions. Also in this case, we can see that NEARME has received very high evaluations: all the mean values are near or above 2. Participants perceived NEARME as particularly perspicuous (i.e., easy to use, understandable, etc.), efficient and attractive. Moreover, they evaluated it as dependable (i.e., predictable, supportive, etc.) and they appreciated its novelty.

Answers to the Open Question. In the free-text answers, several participants confirmed that they appreciated the system. Moreover, they gave some suggestions. For instance, a user proposed to visualize the scale of the circle, in order to understand the size of the area below it. Moreover, another participant complained that, each time the user creates a lens, a new list of checkboxes representing PoI categories appears. She asked to enable the positioning of the widget aside in the user interface.

As far as predictability is concerned, which received the lowest mean evaluation (question 8), a participant suggested to include in the user interface of the system a video tutorial showing the main functions offered by NearMe, similar to the one we proposed before experimental tasks. This might be interpreted as a need to clarify what can be done when interacting with the lens, and a need for help in learning to use the system.

Discussion. The experimental results suggest that NearMe successfully supports geographic information exploration and can be a useful extension to a Web GIS such as OnToMap, in order to help the user focus on small areas when looking for specific categories of PoIs.

There are however some limitations, which we plan to address in our future work. In particular, we tested our model on a small sample of people, most of whom are University students. In order to acquire more significant results, a larger set of participants has to be involved in the user test. Another limitation is the fact that people tested the system by using laptops and desktop PCs. Another test should be carried out to measure User Experience when the system is used on mobile phones.

6   Conclusions

We presented the NearMe information filtering model, aimed at supporting the exploration of geographical maps by focusing on small size areas. In this model, we provide the user with a lens that can be positioned on an area in order to discover which categories of PoIs are available, and to select those to be visualized. Basically, the lens offers a preview of the available options, in that it only proposes categories having at least one available PoI located in the area under the lens. Thus, it guides the user in an informed navigation of the solution space, preventing the zero-results effect. A user test involving 25 participants showed that NearMe is perceived as easy to use, understandable, attractive and efficient in supporting geographic information exploration.

Acknowledgments. This work was supported by the European Community through co3project: co-create, co-produce, co-manage (H2020 - CO3 Grant Agreement 822615).

References

1. Andrienko, G.L., Andrienko, N.V.: Interactive maps for visual data exploration. Int. J. Geogr. Inf. Sci. **13**(4), 355–374 (1999). https://doi.org/10.1080/136588199241247
2. Ardissono, L., Goy, A., Petrone, G., Segnan, M., Torasso, P.: INTRIGUE: personalized recommendation of tourist attractions for desktop and handset devices. Appl. Artif. Intell. **17**(8–9), 687–714 (2003). https://doi.org/10.1080/713827254. Special Issue on Artificial Intelligence for Cultural Heritage and Digital Libraries
3. Ardissono, L., Lucenteforte, M., Mauro, N., Savoca, A., Voghera, A., La Riccia, L.: OnToMap: semantic community maps for knowledge sharing. In: Proceedings of the 28th ACM Conference on Hypertext and Social Media (HT 2017), pp. 317–318. Association for Computing Machinery, New York (2017). https://doi.org/10.1145/3078714.3078747
4. Ardissono, L., Lucenteforte, M., Mauro, N., Savoca, A., Voghera, A., La Riccia, L.: Semantic interpretation of search queries for personalization. In: Adjunct Publication of the 25th Conference on User Modeling, Adaptation and Personalization (UMAP 2017), pp. 101–102. Association for Computing Machinery, New York (2017). https://doi.org/10.1145/3099023.3099030
5. Ardito, C., Costabile, M.F., Desolda, G., Matera, M.: Supporting professional guides to create personalized visit experiences. In: Proceedings of the 18th International Conference on Human-Computer Interaction with Mobile Devices and Services Adjunct (MobileHCI 2016), pp. 1010–1015. ACM, New York (2016). https://doi.org/10.1145/2957265.2962650
6. Bier, E.A., Stone, M.C., Pier, K., Buxton, W., DeRose, T.D.: Toolglass and magic lenses: The see-through interface. In: Proceedings of the 20th Annual Conference on Computer Graphics and Interactive Techniques (SIGGRAPH 1993), pp. 73–80. Association for Computing Machinery, New York (1993). https://doi.org/10.1145/166117.166126
7. Braunhofer, M., Ricci, F.: Selective contextual information acquisition in travel recommender systems. Inf. Technol. Tour. **17**(1), 5–29 (2017). https://doi.org/10.1007/s40558-017-0075-6
8. Brown, G., Weber, D.: Measuring change in place values using public participation GIS (PPGIS). Appl. Geogr. **34**, 316–324 (2012). https://doi.org/10.1016/j.apgeog.2011.12.007
9. Canham, M., Hegarty, M.: Effects of knowledge and displays design on comprehension of complex graphics. Learn. Instr. **20**(2), 155–166 (2010). https://doi.org/10.1016/j.learninstruc.2009.02.014
10. Cheverst, K., Davies, N., Mitchell, K., Smith, P.: Providing tailored (context-aware) information to city visitors. In: Brusilovsky, P., Stock, O., Strapparava, C. (eds.) AH 2000. LNCS, vol. 1892, pp. 73–85. Springer, Heidelberg (2000). https://doi.org/10.1007/3-540-44595-1_8
11. Colby, G., Sholl, L.: Transparency and blur as selective cues for complex visual information. In: Proceedings of Image Handling and Reproduction Systems Integration, vol. 1460, pp. 114–125. SPIE Digital Library (1991). https://doi.org/10.1117/12.44415
12. Coulton, C., Chan, T., Mikelbank, K.: Finding place in community change initiatives: using GIS to uncover resident perceptions of their neighborhoods. J. Community Pract. **19**(1), 10–28 (2011). https://doi.org/10.1080/10705422.2011.550258
13. Google: Google maps (2019). https://www.google.com/maps

14. Hearst, M.A.: Design recommendations for hierarchical faceted search interfaces. In: Proceedings of SIGIR 2006, Workshop on Faceted Search, pp. 26–30, August 2006

15. Hoeber, O., Yang, X.D.: A comparative user study of web search interfaces: HotMap, Concept Highlighter, and Google. In: Proceedings of the 2006 IEEE/WIC/ACM International Conference on Web Intelligence (WI 2006), pp. 866–874. IEEE Computer Society, Washington (2006). https://doi.org/10.1109/WI.2006.6

16. Kraak, M.J.: Playing with maps: explore, discover, learn, categorize,analysis, explain, present geographic and non - geographic data: keynote. Presented at Infovis6, London, UK, 5 July 2006, pp. 1–25 (2006)

17. Laugwitz, B., Held, T., Schrepp, M.: Construction and evaluation of a user experience questionnaire. In: Holzinger, A. (ed.) HCI and Usability for Education and Work, pp. 63–76. Springer, Heidelberg(2008). https://doi.org/10.1007/978-3-540-89350-9_6

18. Lieberman, H.: Powers of ten thousand: navigating in large information spaces. In: Proceedings of the 7th Annual ACM Symposium on User Interface Software and Technology (UIST 1994), pp. 15–16. ACM, New York (1994)

19. Lionakis, P., Tzitzikas, Y.: PFSgeo: preference-enriched faceted search for geographical data. In: Panetto, H., et al. (eds.) OTM 2017. LNCS, vol. 10574, pp. 125–143. Springer, Cham (2017). https://doi.org/10.1007/978-3-319-69459-7_9

20. Lobo, M.J., Pietriga, E., Appert, C.: An evaluation of interactive map comparison techniques. In: Proceedings of the 33rd Annual ACM Conference on Human Factors in Computing Systems (CHI 2015), pp. 3573–3582. ACM, New York (2015). https://doi.org/10.1145/2702123.2702130

21. Marchionini, G.: Exploratory search: from finding to understanding. Commun. ACM 49(4), 41–46 (2006). https://doi.org/10.1145/1121949.1121979

22. Mauro, N., Ardissono, L., Cena, F.: Personalized recommendation of PoIs to people with autism. In: Proceedings of the 28th ACM Conference on User Modeling, Adaptation and Personalization (UMAP 2020), pp. 163–172. ACM, New York (2020). https://doi.org/10.1145/3340631.3394845

23. Mauro, N., Ardissono, L., Lucenteforte, M.: Faceted search of heterogeneous geographic information for dynamic map projection. Inf. Process. Manag. 57(4), 102257 (2020). https://doi.org/10.1016/j.ipm.2020.102257

24. OpenStreetMap Contributors: Openstreetmap (2017). https://www.openstreetmap.org

25. OWL Services Coalition: OWL-S: semantic markup for web services (2004). http://www.daml.org/services/owl-s/1.1B/owl-s/owl-s.html

26. Pietriga, E., Bau, O., Appert, C.: Representation-independent in-place magnification with sigma lenses. IEEE Trans. Vis. Comput. Graph. 16(3), 455–467 (2010). https://doi.org/10.1109/TVCG.2009.98

27. Schrepp, M., Hinderks, A., Thomaschewski, J.: User experience questionnaire (2017). www.ueq-online.org/

28. Shneiderman, B.: Tree visualization with tree-maps: 2-D space-filling approach. ACM Trans. Graph. 11(1), 92–99 (1992). https://doi.org/10.1145/102377.115768

29. TomTom International BV: TomTom (2021). https://www.tomtom.com

30. Waze Mobile Ltd.: Waze (2021). https://www.waze.com

Decision Support for Prolonged, and Tactical Combat Casualty Care

Christopher Nemeth[1]([✉])(iD), Adam Amos-Binks[1], Natalie Keeney[1],
Yuliya Pinevich[2](iD), Gregory Rule[1], Dawn Laufersweiler[1], Isaac Flint[1],
and Vitaly Hereasevich[2](iD)

[1] Applied Research Associates, Inc., Albuquerque, NM, USA
{cnemeth,aamosbinks,nkeeney,grule,dlaufersweiler,iflint}@ara.com
[2] Mayo Clinic, Rochester, MN, USA
{Pinevich.Yuliya,Vitaly}@mayo.edu

Abstract. In Tactical Combat Casualty Care (TCCC), evacuation from a battlefield may not be immediately available, resulting in a medic providing Prolonged Field Care (PFC) for one or more casualties over hours or days.

We report on the development and evaluation of the Trauma Triage Treatment and Training Decision Support (4TDS) system, an Android phone and tablet-based application, that is designed to support clinician casualty care decisions, to detect the probability of a casualty experiencing shock, and to provide refresher training in life-critical skills. Refresher training scenarios ensure retention of crucial knowledge and skills needed to perform life-critical skills during extended deployments. As part of 4TDS, we also used machine learning to develop a model that scans vital signs data to detect the probability of a casualty developing shock, which is a life-threatening condition that is more likely to occur during PFC.

Interface development methods included a literature review, rapid prototyping, and agile software development, while evaluation methods included design requirements reviews, subject matter expert reviews, and usability assessments. We developed a shock model using a logistical regression approach, trained the shock model on public Intensive Care Unit (ICU) patient data, then trained it on de-identified ICU patient data provided by our collaborators at Mayo Clinic. We evaluated the model in a silent test at Mayo Clinic over six months to compare model identification of shock with clinician decisions on the same patients.

More timely, accurate decisions will improve battlefield casualty care outcomes and reduce the potential for misadventures.

Keywords: Design requirements review · Usability assessment · Decision support · Shock assessment · Tactical Combat Casualty Care · Prolonged Field Care

1 Introduction

Tactical Combat Casualty Care (TCCC) [1] is the U.S. military's process to retrieve, stabilize, and transport battlefield casualties to either return them to service or to receive

© Springer Nature Switzerland AG 2021
S. Yamamoto and H. Mori (Eds.): HCII 2021, LNCS 12765, pp. 218–226, 2021.
https://doi.org/10.1007/978-3-030-78321-1_17

further care. Until recently, air evacuation by helicopter made it possible to transport a casualty to a treatment site such as a Battalion Air Station (BAS) or Field Hospital (FH) for further care. Evacuation to such sites within 60 min (the "Golden Hour") has for years been considered the standard for battlefield casualty care. However, operations in remote locations with less likely air support increases the likelihood of Prolonged Field Care (PFC) [2] in which evacuation is not immediately available and may result in a medic caring for one or more casualties for hours or even days. This can pose the risk of complications such as infection (sepsis) or require the medic to perform more complex procedures. In addition, the experience and ability of care providers who are assigned to the battlefield can vary. Training while deployed in the field can help to improve recall and performance of skills procedures that are crucial to saving lives but are used infrequently.

2 Background

Over the past two years Applied Research Associates has collaborated with the Mayo Clinic, Ambient Clinical Analytics, and Pacific Northwest National Laboratory to develop a decision support system for battlefield casualty care. Our Trauma Triage Treatment and Training Decision Support (4TDS) project has developed an application for use on an Android smart phone and tablet [3].

Figure 1 shows the 4TDS components. An inexpensive, durable sensor can be placed on the casualty to stream vital signs to the phone in real time: systolic blood pressure, respiration rate, blood oxygen levels, and heart function. A file with vital signs data and identifying information can be transferred to the tablet for use by clinicians at a field treatment facility such as a Battalion Aid Station. Tablet interface design reflects clinician mental models learned during prior project work to develop real time decision support for Burn Intensive Care Unit (ICU) clinicians [4, 5]. Tabs along left and right side are organized according to body system, showing key vital signs data that make it possible to quickly understand patient condition. Choice of a tab opens more detailed information in a center panel that can be viewed using line graphs or tables.

Both the phone and tablet include a model the team developed that scans the vital signs data to detect the probability of shock in a casualty up to 90 min before it is likely to occur. On the tablet, a "shock influences" tab contains a diagram that shows the degree to which each of the vital signs influences algorithm performance.

The phone app also includes scenarios for refresher training in knowledge and skills such as shock identification and management, and performance of life-critical procedures such as a cricothyroidotomy to restore a casualty's airway. The user is presented with a brief casualty care situation, asked "what is the next step?" and offered a number of choices that lead to different outcomes. An Instructor interface on the tablet enables a senior medic in the field to monitor how well medics perform while using the refresher training scenarios. Rather than having to wait for results to be processed elsewhere and returned, the Instructor app makes it possible to identify and recognize strengths and needs for improvement, then correspond with individuals to provide guidance.

3   Methods

The project team sought and received approval to conduct research from each of the military commands to which sample participants are assigned. Institutional Review Boards (IRB) and the Human Research Protection Office (HRPO) approved human subject research for both field assessments and machine learning algorithm evaluation at Mayo Clinic.

Lack of access to actual battlefield conditions compelled the team to rely on subject matter experts (SME) and a highly experienced sample that had cared for battlefield casualties. Lack of access to sample participants due to pandemic restrictions made it necessary for the team to develop alternate means to collect data.

3.1   Design Requirements Reviews

In the research phase, the team conducted design requirements reviews with 17 experienced participants from each military service (U.S. Army, Navy, Air Force) who had provided TCCC in austere conditions during multiple deployments. Based on a review of TCCC literature and manuals, the team developed materials that described how the 4TDS system could be expected to work.

Flow Chart Diagram. A flow chart (Fig. 1) illustrated the steps medics would take from point of injury through prolonged field care, to evacuation while using 4TDS.

Use Cases. Narratives a few paragraphs long described how the team expected the app would be used by medics during TCCC and by clinicians at a Battalion Aid Station.

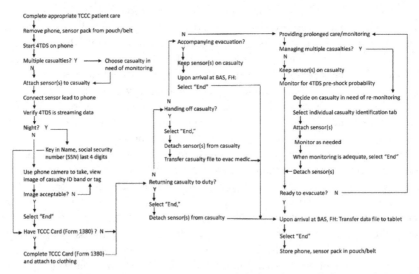

Fig. 1. Notional medic workflow using 4TDS (Copyright © 2021, Applied Research Associates, Inc. Used by permission)

Interface Rough Prototypes. Images of interfaces including those in Fig. 2 illustrated how the team proposed the phone and tablet interfaces would look and operate. Images also showed multiple ways to present data such as shock probability, making it possible for participants to express their preference on how the information should be displayed.

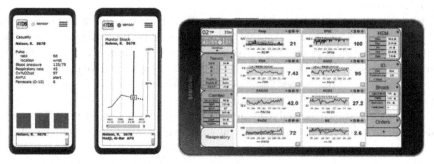

Fig. 2. Initial 4TDS Android phone, tablet interface prototypes (Copyright © 2021, Applied Research Associates, Inc. Used by permission)

In the development phase, pandemic restrictions blocked access to members of the sample we had recruited. As an interim measure, the team reviewed prototype materials with 7 casualty care SME s from the U.S. Army, Navy, and Air Force. With no access to participants at the base where they were assigned, a colleague managed apparatus in person and the team conducted evaluations using the Zoom (Zoom Video Communication, San Jose, CA) videoconferencing platform. Each participant was invited to speak aloud to provide a verbal protocol that indicated his or her thoughts in real time.

Scripts the session facilitator used outlined scenarios that included task assignments. Each script was organized to collect non-identifiable background information to determine experience and point of view, responses to statements about the scenario and the interface, and comments on open-ended questions. One script, for Tactical Combat Casualty Care-Medic, was designed to assess the phone and sensor as well as two training scenarios. Three scripts were designed to assess the tablet app interface: Battalion Aid Station-Medic, Battalion Aid Station-Clinician, and Training Instructor. Responses on the interface and scenarios were: 71% on the TCCC smart phone app, 72% on Battalion Aid Station, 64% on Training tablet apps.

3.2 Field Usability Assessments

Usability assessment makes it possible to determine the fit between a decision support system and the work of those it is intended to assist [6]. In the evaluation phase, the team was finally able to gain access to the sample members when pandemic restrictions were eased. The team conducted two assessments in parallel. Under the circumstances, one of them was managed via the Zoom software platform while the other was conducted in person.

The team performed assessments with 28 exceptionally qualified and experienced participants from the three military services at Joint Base San Antonio, TX. Eighteen of the sample had provided TCCC for over 10 years. Thirteen of the sample had diagnosed casualties with hypovolemic (loss of fluid volume) shock. Many others had diagnosed it in addition to other types, including sepsis (infection), obstructive, and anaphylactic (allergic) shock.

The sessions sought to determine whether the application on the Android phone with a vital signs sensor, and an Android tablet, supported decision making and whether participants found the apps acceptable. We also wanted to determine whether the training scenarios were accurate and suited to refresher training. Scripts used for the session were similar to the prior SME reviews. Participants chose the scripts for which they were most qualified, yielding a total of 51 evaluations.

Assessment results indicated predominantly positive evaluations of the 4TDS applications and provided recommendations to align them even more closely with battlefield casualty care.

3.3 Shock Model Silent Test

In a parallel evaluation, the team conducted a "silent test" that compared our shock detection algorithm's performance with actual clinician diagnoses from May to October 2020. Clinicians were unaware of the algorithm, which is what makes the test "silent". Our Predictive Model Markup Language (PMML) model ran on the 510k-approved AWARE system installed at the Mayo Clinic for six months. The sample was comprised of patients who were admitted to Mayo Clinic ICUs: MICU, surgical, neurosurgical, cardiac surgery ICU; and cardiac progressive units. Each day a research fellow MD reviewed clinician notes assessing ICU patient condition and compared them to shock alerts that the algorithm generated with these results:

Area Under Curve. The team used Receiver Operator Characteristics (ROC). Tradeoff between false positive and false negatives. Anything near 0.80 is considered progress, and the test yielded a result of 0.85.

Sensitivity. Generated an alert for 78% of shock patients (287 out of 365).

Specificity. Generated an alert for 7% of few non-shock patients (345 out of 4785).

Positive Predictive Value. Approximately 1 out of 2 alerts were legitimate (287 legitimate vs. 345 non-legitimate).

Negative Predictive Value. (for shock detection model). Only 78 shock patients were missed out of 365.

Prior published research has used a rich data set including laboratory test results to predict shock probability. The 4TDS model used only vital signs to detect shock probability, as care for casualties in an austere setting would not have access to laboratory tests. Silent test results were noteworthy considering that the model performed in a manner similar to studies using a much richer data set.

4   Discussion

Methods that have been proven in the study of individual and team cognitive work are essential to understand goals, barriers, and operator strategies to accomplish their goals. The development team relied on Cognitive Systems Engineering (CSE) [7] methods to know what data are most important, and how to find and present them in a manner that minimizes demands on the user to make sense of it. The CSE methodology fits within the Naturalistic Decision Making (NDM) approach. The NDM approach differs from other traditions such as heuristic and biases (HB), because "The NDM mindset helps researchers to conduct effective observations and interviews. It enables them to capture the way people use their experience to handle uncertainty and vague goals and high stakes... [8]". NDM is suited to the development of a decision support system for TCCC because "the NDM orientation is to be inclusive and curious about different aspects of cognition that will affect the way people handle the conditions such as limited time, uncertainty, high stakes, vague goals, and instability [9]".

The 4TDS project results show that the NDM approach can develop healthcare IT that is designed to improve actual practice under conditions that are uncertain, time-pressured, and high risk. While 4TDS is a "decision support system," TCCC and PFC care providers perform a range of activities beyond decisions that comprise macro-cognitive activity [10, 11]. For example, the cognitive work of TCCC and PFC also compel individuals and teams to develop treatment plans, assess patients and evaluate their progress, and refine plans to manage care. Medic and clinician cognitive work we discovered through data collection and analysis reflect the kinds of macro-cognitive activity that Table 1 lists.

Alignment with clinician needs and procedures is essential to the development of effective decision support. Decisions about diagnosis and treatment rely on access to the most important, or salient, information about the patient in real time. Use of the CSE methodology reveals the decision making and associated cognitive work of actual casualty care, and the data/information that will effectively support it [13].

While 4TDS does not currently exist as an operational product, it could exist and its introduction into TCCC would change its context substantially. This need to see into the future is termed an "envisioned world problem [14]". Participants in data collection methods such as interviews needed to foresee a condition such as the use of decision support that they have not experienced. Creation of prototype interfaces and workflows invited them into this future setting to consider it.

Frequent and intense interactions between the research team and TCCC care providers identified near term opportunities to improve decisions, as well as practical considerations such as the sensor remaining with the casualty through the continuum of care.

Limitations. As mentioned in the Methods section, the project team did not have access to observe battlefield casualty care. The project was conducted during restrictions imposed during the COVID-19 pandemic which limited access to, and the number of, sample participants.

Table 1. Macro-Cognitive Activities (adapted from [12])

Macro-cognitive activity	Description
Naturalistic decision making	Reliance on experience to identify a plausible course of action and use of mental simulation to evaluate it
Sense making/situation assessment	Diagnosis of how current state came about and anticipation of how it will develop
Planning	Changing action in order to transform a current state into a desired state
Adaptation/Re-planning	Modification, adjustment or replacement of a plan already implemented
Problem detection	Ability to notice potential problems at an early stage
Coordination	How team members sequence actions to perform a task
Developing mental models	Mental imagery and event comprehension, based on abstract knowledge and domain concepts and principles
Mental simulation and storyboarding	Use of mental models to consider the future, enact a series of events, and ponder them as they lead to possible futures
Maintaining common ground	Ongoing maintenance and repair of a calibrated understanding among team members
Managing uncertainty and risk	Coping with a state or feeling in which something is unknown or not understood
Turning leverage points into courses of action	Ability to identify opportunities and turn them into courses of action
Managing attention	Use of perceptual filters to determine the information a person will seek and notice

5 Summary

Automated decision support promises a number of benefits. Use of a durable, simple monitor spares a medic from having to frequently document vital signs, allowing for more attention to be paid to the casualty. Shock monitoring and detection using only vital signs enables a medic to anticipate, identify, and prevent potential shock with enough time before it occurs. Vital signs collection and casualty identity documentation from point of injury improves continuity across echelons of care.

Improved knowledge retention from frequent refresher training on life-critical skills improves the ability for medics to perform life-critical skills more reliably in austere settings. It also enables instructors to give timely guidance to medics based on refresher training performance.

Faster, more accurate decisions can improve TCCC and PFC patient care under austere conditions in which delays can increase morbidity and mortality.

Acknowledgments. This work is supported by the US Army Medical Research and Materiel Command under Contract No. W81XWH-15-9-0001. The views, opinions and/or findings contained in this report are those of the author(s) and should not be construed as an official Department of the Army position, policy or decision unless so designated by other documentation. In the conduct of research where humans are the subjects, the investigator(s) adhered to the policies regarding the protection of human subjects as prescribed by Code of Federal Regulations (CFR) Title 45, Volume 1, Part 46; Title 32, Chapter 1, Part 219; and Title 21, Chapter 1, Part 50 (Protection of Human Subjects).The authors wish to acknowledge the generous support and guidance provided by Dr. Matt D'Angelo (LTC, USAR) (Uniformed Services University of the Health Sciences), Mr. Bret Smith (MAJ, USA ret.) (Tactical Combat Medical Course, Joint Base Ft. Sam Houston), and Dr. Mei Sun, our project sponsor at the U.S. Army Medical Research and Development Command.

References

1. Tactical Combat Casualty Care Handbook. No. 12-10. Center for Army Lessons Learned (CALL) (2012). CALL web site. https://call2.army.mil/toc.aspx?document=6851& filename=/docs/doc6851/12-10.pdf. Accessed 15 Dec 2018
2. PFC WG. Prolonged Field Care Teaching and Training Recommendations. USSOCOM Prolonged Field Care working group (PFC WG), March 2015. PFC Working Group web site. http://prolongedfieldcare.org/about/. Accessed 14 Oct 2015
3. Nemeth, C., et al.: Decision support for tactical combat casualty care using machine learning to detect shock. Mil Med. **186**(Suppl._1), 273–280 (2021). Society of Federal Health Officials (AMSUS). https://doi.org/10.1093/milmed/usaa275
4. Nemeth, C., Blomberg, J., Argenta, C., Serio-Melvin, M., Salinas, J., Pamplin, J.: Revealing ICU cognitive work using NDM methods. special issue on expanding naturalistic decision making. J. Cognit. Eng. Decis. Mak. Hum. Factors Ergon. Soc. **10**(4), 350–368 (2016). https://doi.org/10.1177/1555343416664845
5. Pamplin, J., et al.: Improving burn ICU clinician decision and communication through IT support. Mil. Med. **185**(1-2), e254–e261 (2020). Society of Federal Health Officials (AMSUS). https://doi.org/10.1093/milmed/usz151
6. Nemeth, C.: Human Factors Methods for Design: Making Systems Human-Centered. Taylor and Francis/CRC Press, Boca Raton (2004)
7. Woods, D., Roth, E.: Cognitive systems engineering. In: Helander, M. (ed.) Handbook of Human-Computer Interaction, pp. 3–43. North-Holland, Amsterdam (1988)
8. Nemeth, C., Klein, G.: The naturalistic decision making perspective. In: Cochran, J.J. (ed.) Wiley Encyclopedia of Operations Research and Management Science. Wiley, New York (2011)
9. Orasanu, J and Connolly, T.: The reinvention of decision making. In: Klein, G.A., Orasanu, J., Calderwood, R., Zsambok, C.E. (eds.) Decision Making in Action: Models and Methods, pp. 3–20. Ablex, Norwood (1993)
10. Cacciabue, P.C. Hollnagel, E.: Simulation of cognition: applications. In: Hoc, J.M., Cacciabue, P.C., Hollnagel, E. (eds.) Expertise and Technology: Cognition and Human-Computer Cooperation, pp. 55–73. Lawrence Erlbaum Associates, Mahwah (1995)

11. Klein, G., Ross, K.G., Moon, B.M., Klein, D.E., Hoffman, R.R., Hollnagel, E.: Macro-cognition. IEEE Intell. Syst. **18**(3), 81–85 (2003)
12. Crandall, B., Klein, G., Hoffman, R.R.: Working Minds: A Practitioner's Guide to Cognitive Task Analysis. The MIT Press, Cambridge (2006)
13. Nemeth, C., Anders, S., Brown, J., Grome, A., Crandall, B., Pamplin, J.: Support for ICU clinician cognitive work through CSE. In: Bisantz, A., Burns, C., Fairbanks, T. (eds.) Cognitive Engineering Applications in Health Care, pp. 127–152. Taylor and Francis/CRC Press, Boca Raton (2015)
14. Hoffman, R., Deal, S., Potter, S., Roth, E.: The practitioner's cycles: solving envisioned world problems. IEEE Intell. Syst. 6–11 (2010)

Lessons Learned from Applying Requirements and Design Techniques in the Development of a Machine Learning System for Predicting Lawsuits Against Power Companies

Luis Rivero[1]([📧]), Carlos Portela[2], José Boaro[2], Pedro Santos[2], Venicius Rego[2], Geraldo Braz Junior[1], Anselmo Paiva[1], Erika Alves[3], Milton Oliveira[3], Renato Moraes[3], and Marina Mendes[3]

[1] Programa de Pós-Graduação em Ciência da Computação, Universidade Federal do Maranhão, São Luis, Brazil
{luisrivero,geraldo,paiva}@nca.ufma.br
[2] Departamento de Informática, Universidade Federal do Maranhão, São Luis, Brazil
{portela,boaro,thiagocutrim,venicius.gr}@nca.ufma.br
[3] Equatorial Energia S/A, São Luis, Brazil
{erika.assis,milton.oliveira,renato.moraes,
marina.mendes}@equatorialenergia.com.br

Abstract. Machine Learning (ML) has shown great potential for automating several aspects of everyday life and business. Experiences reports on developing ML systems, however, focus mainly on how to install, configure and maintain an ML system, but do not focus on the requirements engineering process and the design of the user interface that will present the data to the user. In this paper, we report the lessons learned from applying different requirements and design methods in one of the stages of a large development project (18 months). Its goal was to develop and evaluate a Web based application embedding an ML model that would produce data on the probability of lawsuits based on features such as: number of complaints, electrical damage to appliances and power shutdowns. At all, the design team applied the following techniques: (a) Interviews and document analysis, to identify the major reasons for filling a lawsuit; (b) Personas, to define which type of clients could file a lawsuit; (c) Scenarios, to define which interaction (conversations) would be automatically triggered by a chatbot to try to solve client problems, avoiding a lawsuit; and (d) Prototype evaluation, to define the interaction and type of data that would be available to the lawyers through a Web application. Through the lessons learned within this paper and by providing details for its replication, we intend to encourage software companies to combine requirements and design approaches for cost-effective user centered design, especially in decision support intelligent systems.

Supported by Equatorial Energia.

© Springer Nature Switzerland AG 2021
S. Yamamoto and H. Mori (Eds.): HCII 2021, LNCS 12765, pp. 227–243, 2021.
https://doi.org/10.1007/978-3-030-78321-1_18

Keywords: Decision support systems · Machine learning system ·
Power company · Combined design approaches

1   Introduction

Machine learning (ML) algorithms and tools are now available to allow the development of ML-based software solutions for real-world problems. ML based systems have gained popularity as they produce usable data or provide suggestions about items that are most likely to represent interest for a specific user [18]. As several companies are still beginning to apply ML solutions, several challenges arise to guarantee the quality of the ML systems so that they are usable to the managers that require the data to make informed decisions [11].

According to Menzies [22], the software industry is searching for new alternatives on how to better develop and maintain ML systems. Kumeno [16] states that new technical and non-technical challenges that complicate ML systems development activities arose since software development teams must focus on the specifics of designing an appropriate interaction and presentation of the data. Therefore, there is a need to develop new and more effective types of computer interfaces addressing the emerging challenges, and evaluating their effectiveness.

In their survey, Dove et al. [6] identified challenges regarding the design of ML systems, such as the difficulties in prototyping due to their dynamic interactions and potentially unpredictable outcomes. However, experience reports on developing ML systems focus mainly on how to install, configure and maintain ML systems [22], but do not focus on the requirements engineering process and the design of the user interface that will present the data to the user. Lessons learned from real case studies can provide software development teams with ideas on how to capture user requirements and propose appropriate user interfaces to meet the users' expectations, while also discussing what works when considering the applied software development methodology (e.g. agile, model-based, test-driven and others). As software engineers developing ML systems need to meet customers' expectations in terms of schedule and quality, it is necessary that lessons learned from the application of requirements and design approaches in ML contexts are shared to improve productivity and avoid pitfalls.

In this paper, we explore the obtained results of applying a combined methodology for identifying user needs and proposing user interfaces in the context of an ML system. The system was developed within a partnership between the Equatorial S/A Brazilian power company that provides around 22% of the overall power in Brazil, and the Applied Computing Group (Núcleo de Computação Aplicada - NCA) from Federal University of Maranhão in Brazil. The purpose of the system was to analyze the degree to which a client could file lawsuits against a power company based on features such as: number of complaints, electrical damage to appliances and power shutdowns. Also, based on such data and using a chatbot, the application would automatically approach a client with a high risk of filing a lawsuit and try to negotiate with him/her. Furthermore, the application would provide information to lawyers to support their decision on what actions to take when proposing deals to clients. During the six-month duration of the requirements and design stage of the development process, we applied an iterative approach through the SCRUM methodology [4] using the following techniques: (a) Interviews and document analysis, to identify the major reasons for filling a lawsuit; (b) Personas, to define which type of clients could file a lawsuit; (c) Scenarios, to define which interaction (conversations) would be automatically triggered by the chatbot to try to solve client problems, avoiding a lawsuit; and (d) Prototype evaluation, to define the interaction and type of data that would be available to the lawyers through a Web application.

The remainder of this paper is organized as follows. Section 2 shows related work on the requirements engineering and design processes of ML systems, while presenting an overview of the methods we applied in such context. Section 3 shows the context of the developed ML system for dealing with lawsuits against power companies. Then, Sect. 4 shows the requirements engineering and design approaches employed for guaranteeing the quality of the ML system. Moreover, Sect. 5 discusses the lessons learned from this experience report. Finally, Sect. 6 concludes the paper indicating future works.

2 Background

2.1 ML Systems Development Process

The Artificial Intelligence (AI) industry is searching for new alternatives on how to better develop and maintain ML systems [22]. Kumeno [16] states that new technical and non-technical challenges that complicate ML systems development activities arose since, besides employing traditional software engineering practices, software development teams working with ML must also integrate the AI model lifecycle (training, testing, deploying, evolving, and so on) into their software process. In this sense, when developing an ML based system, human computer interaction approaches are important to improve the quality of the application under development.

To verify the extent to which a repeatable development process was available for ML systems, Hill et al. [9] carried out a set of field interviews. In their paper, the authors identified that development teams faced difficulties to establish a repeatable process. Thus, due to the popularization of ML in the development of artificial intelligent systems, several authors are reporting their experiences to support novice software engineering teams in this research area.

In their paper, Hirsch et al. [10] describe the design of an automated assessment and training tool for psychotherapists to illustrate challenges when creating interactive machine learning systems. The authors propose strategies to improve the UX of the system, such as providing contestability, specially in applications that affect human life, livelihood, and well-being. On the other hand, the authors do not provide examples of requirements elicitation techniques applied to the context of ML systems or related systems.

In another research, Lytvyn et al. [18] described the design of a recommendation system considering personal needs of the users. The proposed decision-making system aimed at developing recommendations on content based on the collaborative filtering and Machine Learning considering the user's personal needs. Although the authors provide information on how to develop the ML model and the decisions taken during the development process, the process for developing the interface of the web application embedding the ML model was not discussed.

With regards to improving the user experience of machine learning systems, Yang [26] discussed how to design ML systems considering his experience in practice. The author shows an example of a clinical decision support system, which mines thousands of patient records, bringing the collective intelligence of many physicians to each implant decision of whether and when to implant an artificial heart into an end-stage heart failure patient. Although the author explores problem setting and problem solving in the development process, there are not many details on the use of traditional human computer interaction approaches for the identification of requirements and proposal of a solution.

A further analysis of the papers described above shows that most papers reporting the development process of ML systems focus on the development of the statistical model for reasoning, problem solving, planning, and learning in several marketplace contexts. These papers, however, do not focus on how to identify the requirements of the final end-user application embedding the ML model. Furthermore, when designing the proposed solution of the user interface or interaction related to the final application of the ML system, few details are provided, which are insufficient for inexperienced software development teams willing to improve the quality of the ML systems they are developing. Moreover, when considering the context of agile software development for ML systems, few studies report the results of such experiences [16]. Instead, authors report the use of ML for supporting agile development processes [22]. For instance, Arora et al. [3] discuss how machine learning is employed for estimation approaches in scrum projects, while Kumeno [16] indicates that one of the main questions in agile development of ML systems is related to the new roles, artifacts, and activities that come into play into the agile development processes.

Considering the above, there is a need to report experiences in ML systems development that share experiences in the agile context in order to support software engineers facing difficulties in the application of methodologies and approaches for guaranteeing the quality of ML systems under development. In our context, sharing our experience of applying requirements elicitation and design techniques for different stages of the development of an ML application (i.e. Model and User Interface) can be useful to software engineers. There is a need for more examples of ML system requirements elicitation and design that provide information on how to propose solutions in terms of user interfaces and interactions. Also, we did not find a report on how the requirements elicitation and design could occur within an agile development process. In this paper, we intend to present such example, by showing how we applied some of these techniques in the context of the development of a real ML system for power companies using the Scrum development life cycle. In the following section we provide further details on the applied techniques within this experience report.

2.2   Techniques for Requirements Elicitation and Design

Meeting the users expectations is fundamental in any development process that seeks to produce an interactive system with high-quality of use. Thus, the use of requirements elicitation techniques and design techniques is useful for systematically achieve an understanding of the users' needs and propose solutions that meet their expectations [23].

The methods we applied during the project were: document analysis, scenarios production, personas, prototype development and interviews. These methods can gather insightful information on the requirements of a product, as well as the opinion of the users on the proposed solutions and ideas [5]. Additionally, each method can contribute to gathering data from different sources, while producing different artifacts documenting the proposed solution. We briefly summarize each method as follows.

Document Analysis: This technique is useful for starting the requirements elicitation phase [20]. In this technique, analysts study the available documentation on the problem and/or existing solution to identify information relevant to the development of a new solution. This technique can be useful for gathering information before scheduling interviews or other elicitation sessions with stakeholders.

Scenarios: According to Holbrook [12], a scenario-based requirements elicitation approach structures the early interaction between users and designers in order to quickly develop a set of initial requirements. Scenarios are stories that illustrate how a perceived system will satisfy a user's needs. Thinking of scenarios as stories is relevant as they constitute an almost universal way for the organization and dissemination of an experience. Scenarios are an idealized but detailed description of a specific instance of human computer interaction.

Personas: A persona is a fictional character or hypothetical model of a group of real users, created to describe a typical user [14]. It is based on data from user profiles that present a detailed description of the characteristics of users, their relationship with technology, their knowledge of the domain and tasks. As personas define expectations, concerns and motivations, they help design teams to understand how to design a product that will satisfy users needs.

Prototype Development: According to Camburn [4], prototype is a pre-production representation of some aspect of a concept or final design. Thus, each prototyping effort requires a certain unique strategy to resolve a design problem or opportunity. In this sense, prototyping can provide a demonstration of how stakeholders and end-users can carry out tasks and interact with the proposed solution.

Interviews: They are guided conversation that may or not follow a script of questions or topics, in which an interviewer seeks to obtain information from an interviewee. This method allows to collect detailed and in-depth information from individual users. According to Rafiq et al. [24], interviews are still popular among small software development companies and startups due to their low costs and duration.

3   The ML System for Predicting and Dealing Lawsuits

3.1   Application Context

Legal and managerial mechanisms for consumer protection have brought changes to the energy market in several countries [13]. In the context of Brazil, the need for changes in the market for electric energy companies has become more evident due to the appearance of lawsuits against those companies. The main reasons for lawsuits are related to client dissatisfaction with the treatment and/or service provided by the company. Consequently, identifying customers who are likely to file a lawsuit against the company is a way to identify customer dissatisfaction [8]. Since dissatisfaction can lead to considerable monetary loss when clients leave the company or file lawsuits, improving the quality of the services by meeting the needs and addressing the complaint of clients is crucial.

Companies that provide services are employing ML systems to predict lawsuits [2,15]. In the context of Brazilian power provision, ML solutions have been proposed to investigate the relationship between customer satisfaction and filed lawsuits. For instance, Marchetti and Prado [19] developed a customer satisfaction model using structural equation models. Another work proposed a decision-making technique based on the Approximate Set Theory (AST) to extract important decision rules related to customer turnover [2].

Although the works above investigate ways to predict possible actions from customers, they do not provide a tool for presenting details on the reasons for the clients' actions. Also, they do not automate the process of approaching clients who require special assistance and, if left unattended, could file lawsuits. The ML system discussed in this paper intends to do that, by predicting lawsuits related to unregistered consumption and power outage, and approaching the clients to solve problems from their point of view. In the following subsection, we present further details on how the application was developed.

3.2   Characterization of the Project

The development project was called "Intelligent Legal System for Analysis and Prediction of Lawsuits and Deals" (SIJURI). This project had a duration of 24 months (from November 2019 until October 2021). At all, the development team had 20 software engineers, including: project managers, ML specialists, software developers, front-end developers, software analysts, data base administrators and software testers.

To develop the ML system, the development team applied the Scrum agile methodology, which is one of the most popular agile development processes in the Brazilian development community [21]. Scrum uses iteration and incrementation to manage rapidly-changing project requirements by improving communication between [17]: the project owner, responsible to be the voice of business inside a Scrum project; Scrum Team, formed by its developers, testers and other roles within the project; and Scrum Master, who is responsible to keep the team focused on the practices and values that are needed to be applied inside the project and is also responsible to help the team whenever they face some problem during the development process. In our context, the product owner was a manager inside the Equatorial Energia power company. She would have access to several sectors within the company and would enable meetings with future users of the system, so that system requirements could be discussed and defined. The Scrum master was an ML experienced developer with more than 5 years of experience developing ML models. Within the project, there was also an experienced (more than 5 years) usability and UX consultant with experience in system analysis, user interface design and user experience evaluation. Finally, the Scrum team was composed of both experienced (more than 3 years of experience) and inexperienced (graduate students) software engineers.

According to Lei et al. [17] the Scrum team is responsible for implementing the deliverables of the project at the end of each "Sprint", which is a period of time (in our case, 1 month) to create a usable increment of the product. In the SIJURI project, the following activities were set: (a) Analysis and Design (Sprints 1–9); (b) Development of Data Base and Queries (Sprints 4–12); (c) Development of user interface for Web ML application (Sprints 4–18); (d) Development of chatbot to approach clients (Sprints 4–18); (e) Integration with an ML model, which was developed in another project (see [25]) (Sprints 4–18); and (f) Testing, Integration and Release (Sprints 7–24). After defining the main activities to be performed in the project, the development team followed Scrum practices. The sprint planning meeting occurred in the beginning of each Sprint. During the sprint, the team worked in several activities according to the project planning. To keep track of the design decisions and requirements for each sprint, the team would use an online Kanban board, which had columns to categorize tasks in: "waiting", "in progress" and "completed" or "to-do", "doing", and "done". Also, daily meetings were performed to assess problems during software development.

As described above, the Analysis and Design activity began on month 1 of the project and ended in month 9. The long duration of this stage was defined due to the need to define the requirements of the Web application embedding the information of the ML model, while also defining the scenarios and interaction of the chatbot, which would approach the clients with high risk of filing lawsuits. We will focus on the lessons learned from the use of requirements and design approaches of the Web application and dialogues definition. We decided to focus on these examples so that we can show how to combine different requirements engineering and design approaches to gain a better understanding of the problem to be solved, while proposing solutions to this problem. Below, we present the execution process of these approaches and their results.

4   The Requirements Engineering and Design Processes

In this project, we had two major goals: (1) to develop a user interface capable of providing information to lawyers regarding the clients that had a high risk of filing a lawsuit; and (2) to develop a chatbot to approach clients whit a high risk of filing a lawsuit to try to solve their problems, aiming at avoiding lawsuits. For each goal, complementary requirements and design techniques were employed to gain insight into the clients' problem and proposing a solution that would meet their expectations. In the following subsections, we present how these techniques were applied and how they contributed to the development process.

4.1   Document Analysis and Scenarios

The document analysis was performed to understand which data was necessary to identify if a client was going to file a lawsuit. After building the ML model considering several factors such as power blackout, non-registered consumption, increased consumption, incorrect power cut and incorrect billing, we identified which were the variables that highly correlated to a high risk of filing a lawsuit. The categorization of the data was performed automatically through ML algorithms and is described in a previous paper [7]. After that, we analyzed which were the main complaints of the users according to the factors described above. To gather the data, we manually searched one of the most popular Brazilian complaint websites[1]. Also, we analyzed laws and rules related to the Equatorial Power Company by searching its official website[2], specifically its technical standards and frequently asked questions web pages.

Through the analysis of the frequently asked questions and the complaints of the clients, we were able to identify: (a) data related to cases in which a client would complain against the power company; (b) measures taken by the power company to either justify a legal action or solve a problem related to customer service; and (c) data required or produced within a costumer attendance. The identified data was employed to define scenarios in which a client could be approached by the chatbot to: (a) receive information that would be useful to make the client aware of any problems and their solution; and (b) try to handle specific issues or enable the communication with a human attendant. For each developed scenario, we created a text description and a diagram to represent the flow of actions, the required data and the messages that would be provided. Such scenarios would be useful in the process of developing the chatbot from the ML solution. With regards to the development of the Web application for supporting Lawyers in the decision making process of dealing with lawsuits, the data from the chatbot interaction was used as basis for the generation of reports. The process of developing this prototype will be explained in Subsect. 4.2.

Below, we present an example of the scenarios that were developed in their text and diagram form (see Fig. 1). The text format was employed to provide an introduction of the problem being handled in the scenario, while the diagram would provide the steps necessary for it to be implemented in the chatbot. Also, the diagram contained information on the type of data that would be required and the messages exchanged between the client and the chatbot.

---

[1] Brazilian Website "Complain Here" (*"Reclame Aqui"* in Portuguese) - https://www.reclameaqui.com.br/.

[2] Equatorial Power Company Brazilian Website - https://ma.equatorialenergia.com.br/.

Scenario 1: After a power blackout, a client verifies that some of his appliances have suffered an electrical damage. If this client has had several blackout complaints or repair requests, after the identification of a Blackout, our chatbot can approach the client, asking if everything is fine, or if he requires assistance due to damaged appliances.

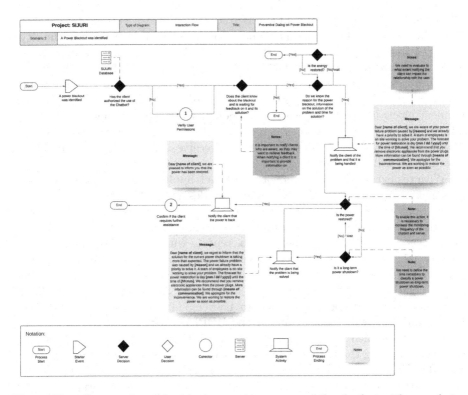

Fig. 1. Flow diagram describing the interaction process of the chatbot with regards to a Power Blackout Scenario

4.2 Personas and Prototype Proposal

Once we gathered the information regarding the processes and data necessary to approach clients and track if they required any support, we developed personas as a complement of the scenarios described above. Each persona had specific attributes or problems related to the five reasons for filing lawsuits. Below, we present and example of two personas and different issues they faced with regards to Power Blackouts.

Persona 1: Jane is 40 years old and has 3 children. A power blackout caused her refrigerator to break, and also two fans, which she and her family use during hot days. She reported the problem to the power company, aiming at receiving the corresponding refund. However, she received no response. As a result, she files for a lawsuit against the power company due to lack of or late attendance. Our chatbot could ask Jane if, after a blackout, there is a need to request repairs or request refunds. Also, if a request is being processed, the chatbot can inform Jane how long it will take for her to receive an answer and what are the stages of her request within the company.

Persona 2: Anne is 25 years old and is studying online to get a masters degree. Power blackouts are becoming more common due to the rainy season in the city. A proper analysis of the electric structure could avoid blackout problems, so Anne decides to request an electric infrastructure check-up. The fact that there is no feedback on the repair request makes her very angry, which causes her to file a lawsuit against the power company. The chatbot could analyze that Anne has frequent blackouts and ask, before she does, if she wants to request an electric infrastructure check-up. Also, if a request is being processed, the chatbot can inform Anne how long it will take for her to receive a reply or the date in which a technician will go to her house.

With regards to the development of the Web application for supporting decision making of lawyers/company workers, we designed a high-fidelity prototype of the application, showing the structure of the different reports. We considered that lawyers could use the information of conversations with the chatbot to understand the history of a client. Also, reports with data on the main events and variables highly correlated to lawsuits could be useful. This information was extracted though the document analysis performed in Sect. 4.1, and verifying how a company worker dealt with clients making complaints. Figure 2 shows an example of the reports provided to lawyers, in which the presented data is fictitious (following suggestions of the BDPL). First, a lawyer can view to which degree a client requires assistance: (a) in terms of avoiding a lawsuit, or (b) understanding the reasons why (s)he is filing a lawsuit. Clients are grouped according to the risk they represent based on the main reasons for filing lawsuits. After selecting a client, a lawyer can view his/her data and graphs on how his/her degree of filing a lawsuit has evolved. For each point in time, lawyers can view which events impacted positively or negatively in the risk of a client. Also, they can view the conversations of the client with the chatbot, to draw further conclusions, or understand the expectations/frustrations of the clients.

4.3 Validation Interviews

After defining the scenarios, personas and developing the high fidelity prototypes, we carried out interviews to identify which requirements and scenarios should be explored, as well as validating the data necessary to perform tasks within the Web system. The main purpose of this step, was to understand, to which extend the data could be retrieved and if the web system and chatbot could be implemented along with the other existing solutions from the power company,

With regards to goal 1 (web application for providing data to lawyers), we carried out online meetings with possible users of the systems (e.g. lawyers, customer

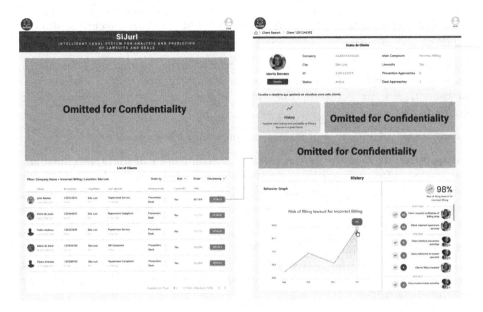

Fig. 2. Part of the Prototype of the web information system embedding the ML model

assistants, managers, others). In these meetings, we would discuss which was the necessary data to properly allow a user to track the behavior of a client and decide which measures to adopt when avoiding lawsuits. Since this project was being developed remotely due to the pandemic of Covid-19, the validation meetings were online, using prototype interaction simulation software. During these meetings we asked questions related to the adequacy and feasibility of the proposed solutions, as well as improvement opportunities and limitations regarding data access and systems integration. Below, we present an extract of part of the interviews and some of the questions that were asked.

Interviewer: *Do you think that there are further functionalities besides the ones explored in this prototype?*

Interviewee 1: *Since we are thinking about a system that allows shedding light on doubts regarding preparing a defence so that this system is a support to lawyers, I believe that providing a history of the messages about the topic and what treatments were made. Also, we can verify with the corresponding area, which are the documents that proof the handling of the situation, perhaps an icon that could show the documents or deals. Maybe, juridically speaking, we need the proof. And for the lawyer, it is important that he has the documents that proof that information. We are worried about the BDPL (Brazilian Data Protection Law), since we are providing a tool that provides data from our clients that we need to consider.*

Interviewee 2: *We need to verify which data is not relevant and should not be presented in the screen.*

Interviewer: *Regarding the mentioned documentation, which documentation is available and where?*

Interviewee 1: *We need to talk to the related areas, so that we know which data can be provided. Sometimes a document is inside the system, and we need to know if you can access it. Each area is going to inform if we can provide access to that data.*

Interviewer: *Are there any further comments on the proposal?*

Interviewee 1: *Perhaps we can think of different types of users who have different types of access and the data. For instance, the lawyer who work at our company, have access to sensitive data. Lawyers who are from third parties may have less access. We need to think about signing confidentiality forms. We need to think about the implications about such access.*

Interviewer: *We have questions about the user management.* [Confidential Information] *What is the access that these lawyers have? Having access related to the firm's lawsuits is common? Or do we need to verify, to which extent (s)he can access lawsuits?*

Interviewee 1: *Our lawyers have a specific registration. But third party lawyers must also have access to the data. However, we need to limit such access. We need to verify this issue with the BDPL sector.*

Interviewer: *Do you identify further profiles that need to be considered, besides internal and external lawyers?*

Interviewee 1: *If we limit too much, we can complicate the use of the system. [Confidential Information] I don't see a problem that some data could be provided to related users. For instance, consumer assistance.*

Interviewer: *Wouldn't it be possible to export data reports, so that people who should not access the system are not able to access the system, but still can use the data?*

Interviewee 1: *It would be interesting to define a set of degree of access in terms of profile, related to the BDPL. Such and such information are degree 1 of secrecy, thus, only type X of users can access it. And so on.*

Similar questions were carried out with regards to the scenarios that were prepared. As a result we were able to validate which scenarios were correlated and which data was available for analysis. By ending the requirements elicitation and design processes with the interviews, we were able to produce and validate several artifacts that were useful in further stages of the development process. In the following section, we present the lessons learned from this experience and how we managed problems thought the requirements elicitation and design processes.

5   Discussion and Lessons Learned

Based on this experience of designing the Web application for decision making support and the chatbot for approaching clients, we were able to obtain lessons learned regarding the feasibility of combining different requirements and design approaches. We will present these lessons learnt reporting the decisions we made and how they were positive or negative for the project. These lessons are mainly based on the opinion of the design team as well as the managers of the project.

5.1   How the Applied Methods Contributed to the Design of the Solution

With regards to the requirements identification, through the document analysis, we were able to understand the process that attendants follow to deal with complaints. Thus, the document analysis was useful for understanding why users complained and what were the courses of action that the company suggested for different types of complaints. As the "Complain Here" website is a popular communication platform between customers and companies with more than a thousand companies in Brazil [1], we were able to identify messages ideas from real customer/attendant conversations. Also, we were able to identify which data was required by the attendant, and how this data was provided by the user, either when explaining an issue, or during the solution of a problem, by the request of an attendant. Furthermore, with regards to the power company official website, we were able to identify the answer to possible questions that were not clear to the development team. Also, the ML model was useful to define which were the factors that impacted the degree of risk of a client filing a lawsuit. In this sense, the identified factors and their related events were employed as basis for the development of Scenarios and Personas.

The Personas technique allowed us to identify user profiles and understand their pains. This is important since, in our case, creating empathy with the customers makes the development team understand why they are upset with the company, and what would be expected in terms of the company's action. In this sense, we were able to propose messages and dialogues according to different scenarios that could trigger a lawsuit from a user. In turn, the scenarios provided examples of cases in which the chatbot could be useful to mitigate lawsuits. When combining the personas, scenarios and modeling of the scenarios, we were able to create a robust documentation and basis for the development of interactions with the chatbot.

With regards to the prototype, based on the data collected from documents analysis, we were able to identify which data was necessary for an attendant to analyze the profile and problems of a customer. This information was considered in the generation of reports. Also, the different factors and events that impacted the risk of a client filing lawsuits were considered in the development of graphs showing the history of the client. Through the integration of the Scenarios, Personas and Document Analysis, the input for developing the prototype was more robust. Additionally, to develop the prototype, by analyzing similar applications showing graphs and reports of customers, we were able to propose an application that may be more useful. As such, it would provide functionalities found in reports, such as search, zoom in and out in graphs, filter data, among others.

Finally, the interviews were applied to validate the obtained data and evaluate the proposal of scenarios for the chatbot and the prototype of the web application. At all, several scenarios and questions related to the interaction of the chatbot were discussed. Also, some reports were validated with the power company team. Furthermore, we were able to verify if an interaction was possible or if the implementation of a scenario was nos feasible.

5.2 Difficulties in the Application of the Methods

Although the application of the methods was easy from the point of view of the design team, some difficulties arose from the different jargon employed by the development team, clients and attendants from the power company. One of the main problems in the document analysis was related to the view of clients when complaining in the "Complain Here" website. In the view of the company, the Incorrect Billing factor and the Increased Consumption factor are very different. While the first one is related to problems when charging clients with items that were not accepted by them, the second one is related to a problem in the measure of the client's power consumption. In the view of clients, both are the same thing, since they usually don't understand when a bill is higher due to an increase in power consumption or for further items that were charged. Therefore, when analyzing these problems in the complaints from clients in the website, some scenarios were mixed or, in some cases, categorized in the wrong factor. Fortunately, the validation sessions allowed the power company attendants to point out these issues and describe the correct scenario or its categorization. Also, it allowed correcting terms that were misused in the scenarios, personas and prototype.

With regards to the difficulties in producing the personas, scenarios and prototype, the main issues were related to the variability of the interaction possibilities. At the beginning of the analysis, we identified around 13 scenarios considering the data from the ML system, indicating events that could trigger a lawsuit. However, as the analysis progressed, new events and data were identified, and some of them were intertwined. For instance, we noticed that a scenario of incorrect billing related to increased bill value, could also be related to an increase in power consumption. Thus, there was a need to analyze which events and in which factors would activate the chatbot, and how to separate the scenarios, identifying a total of 24 main scenarios and 4 common dialogues (i.e. communication acceptance, feedback analysis, proof providing and payment management). The changes in the scenarios significantly increased rework and verification. However, the documentation in terms of modeling became useful for discussing the implementation of the chatbot, while also allowing to discuss which data was available for use, and which required further access from the databases at the power company.

Finally, although the development of the prototype did not suffer major changes throughout the validation process, an analysis of similar applications was necessary to provide ideas on how to implement similar reports. Furthermore, in order to meet the needs of the BDPL, it was necessary to think of a user management system, guaranteeing that attendants would only have access to the required information of clients. In this sense, the validation process of the prototype was useful for analyzing pending decisions regarding both the prototype and chatbot interaction scenarios.

6   Conclusions and Future Work

This paper described the requirements engineering process and design proposal of an ML based web application and chatbot for managing and approaching clients with strong chances of filing lawsuits on unregistered consumption and power outages. We were able to integrate several requirements elicitation and design approaches to guarantee the quality of the proposed solutions. After applying the interviews and document analysis, we identified the data that needed to be collected. Also, we identified the five main reasons why a client would file a lawsuit, being: power blackout, non-registered consumption, increased consumption, incorrect power cut and incorrect billing. Additionally, through the analysis of company related documents and customer service logs in websites, we managed to identify which processes were applied when dealing with clients' complaints before they filed a lawsuit. With this information, we defined personas and scenarios for the chatbot interaction. With the information regarding the possible required data and type of interaction of the clients with the lawyers/ company workers, we designed a high-fidelity prototype of the application, showing the structure of the different reports that would be presented to the lawyers, including interaction of clients with the chatbot.

With regards to the lessons learned from this experience, we managed to connect the output of the interviews and document analysis as the input for the Personas, which in turn, were the input of the scenarios and prototype. By searching for real cases of customer services in websites for complaints, we were able to obtain accurate data of how the company handles clients in order to avoid lawsuits, which can be useful for designing the interaction of chatbots, mirroring the interaction with company attendants. The use of interviews along with prototype creation to gain feedback was useful for identifying further data and functionalities, as well as organizing the presentation of the data respecting legal Brazilian requirements on the display of sensitive data.

As future work, we intend to use further evaluation techniques to identify improvement opportunities in the proposed solutions, once they are implemented. We intend to: (a) evaluate the perception of the users (i.e. power company managers, lawyers and clients) in real life scenarios; and (b) improve the usefulness and ease of use of proposed solution by correcting the identified problems. Through the lessons learned within this paper and providing details for its replication, we intend to encourage software companies to combine requirements and design approaches for cost-effective user centered design, especially in decision support intelligent systems.

Acknowledgments. This work was supported by SijurI project funded by Equatorial Energy under the Brazilian Electricity Regulatory Agency (ANEEL) P&D Program Grant number APLPED00044_PROJETOPED_0036_S01. Additionally, this work was supported by the Foundation for the Support of Research and Scientific Development of Maranho (FAPEMA), the Coordination for the Improvement of Higher Education Personnel (CAPES) and the National Council for Scientific and Technological Development (CNPq).

References

1. de Almeida, G.R., Cirqueira, D.R., Lobato, F.M.: Improving social CRM through electronic word-of-mouth: a case study of ReclameAqui. In: Anais Estendidos do XXIII Simpósio Brasileiro de Sistemas Multimídia e Web, pp. 107–110. SBC (2017)
2. Amin, A., et al.: Customer churn prediction in the telecommunication sector using a rough set approach. Neurocomputing **237**, 242–254 (2017)
3. Arora, M., Verma, S., Kavita, Chopra S: A systematic literature review of machine learning estimation approaches in scrum projects. In: Mallick, P., Balas, V., Bhoi, A., Chae, G.S. (eds.) Cognitive Informatics and Soft Computing. Advances in Intelligent Systems and Computing, vol. 1040, pp. 573–586. Springer, Singapore (2020). https://doi.org/10.1007/978-981-15-1451-7_59
4. Camburn, B., et al.: Design prototyping methods: state of the art in strategies, techniques, and guidelines. Des. Sci. **3**, e13 (2017)
5. Davis, A., Dieste, O., Hickey, A., Juristo, N., Moreno, A.M.: Effectiveness of requirements elicitation techniques: empirical results derived from a systematic review. In: 14th IEEE International Requirements Engineering Conference (RE 2006), pp. 179–188. IEEE (2006)
6. Dove, G., Halskov, K., Forlizzi, J., Zimmerman, J.: UX design innovation: challenges for working with machine learning as a design material. In: Proceedings of the 2017 CHI Conference on Human Factors in Computing Systems, pp. 278–288 (2017)
7. França, J.V., et al.: Legal judgment prediction in the context of energy market using gradient boosting. In: 2020 IEEE International Conference on Systems, Man, and Cybernetics (SMC), pp. 875–880. IEEE (2020)
8. Gruginskie, L., Vaccaro, G.L.R.: Lawsuit lead time prediction: Comparison of data mining techniques based on categorical response variable. PLoS One **13**(6), e0198122–e0198122 (2018)
9. Hill, C., Bellamy, R., Erickson, T., Burnett, M.: Trials and tribulations of developers of intelligent systems: a field study. In: 2016 IEEE Symposium on Visual Languages and Human-Centric Computing (VL/HCC), pp. 162–170. IEEE (2016)
10. Hirsch, T., Merced, K., Narayanan, S., Imel, Z.E., Atkins, D.C.: Designing contestability: interaction design, machine learning, and mental health. In: Proceedings of the 2017 Conference on Designing Interactive Systems, pp. 95–99 (2017)
11. Hohman, F., Wongsuphasawat, K., Kery, M.B., Patel, K.: Understanding and visualizing data iteration in machine learning. In: Proceedings of the 2020 CHI Conference on Human Factors in Computing Systems, pp. 1–13 (2020)

12. Holbrook III, H.: A scenario-based methodology for conducting requirements elicitation. ACM SIGSOFT Softw. Eng. Notes **15**(1), 95–104 (1990)
13. Ibáñez, V.A., Hartmann, P., Calvo, P.Z.: Antecedents of customer loyalty in residential energy markets: Service quality, satisfaction, trust and switching costs. Serv. Ind. J. **26**(6), 633–650 (2006)
14. Jain, P., Djamasbi, S., Wyatt, J.: Creating value with proto-research persona development. In: Nah, F.F.-H., Siau, K. (eds.) HCII 2019. LNCS, vol. 11589, pp. 72–82. Springer, Cham (2019). https://doi.org/10.1007/978-3-030-22338-0_6
15. Keramati, A., Ghaneei, H., Mirmohammadi, S.M.: Developing a prediction model for customer churn from electronic banking services using data mining. Financ. Innov. **2**(1), 1–13 (2016). https://doi.org/10.1186/s40854-016-0029-6
16. Kumeno, F.: Software engineering challenges for machine learning applications: a literature review. Intell. Decis. Technol. **13**(4), 463–476 (2019)
17. Lei, H., Ganjeizadeh, F., Jayachandran, P.K., Ozcan, P.: A statistical analysis of the effects of Scrum and Kanban on software development projects. Robot. Comput.-Integr. Manuf. **43**, 59–67 (2017)
18. Lytvyn, V., et al.: Design of a recommendation system based on collaborative filtering and machine learning considering personal needs of the user, vol. 4, no. 2, pp. 6–28 (2019)
19. Marchetti, R., Prado, P.H.: Avaliação da satisfação do consumidor utilizando o método de equações estruturais: um modelo aplicado ao setor elétrico brasileiro. Revista de Administração Contemporânea **8**(4), 9–32 (2004)
20. Masuda, S., Matsuodani, T., Tsuda, K.: Automatic generation of test cases using document analysis techniques. Int. J. New Technol. Res. **2**(7), 59–64 (2016)
21. de O. Melo, C., et al.: The evolution of agile software development in Brazil. J. Braz. Comput. Soc. **19**(4), 523–552 (2013). https://doi.org/10.1007/s13173-013-0114-x
22. Menzies, T.: The five laws of se for AI. IEEE Softw. **37**(1), 81–85 (2019)
23. Poth, A., Riel, A.: Quality requirements elicitation by ideation of product quality risks with design thinking. In: 2020 IEEE 28th International Requirements Engineering Conference (RE), pp. 238–249. IEEE (2020)
24. Rafiq, U., Bajwa, S.S., Wang, X., Lunesu, I.: Requirements elicitation techniques applied in software startups. In: 2017 43rd Euromicro Conference on Software Engineering and Advanced Applications (SEAA), pp. 141–144. IEEE (2017)
25. Rivero, L., et al.: Deployment of a machine learning system for predicting lawsuits against power companies: lessons learned from an agile testing experience for improving software quality. In: Brazilian Symposium on Software Quality. ACM (2020)
26. Yang, Q.: The role of design in creating machine-learning-enhanced user experience. In: 2017 AAAI Spring Symposium Series (2017)

Information in VR and Multimodal User Interfaces

Asymmetric Gravitational Oscillation on Fingertips Increased the Perceived Heaviness of a Pinched Object

Tomohiro Amemiya[1,2]([✉]) [iD]

[1] Virtual Reality Educational Research Center, The University of Tokyo, Tokyo, Japan
amemiya@vr.u-tokyo.ac.jp
[2] Graduate School of Information Science and Technology, The University of Tokyo,
7-3-1 Hongo, Bunkyo, Tokyo 113-8656, Japan

Abstract. Studies have shown that changes in shear or friction force at the fingertips contribute to weight perception. Pinching an actively moving object induces a similar change in shear or friction force at the fingertips, and humans may perceive the object as feeling heavier than a motionless object. However, little is known about weight perception when pinching an oscillating object. This study aimed to investigate the role of an object's oscillation in weight perception, with participants assessing heaviness when pinching a pair of boxes. The results showed that a vibrating box felt heavier than a static box, and an asymmetrically vibrating box felt heavier than a symmetrical box. These findings are similar to those of our previous study, when participants grasped a larger and heavier box that vibrated asymmetrically in the vertical direction.

Keywords: Haptics · Force sensation · Sensory illusion

1 Introduction

Many studies have reported that the impression of an object's heaviness changes in relation to surface roughness [1], size [2, 3], width of grip [3], temperature [4], visual effect [5, 6], or cognitive factors [7, 8]. Most research concerning heaviness perception has been conducted in relation to the weight of static objects. Our previous study examined the perceived weight of non-static objects, such as vibrating objects, and our results showed a perceptual change in heaviness when grasping a hand-held box in which a small mass oscillated with an asymmetric acceleration in the gravity direction, at a frequency of 5 or 9 cycles per second [9]. Stemming from the nonlinearity of human perception, a weak force stimulus is not clearly perceived, even if presented for a long period of time. Thus, vibrations with asymmetric acceleration, consisting of intense pulses with longer periods of low-amplitude recovery, induce a sensation of being pulled or pushed in a particular direction [10–12]. Recently, we succeeded in markedly reducing the size and weight of a force display through using a voice-coil vibrator [13, 14].

This study focused on the effect of oscillations on perceived heaviness of a pinched object using cutaneous information derived from only the index finger and the thumb.

© Springer Nature Switzerland AG 2021
S. Yamamoto and H. Mori (Eds.): HCII 2021, LNCS 12765, pp. 247–256, 2021.
https://doi.org/10.1007/978-3-030-78321-1_19

This study aimed to elicit distorted heaviness perception using asymmetric and symmetric oscillations in the gravity direction without altering the physical mass of a pinched object. The point of subjective equality (PSE) of distorted heaviness perception was calculated through comparing vibrating and static box weights. Moreover, changes in heaviness perception were compared between asymmetric and symmetric oscillations in the gravitational direction.

2 Weight Perception Using Asymmetric Oscillation

Conventional force displays use mechanical linkages to establish a fulcrum relative to the ground. The fulcrum, or grounding support, is required based on the action-reaction principle. However, most conventional mobile force display systems can produce neither a constant nor a translational force, but only a short-term rotational force, as they lack a fulcrum. The use of asymmetric vibration to create a force sensation of being pulled or pushed, using various types of mobile devices, has previously been proposed [9–14]. Asymmetric vibration involves different acceleration patterns in two directions that create a perceived force imbalance. Although the forces' averages in each direction are identical, someone holding a device that is vibrating using an acceleration pattern feels pulled in one direction, as the amplitude of the weaker force has been adjusted to be below the sensory threshold [15]. Previously, we developed a prototype to create an asymmetric vibration using a crank-slider mechanism and successfully induced a force sensation using asymmetric vibration with frequencies ranging from 5 to 10 Hz [9–11].

We subsequently developed a prototype with a voice-coil vibrator, designed to be pinched by the fingers, and we successfully induced a force sensation through asymmetric vibration with higher frequencies of approximately 40 Hz [13, 14]. The Meissner corpuscles and Ruffini endings are considered to contribute to this illusion, as they are sensitive to a force tangential to the skin [16, 17]. Moreover, Pacinian corpuscles do not seem to contribute to the illusion because they cannot sense direction [18, 19]. Furthermore, we found that an asymmetric oscillation of approximately 40 Hz was effective for the prototype with a voice-coil vibrator. In contrast, Merkel disks and Meissner corpuscles are sensitive to vibrations with lower frequencies of < 100 Hz and can sometimes clearly code the sliding or tangential force direction [20]. The SA-II fiber, innervating the Ruffini endings, is also sensitive to skin stretch, particularly to tangential forces generated in the skin. Therefore, to create an illusory force sensation of being pulled, an asymmetrical oscillation pattern should be designed that contains the frequency components stimulating these receptors (i.e., < 100 Hz) and an asymmetric magnitude that exceeds the thresholds of shearing displacement in one direction, but not in the other.

One study undertook psychophysical experiments on force direction discrimination of asymmetric oscillation and confirmed that almost all participants felt an apparent illusory force of being pulled or pushed persistently and that they were able to distinguish the force direction correctly [15]. In terms of perceived intensity of the force sensation, two studies have reported perceived intensity to be in the order of 10^{-1} N [21, 22]. In terms of gravity direction, our previous study showed a perceived weight change generated using asymmetric versus symmetric acceleration when grasping a box (size, 56 mm × 175 mm × 27 mm; standard weight, 400 g) with asymmetrically oscillating frequencies of 5 or 9 cycles per second [9].

In this study, we used a pinchable (20 mm × 20 mm × 40 mm, excluding soft foam as described below) and lighter box (standard weight, 50 g) to examine the effect of a higher frequency of asymmetric oscillation (40 cycles per second) on heaviness perception. When a small object is picked up using the index finger and the thumb in a pinching motion, its weight is implicitly detected and grip and load force change accordingly [23, 24]. Several studies have shown that changes in shear or friction forces at the fingertips contribute to weight perception [25–27]. Thus, a pinchable object was selected to examine changes in perception of heaviness using asymmetric oscillation.

3 Methods

3.1 Participants

Thirteen right-handed participants (nine females) aged 21–38 (average, 32.3) years participated in the main experiment. For further control experiments, eight participants (six females) were selected from this pool. The participants had no known abnormalities in tactile sensory systems or upper limb motor systems. The participants were paid for their time and provided informed consent prior to their inclusion in the study. Recruitment of the participants and experimental procedures were approved by the local Ethics Committee, and the study was conducted in accordance with the Declaration of Helsinki.

3.2 Vibrating Device and Stimuli

The weight stimulus involved gravitational force, in terms of gross weight and vibration in the gravitational direction. A vibration stimulus was created using a linearly vibrating actuator (Haptuator Original, Tactile Labs, Canada). Each actuator was covered with a box (dimensions, 20 mm × 20 mm × 40 mm), which was made of ABS resin. The appearance of the boxes was identical in order to eliminate any effect of visual differences [5, 6]. A piece of sandpaper (#1000 grit) was pasted onto its surface to control surface roughness [28]. Between the sandpaper and the device, a film-shaped pressure sensor (FlexiForce, Nitta Corporation, Japan) was attached to one side of the box to measure the grip force. A soft foam was attached at the bottom of the box to absorb vibration when it was placed on the table. The box was connected to a weight pan with a metallic wire to change its gross weight.

The weight of the vibrating device with a metallic wire was 25 g, and that of the weight pan was 25 g. The standard stimuli were static and were all identical (50 g), whereas the comparison stimuli were vibrating and varied (25, 37.5, 50, 62.5, 75, 87.5, 100, 125, or 150 g) to allow construction of a psychometric function indicating the likelihood of reporting the comparison stimuli as heavier. In this experiment, the trials were divided into six sets to reduce the effect of fatigue. Including a practice set, each participant performed 900 trials. In order to eliminate the influence of adaptation with long-term vibration, a participant was given a break of at least 5 min after every set. The total duration of the experiment was approximately 3 h.

3.3 Procedure

Participants were seated in front of a monitor. Three boxes labeled "A," "B," and "C" were placed on a table with 10 cm spacing (Fig. 1). An experimenter, seated behind the monitor, changed the weight of the weight pan for every trial. A trial consisted of two lifts, one with the standard stimuli and one with the comparison stimuli, in which participants pinched one of the three boxes, as instructed on the monitor. Participants were required to pinch and lift the box in a precise and firm grip that involved using the sides of the distal pads of the thumb and index finger to avoid any slipping out of the hand. They were instructed not to twist the box, as the cue of inertia might have influenced their report of weight perception (cf. [29, 30] but see also [31]). On hearing a tone, the participants lifted the box vertically approximately 1 cm with their hypothenar eminences on a wrist rest on the table. After three seconds, they placed the box on the table on hearing a second tone (Fig. 2). The participants then pinched an-other box, as instructed on the monitor, and repeated the same procedure as with the first lift. After two lifts, they pressed a button to indicate which box felt heavier (Fig. 3). The presentation order was counterbalanced to minimize a known cognitive bias toward judging the second of the two weights as heavier [32].

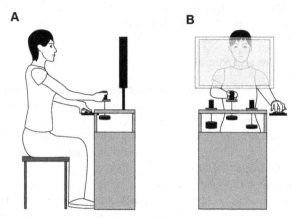

Fig. 1. An illustration of the experimental setup. Three boxes were placed in front of the participants. The monitor in front showed which box should be pinched and picked up next. Participants responded by pressing a key with the left hand.

This experiment was designed to quantify the amplitude of perceived heaviness due to asymmetric oscillation. For this purpose, we used a two-interval two-alternative forced choice (2I-2AFC) discrimination paradigm. In each trial, participants were asked to determine differences in heaviness between two vibrating boxes whose gross weight had been changed by the experimenter. We generated a psychometric function for the heaviness evaluation through varying the relative gross weights of the stimulations.

Our previous study confirmed that precision, in terms of the perceived force direction induced through asymmetric vibration, was significantly better when the hand or arm was actively moved in the horizontal plane [33]. Furthermore, humans have been reported to

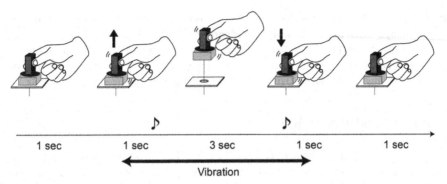

1 sec   1 sec   3 sec   1 sec   1 sec

Vibration

Fig. 2. A temporal sequence of one lift. On hearing a tone, the participants lifted the box vertically approximately 1 cm from the table. After three seconds, they placed the box on the table on hearing a second tone. The vibrating box was connected to a weight via a metallic wire.

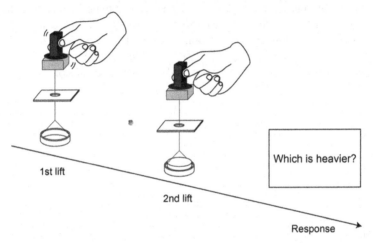

1st lift

2nd lift

Which is heavier?

Response

Fig. 3. Experimental procedure of one trial involving two lifts. The task of the participants was to report which interval contained the heavier stimuli.

frequently seek the determination of an object's inertia through actively moving it in an oscillatory manner [34]. To avoid interference between stimulus vibrations and manual movements, we asked the participants to lift the box slowly. In addition, to eliminate any effect of visual difference, the three boxes were identical in terms of appearance and material.

3.4 Data Analysis

Data obtained using the method of constant stimuli, that is, the proportion of trials in which "the test stimulus (vibration box) is heavier" were reported and were fitted with a cumulative Gaussian function using the maximum likelihood method. The PSE and just noticeable difference (JND) values were calculated using the fitted functions. The 50%

point on the psychometric function was defined as the PSE, which provided a perceived weight change when the tested weight was perceived subjectively as being equal to that of the compared weight. Half the interquartile range of the fitted function was defined as the JND value. JND is a measure of weight change sensitivity. All statistical tests were performed using MATLAB R2020a (The MathWorks, Massachusetts, USA).

4   Results and Discussion

Here, we estimated the intensity of perceived weight change, due to an oscillating object, by asking participants to indicate which of two boxes felt heavier: one vibrating and one not vibrating. Figure 4 illustrates the mean proportion of heavy evaluation for each stimulus and fitted psychometric functions determined through plotting a cumulative Gaussian model. Statistical analyses were conducted involving PSE and JND, obtained through fitting a psychometric function to data for each individual participant. If discrimination was not affected through vibration, PSE would be expected to be 50 g, namely, the gross weight of the standard stimulus. One-sample t-test results indicated a significant PSE shift from 50 g in the asymmetric oscillation condition [$t(12) = 5.80, p < 0.0001$] and in the symmetric oscillation condition [$t(12) = 3.29, p < 0.01$], suggesting that these vibrations increased perceived weight compared to the static condition.

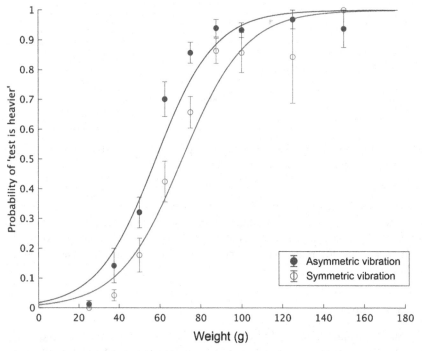

Fig. 4. Psychometric functions for perceived weight change generated using stimuli with asymmetric and symmetric accelerations compared to a static stimulus. The weight of the standard stimulus was 50 g. Solid lines show the psychometric functions fitted using mean data.

Pairwise t-tests were conducted to compare the effect of vibration symmetry on the perception of heaviness. The results indicated a significant PSE for the asymmetric oscillation condition compared to the symmetric oscillation condition [$t(12) = 6.49$, $p < 0.0001$, Cohen's $d = 1.06$), showing that the participants felt that the asymmetric vibrating box was heavier than the symmetric vibrating box (Fig. 5A). However, this analysis showed no significant difference in the JND between asymmetric and symmetric vibration conditions [$t(12) = 1.33$, $p = 0.208$, Cohen's $d = 0.09$), suggesting that differences in the sensitivity of weight perception between the asymmetric and symmetric vibrations are not readily determinable (Fig. 5B).

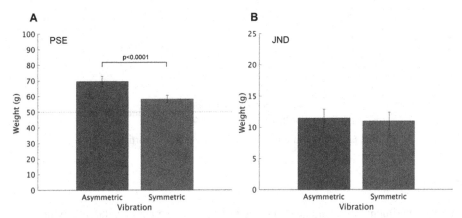

Fig. 5. Average perceived weight change using vibration with asymmetric or symmetric acceleration for PSE (A) and JND (B). The horizontal dotted line indicates the weight of the standard stimuli (50 g). Error bars show standard errors of the mean (SEM).

Our results showed that the perceived weight of an object vibrating with asymmetric acceleration increased compared to that with symmetric acceleration, when the acceleration peaked in the direction of gravity. This finding was in line with our previous study results that showed an object with asymmetric oscillation in the gravitational direction modulated heaviness perception when a 250 g box was grasped [9]. Here, we found that the perception of heaviness changed when a pinched object oscillated. An asymmetric oscillation, presented only at the fingertips, created a perception of increased heaviness.

Figure 6 shows the grip force during the experiment for a representative participant. There were no clear differences between the vibration and static conditions. This could imply that the change in the perception of heaviness did not stem from the change in grip force. However, the pressure sensor used to measure the grip force in this experiment could not detect any slippage at the fingertips. A higher frequency of vibration may cause a stick-and-slip phenomenon on the contact surface between the fingertips and the box. Further investigation will be required using other force sensors or an adhesive to avoid slippage between the fingertips and the box.

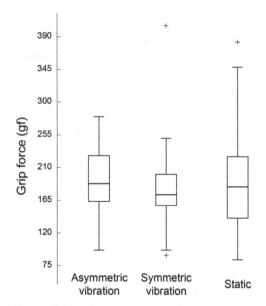

Fig. 6. Maximum grip force for a representative participant.

5 Conclusion

In this experiment, we evaluated changes in heaviness perception due to asymmetric oscillation and we examined the PSE of the amplitude of these changes. Our results suggested that the perceived weight of a vibrating object increased compared to that of a static object. Moreover, an object vibrating with asymmetric acceleration led to an increased perception of heaviness compared to that of an object vibrating with symmetric acceleration, when a high acceleration peak was generated in the direction of gravity. Future studies are required to examine the frequency and amplitude of asymmetric vibration in relation to heaviness perception.

Acknowledgements. This work was supported in part by JSPS KAKENHI Grant-in-Aid for Scientific Research (B) 18H03283.

References

1. Flanagan, J.R., Wing, A.M.: Effects of surface texture and grip force on the discrimination of hand-held loads. Percept. Psychophys. **59**(1), 111–118 (1997)
2. Charpentier, A.: Analyse experimentale de quelques elements de la sensation de poids. Archive de Physiologie Normale et Pathologiques **3**, 122–135 (1891)
3. Flanagan, J.R., Beltzner, M.A.: Independence of perceptual and sensorimotor predictions in the size-weight illusions. Nat. Neurosci. **3**(7), 737–741 (2000)
4. Boff, K.R., Kauffman, L., Thomas, J.P.: Handbook of Perception and Human Performance. Wiley, New York (1986)

5. De Camp, J.E.: The influence of color on apparent weight. a preliminary study. J. Exp. Psychol. **2**(5), 347–370 (1917)
6. Walker, P., Francis, B.J., Walker, L.: The brightness-weight illusion: darker objects look heavier but feel lighter. Exp. Psychol. **57**(6), 462–469 (2010)
7. Ellis, R.R., Lederman, S.J.: The golf-ball illusion: evidence for top-down processing in weight perception. Perception **27**(2), 193–201 (1998)
8. Reiner, M., Hecht, D., Halevy, G., Furman, M.: Semantic interference and facilitation in haptic perception. In: Proceedings of Eurohaptics, pp. 31–35 (2006)
9. Amemiya, T., Maeda, T.: Asymmetric oscillation distorts the perceived heaviness of handheld objects. IEEE Trans. Hapt. **1**(1), 9–18 (2008)
10. Amemiya, T., Ando, H., Maeda, T.: Virtual force display: direction guidance using asymmetric acceleration via periodic translational motion. In: Proceedings of World Haptics Conference, pp. 619–622 (2005)
11. Amemiya, T., Ando, H., Maeda, T.: Lead-me interface for a pulling sensation from hand-held devices. ACM Trans. Appl. Percept. **5**(3), 1–17, article 15 (2008)
12. Amemiya, T.: Haptic interface technologies using perceptual illusions. In: Proceedings of 20th International Conference on Human-Computer Interaction, Las Vegas, NV, pp. 168–174 (2018)
13. Amemiya, T., Gomi, H.: Distinct pseudo-attraction force sensation by a thumb-sized vibrator that oscillates asymmetrically. In: Proceedings of Eurohaptics 2014 Versailles, France, II, pp. 88–95 (2014)
14. Amemiya, T.: Perceptual illusions for multisensory displays. Invited talk. In: Proceedings of 22nd International Display Workshops (IDW 2015), vol. 22, Otsu, Japan, pp. 1276–1279 (2015)
15. Amemiya, T.: Virtual reality applications using pseudo-attraction force by asymmetric oscillation. Proc. HCI Int. **2020**, 331–340 (2020)
16. Bolanowski, Jr. S.J., Gescheider, G.A., Verrillo, R.T., Checkosky, C.M.: Four channels mediate the mechanical aspects of touch. J. Acoust. Soc. Am. **84**(5), 1680–1694 (1988)
17. Srinivasan, M.A., Whitehouse, J.M., Lamotte, R.H.: Tactile detection of slip: surface microgeometry and peripheral neural codes. J. Neurophysiol. **63**(6), 1323–1332 (1990)
18. Culbertson, H., Walker, J.M., Okamura, A.M.: Modeling and design of asymmetric vibrations to induce ungrounded pulling sensation through asymmetric skin displacement. In: Proceedings of IEEE Haptics Symposium (HAPTICS), Philadelphia, PA, pp. 27–33 (2016)
19. Bell, J., Bolanowski, S., Holmes, M.H.: The structure and function of Pacinian corpuscles: a review. Prog. Neurobiol. **42**(1), 79–128 (1994)
20. Maeno, T., Kobayashi, K., Yamazaki, N.: Relationship between the structure of human finger tissue and the location of tactile receptors. JSME Int J., Ser. C **41**(1), 94–100 (1998)
21. Tanabe, T., Yano, H., Iwata, H.: Evaluation of the perceptual characteristics of a force induced by asymmetric vibrations. IEEE Trans. Hapt. **11**(2), 220–231 (2018)
22. Rekimoto, J.: Traxion: a tactile interaction device with virtual force sensation. In: Proceedings of the 26th Annuals ACM Symposium User Interface Software & Technology, pp. 427–431 (2013)
23. Flanagan, J.R., Wing, A.M.: Modulation of grip force with load force during point-to-point arm movements. Exp. Brain Res. **95**(1), 131–143 (1993)
24. Johansson, R.S., Westling, G.: Roles of glabrous skin receptors and sensorimotor memory in automatic control of precision grip when lifting rougher or more slippery objects. Exp. Brain Res. **56**(3), 550–564 (1984)
25. Augurelle, A.S., Smith, A.M., Lejeune, T., Thonnard, J.L.: Importance of cutaneous feedback in maintaining a secure grip during manipulation of hand-held objects. J. Neurophysiol. **89**(2), 665–671 (2003)

26. Kilbreath, S.L., Refshauge, K., Gandevia, S.C.: Differential control of the digits of the human hand: evidence from digital anaesthesia and weight matching. Exp. Brain Res. **117**(3), 507–511 (1997)

27. Choi, I., Culbertson, H., Miller, M., Alex Olwal, A., Follmer, S.: Grabity: a wearable haptic interface for simulating weight and grasping in virtual reality. In: Proceedings of the 30th Annual ACM Symposium on User Interface Software and Technology, pp. 119–130 (2017)

28. Flanagan, J.R., Wing, A.M., Allison, S., Spenceley, A.: Effects of surface texture on weight perception when lifting objects with a precision grip. Percept. Psychophys. **57**(3), 282–290 (1995)

29. Brodie, E.E., Ross, H.E.: Jiggling a lifted weight does aid discrimination. Am. J. Psychol. **98**(3), 469–471 (1985)

30. Amazeen, E.L., Turvey, M.T.: Weight perception and the haptic size-weight illusion are functions of the inertia tensor. J. Exp. Psychol. Hum. Percept. Perform. **22**(1), 213–232 (1996)

31. Sekuler, R.W., Hartings, M.F., Bauer, J.A.: 'Jiggling' a lifted object does not aid judgment of its perceived weight. Am. J. Psychol. **87**(1/2), 255 (1974)

32. Ross, H.E.: Consistent errors in weight judgements as a function of the differential threshold. Br. J. Psychol. **55**(2), 133–141 (1964)

33. Amemiya, T., Gomi, H.: Active manual movement improves directional perception of illusory force. IEEE Trans. Hapt. **9**(4), 465–473 (2016)

34. Gibson, J.J.: Observations on active touch. Psychol. Rev. **69**, 477–491 (1962)

Thematic Units Comparisons Between Analog and Digital Brainstorming

Shannon Briggs[1]([✉]), Matthew Peveler[1], Jaimie Drozdal[2], Hui Su[2], and Jonas Braasch[1]

[1] Rensselear Polytechnic Institute, Troy, NY 12180, USA
briggs4@rpi.edu
[2] Cognitive Immersive Systems Lab, Rensselear Polytechnic Institute, Troy, NY 12180, USA

Abstract. This paper discusses findings in a user study evaluation of a novel digital brainstorming tool situated in a cognitive immersive environment. We compare how the brainstorming tools enable sensemaking by examining thematic units produced by users. The digital tool integrates multimodal inputs that are novel for brainstorming applications, and allows for multiuser collaboration. This user study examines how differences between the analog and digital brainstorming tool formats impacts sensemaking metrics. We find that the digital brainstorming tool allows users to reflect content from a source material more accurately than the analog brainstorming tool. Future work needs to identify the reason for increased idea cohesion within the cognitive immersive room's digital brainstorming tool.

Keywords: HCI methods and theories · Evaluation methods and techniques · Qualitative and quantitative measurement and evaluation · User experience

1 Digital Brainstorming Environment

The Cognitive Immersive Room is a suite of technologies within which the digital brainstorming tool discusses within this paper is situated. The Cognitive Immersive Room Architecture is a smart-room, the original design of which was meant to enable three areas of human action and interaction. The elements of the room consist of: 1. technology which captures, recognizes, and organizes information via microphones, cameras, Kinect® and other sensors; 2. natural language processing algorithms bringing in and interpreting the data through the collection of automated agents, such as Watson; and finally 3. technology that displays the system output to users through projectors and speakers [1]. The benefit of high fidelity, high clarity technology in the system, such as the set-up of the Cognitive Immersive Room, leads to higher immersion and higher presence in users. This increases emotional affect and allows for alternative information processing strategies compared to traditional desktop software [2]. Previous use cases of the Cognitive Immersive Room include corporate boardroom meeting; classrooms to aid in foreign language teaching and cultural immersion; group discussions where integrating and understanding large amounts of data is crucial, and finally a healthcare environment to enable an interdisciplinary group of people to diagnose cancer in patients

S. Yamamoto and H. Mori (Eds.): HCII 2021, LNCS 12765, pp. 257–267, 2021.
https://doi.org/10.1007/978-3-030-78321-1_20

[1]. The intelligence analysis room is one of the newest use cases for the Cognitive Immersive Room. The overall goal of the software is to be robust enough to support this wide variety of difficult cognitive processes across a range of domains, while also being flexible enough to use the same technology framework and AI programming to address each of these areas. Previous work concerning the kind of technology that is currently being enabled for the concept of immersive rooms involves different pieces of technology.

This paper discusses a user study conducted in the sensemaking room within the cognitive immersive environment. The hypotheses we examined for this section of the user study focused on thematic units, which will be explained more fully later. However, we posited that thematic units: the digital tool will have a higher proportion of more relevant and representative thematic units than the analog tool. We assume that the format of the digital tool positively affects the sensemaking process, where we can see if the digital tool produces thematic units that are closer to the thematic units in the text segment given to participants. He work we discuss

2   Related Research

2.1   Sensemaking

Pirolli and Card's sensemaking theory [3] provides the theoretical framework in order to help us understand how to best design and implement a technology that will be most useful and usable to our intended user. Pirolli and Card employ a cognitive model of sensemaking that focuses solely on intelligence analysts. The authors position intelligence analysts as experts within their domain who develop cognitive schema, which they characterize as patterns built around the most important parts of their tasks. The authors cite Ericsson and his theory on performance in expert domains [4], as well as the study from Chase and Simon [5] involving chess masters and how they built master level performance over a 15,000 h engagement with the domain. Therefore, it can be argued that Pirolli and Card [3] suggest that expertise is developed through training and engagement within a given domain.

2.2   Structured Analytic Techniques

Beebe and Pherson's book, Cases in Intelligence Analysis [6], is a collection of intelligence analysis problems previously handled by the U.S. intelligence community, accompanied by a number of structured analytic techniques used by professionals in the intelligence analysis domain. Structured brainstorming is one of the techniques discussed by the authors, which they describe as, "… a group process that follows specific rules and procedures designed to generate new ideas and concepts" [6]. The process involves at least two analysts discussing ideas to generate creativity, and adds value to intelligence products by "expos[ing] an analyst to a greater range of ideas and perspectives than the analyst could generate alone…" [6]. The structured analytic techniques described in intelligence analysis training literature was used as the basis for the digital brainstorming tool in the Cognitive Immersive Room. We felt that techniques already in use by the intended user would be cognitively accessible. The structured analytic technique described by Beebe and Pherson is moderated, with participants working on the same (sometimes, as in our user study, from source material such as a written prompt), starting with generating ideas and concepts separately, creating individual groupings of notes, and then coming together to share their ideas and categories [6]. Different training materials from the intelligence analysis domain differ slightly in the specific number of steps needed to conduct a structured brainstorming session, but usually are completed between 8 to 12 steps.

2.3 Digital Brainstorming

In general, the two most widely covered brainstorming techniques mentioned are individual brainstorming and collective brainstorming. A third type, called nominal brainstorming, is similar to individual brainstorming, but involves the combination of individual ideas after a period of individual brainstorming [7]. Previous studies have found nominal brainstorming to be more effective than brainstorming done alone or in groups [7]. This is important to consider, as the structured analytic technique developed by the intelligence analysis domain would be considered a nominal brainstorming tool.

Many fields utilize brainstorming as a cognitive exercise, which is defined as generating, sharing, and combining ideas about a problem or task by more than one individual [8, 9]. Brainstorming has also long been supported by electronic media. The reported benefits of electronic brainstorming systems (EBS) include cognitive stimulation and synergy, reduced production blocking, and reduced evaluation apprehension in anonymous EBS [8, 10, 11]. For intelligence analysis, being able to understand an intelligence issue from multiple perspectives and identifying where ideas need more evidence and information is particularly important.

The digital brainstorming tool discussed in this paper follows a blended approach, by allowing users to engage in divergent and convergent thinking phases. This entails participants first generating ideas on notes by themselves, creating provisional category types by themselves, and then engaging in a final communal discussion period, where the participants discuss their sticky notes and category types, and merge the categories together. Therefore, we can see the brainstorming technique used in the structed analytic brainstorming uses both nominal and collective types. This type of brainstorming is also set apart by its strict compartmentalized timed units, where participants move from step to step together. The structured analytic brainstorming tool is comprised of about eight to twelve different steps, depending on the source describing the technique.

3 Thematic Units

The brainstorming exercise has a categorization period, where participants collaborate on creating groups of notes that represent large thematic categories. While categories often reflected the major thematic units as discussed above, they often combined or re-chunked semantic units to reflect smaller units that had a semantic relationship [12].

For example, in the analog portion of Session 2, participants grouped debt with Luna's work, as well as work with sexual elements. For the first category, participants demonstrated that they suspected Luna was involved with the missing money from the bank robbery case, as he was in debt for roughly the same amount of money. Examples like this are clear indications of how different groups show different cognitive schemata.

Idea chunks are coded differently from categories and word frequencies; these concept chunks focus on the major theme or themes in each sticky-note per session. Sometimes sticky- notes contain multiple conceptual chunks, according to the participants' analysis of the material, and don't always align precisely with the final category the participants have assigned it to. We see that due to this ambiguity, participants will often vacillate in assigning their sticky-notes to categories.

Therefore, concept chunks are an intermediary step between the high granularity of words in sticky-notes and the low granularity of categories. Concept chunking also allows for coding of mixed concepts; for example, some sticky-notes mentioned that Luna took out a personal loan at the same time money disappeared from his place of work from a case he prosecuted. There are two major concept categories in this example, Luna's financial history, which would be classified as "debt", as well as Luna's work life, which would be classified as "work." Based on each participant and the overall group mentality, the participants would collaborate and discuss which category such a sticky-note might belong to, and might differ between participant groups.

Semantic similarity between the words alone don't capture the full concept chunk—for example, "loan," "personal loan," and "card" or "credit card" might appear in the same sentences and refer to the same thing, but result in different total frequencies across the nine sessions. Therefore, concept chunks allow for a midrange estimation of concept prevalence within each session.

4   User Study Methodology

Our study was a usability study carried across nine sessions, with 2–3 participants per session to study how the change from an analog brainstorming system to a digital brainstorming system affected users' sensemaking abilities. Participants were given text segments before each brainstorming session to encourage them to brainstorm major ideas. The text segments were different between analog and digital brainstorming sessions, in order to control for familiarity of ideas which might otherwise skew the data to be too similar between analog and digital brainstorming sessions. Participants were also instructed to limit a single fact or piece or evidence to a single sticky note, in order to facilitate grouping later. Participants were also instructed to reserve personal judgement statements on the sticky notes, in order to be as representative of the source material as possible. Participants were later able to discuss personal thoughts and judgements about the source material during a dedicated discussion phase.

Our overall goal was to determine if the design of the digital tool, which was based on the analog brainstorming tool, was as useful and usable as the original analog tool. Thematic units were selected as a metric in order to understand how topics are represented between the digital and analog brainstorming formats. Thematic units were first extracted from the text segment that was distributed to users before the brainstorming session.

Then, we extracted from users' sticky notes after all nine user study sessions, and were coded for topic chunks. In the analog brainstorming sessions, we determined there were three major thematic units in the in text segment. Therefore, we would anticipate that users would recognize and demonstrate the same three major thematic units. The digital brainstorming tool was found to have four major thematic units in the text segment as well as the users' sticky notes. We analyzed the differences between the text segment and the users' thematic units, as we expected users would produce a proportionally similar number of thematic units as were displayed in the in-text, taking into account the number of thematic units produced by multiple people.

From this design, we wanted to study how the digital brainstorming tool performs compared to the traditional pen-and-paper brainstorming tool. The analog brainstorming tool is an important part of the study, as the use of the traditional tool is well documented [6]. We anticipated that by comparing the experience between the analog and the digital tools, we will be able to gain insight into how the analog tool can facilitate brainstorming in a way that can be transferred to the digital tool, and we have assumed that a digital tool using the same interactions and technique will be as useful and usable as the analog tool.

We were interested to see if the digital brainstorming tool is as sufficient as the pen-and-paper process in allowing users to make sense of information as they are working through the text segments given to them for the case study. For qualitative measures, we decided to focus on two major aspects of the brainstorming tool that are reflective of the brainstorming process regardless of format: the overall number of sticky-notes created per session, and how many categories participants created. For this user study, we hypothesized that that the thematic units in the digital brainstorming tool sticky will have a higher proportion of more relevant and representative thematic units than the analog tool when compared to their respective source text segments.

5   Results

5.1   Analog

When comparing the separate thematic units with the text samples, we performed a basic frequency analysis on users' text sorted by the three major concept categories, seen in Table 1. Figure 2 shows the relative proportion of thematic units found within users' sticky notes compared to the thematic units found with the text segment. For the analog brainstorming session, we found that there were three major thematic units that could be coded from the text segment and the users' sticky notes. These three thematic units were "debt", "work", and "sex". We found that representation among thematic units would often differ dramatically between sessions. Given that this disparity isn't displayed in the digital brainstorming thematic units, discussed in the next section, we anticipate this variety in representation might be due to the format of the brainstorming session itself. We found that especially in the analog brainstorming sessions, users felt more flexible in assigning or reassigning thematic chunks that have a semantic relation to another chunk, regardless of the similarity of topic of the thematic units. For example, in the same sticky note we would often find a user combining concepts relating to debt with concepts relating to his personal life or work. For example, Fig. 1 shows how a user combined thematic units concerning debt and work.

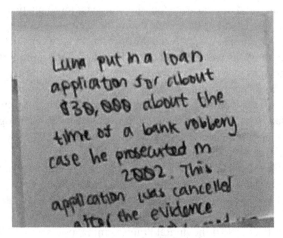

Fig. 1. Analog session sticky note displaying multiple thematic units.

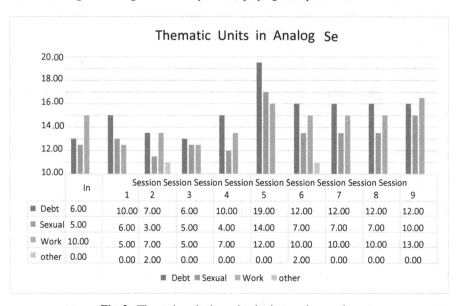

	In	Session 1	Session 2	Session 3	Session 4	Session 5	Session 6	Session 7	Session 8	Session 9
■ Debt	6.00	10.00	7.00	6.00	10.00	19.00	12.00	12.00	12.00	12.00
■ Sexual	5.00	6.00	3.00	5.00	4.00	14.00	7.00	7.00	7.00	10.00
■ Work	10.00	5.00	7.00	5.00	7.00	12.00	10.00	10.00	10.00	13.00
■ other	0.00	0.00	2.00	0.00	0.00	0.00	2.00	0.00	0.00	0.00

Thematic Units in Analog Se

■ Debt ■ Sexual ■ Work ■ other

Fig. 2. Thematic units in analog brainstorming session

Table 1. Descriptive statistics for analog brainstorming thematic units

		Statistics		
		Debt	Sex	Work
N	Valid	9	9	9
Missing		0	0	0
Mean		11.1111	7.0000	8.7778
Median		12.0000	7.0000	10.0000
Mode		12.00	7.00	10.00
Std. Deviation		3.72305	3.31662	2.90593
Variance		13.861	11.000	8.444
Minimum		6.00	3.00	5.00
Maximum		19.00	14.00	13.00
Percentiles	25	8.5000	4.5000	6.0000
	50	12.0000	7.0000	10.0000
	75	12.0000	8.5000	11.0000

Across nine sessions, the mean number of thematic units in debt was 11.11, the mean number of thematic units in the category "sex" was 7, and the mean number of thematic units in "work" was 8.8. These results can be compared against the themes in the sample text that was given to participants in Table 2. In order to understand if the analog brainstorming tool enabled participants to successfully record the thematic units found in the text segment, Table 2 compares the mean number of participants' thematic units across nine sessions to the total number of thematic units in the text. As Table 2 shows, there is a fairly significant gap between the participants' thematic units and the in-text thematic units, with the largest of 5.11, where participants under-reported the "work" concepts. This variability in participant thematic unit means compared to the in-text thematic totals indicates that the analog tool did not successfully support the sensemaking process.

Table 2. Differences between text sample and user thematic units.

Debt concepts	Sex concepts	Work concepts
Participant: 11.11	Participant: 7	Participant: 8.8
In-text: 6	In-text: 5	In-text: 10
Difference: 5.11	Difference: 2	Difference: −1.2

5.2 Digital Thematic Units

We found there were four thematic units in the text segment given to participants for their digital brainstorming sessions. Figure 3 shows the breakdown between the thematic units found in the text segment, compared to the thematic units found in the sticky notes generated by users across the nine digital brainstorming sessions.

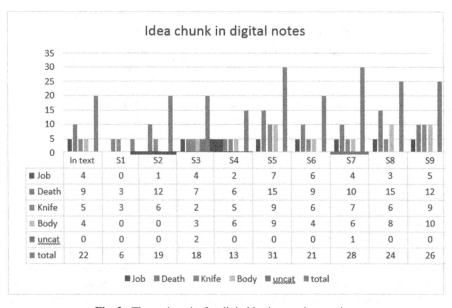

	In text	S1	S2	S3	S4	S5	S6	S7	S8	S9
■ Job	4	0	1	4	2	7	6	4	3	5
■ Death	9	3	12	7	6	15	9	10	15	12
■ Knife	5	3	6	2	5	9	6	7	6	9
▦ Body	4	0	0	3	6	9	4	6	8	10
■ uncat	0	0	0	2	0	0	0	1	0	0
■ total	22	6	19	18	13	31	21	28	24	26

Fig. 3. Thematic units for digital brainstorming sessions

Table 3 shows the descriptive statistics for the thematic units in the digital notes across nine sessions. The mean number of thematic units in the "job" category was 3.5, the mean number of thematic units in the category "death" was 9.8, the mean number of thematic units in "knife" was 5.9, and the mean number of units in the "body" category was 5.11. This compares against the themes in the sample text chunk given to participants in Table 4. In order to understand if these are representative of the text segment given to participants during the digital notes session, Table 4 shows a comparison of the mean number of concepts across nine sessions compared to the total number of thematic units in the text segment. The table shows the mean number of participant thematic units were very close to the text segment, so we infer that the digital tool successfully enabled users to appropriately represent the thematic units in the text segment, and lead to a more cohesive sensemaking session.

Table 3. Digital brainstorming session descriptive statistics

Statistics					
		Job	Death	Knife	Body
N	Valid	9	9	9	9
	Missing	0	0	0	0
Mean		3.5556	9.8889	5.8889	5.1111
Median		4.0000	10.0000	6.0000	6.0000
Mode		4.00	12.00[a]	6.00	.00[a]
Std. Deviation		2.29734	4.07567	2.36878	3.65529
Variance		5.278	16.611	5.611	13.361
Range		7.00	12.00	7.00	10.00
Minimum		.00	3.00	2.00	.00
Maximum		7.00	15.00	9.00	10.00
Percentiles	25	1.5000	6.5000	4.0000	1.5000
	50	4.0000	10.0000	6.0000	6.0000
	75	5.5000	13.5000	8.0000	8.5000

[a]Multiple modes exist. The smallest value is shown

Table 4. Differences between in-text thematic units and user thematic units for digital brainstorming.

Job concepts	Death concepts	Knife concepts	Body concepts
Participant: 3.5	9.8	5.8	5.11
In-text: 4	In-text: 9	In-text: 5	In-text: 4
Difference: 1.5	Difference: .8	Difference: .8	Difference: 1.11

6 Discussion

We performed content analysis on participants' contents from the sticky-notes and categories across the nine user studies for both analog and digital brainstorming tools. We first identified thematic units in participants' sticky-notes that aligned with the thematic units in the concept categories from each of the text segments, and then compared the proportion of relevant words in sticky-notes to their respective text segments from the testing. Our hypothesis for observing thematic units in sticky-notes was that the digital tool will have a higher proportion of more relevant and representative thematic units than the analog tool. When looking at the overall means for nine sessions between the analog and digital sticky-notes tools compared to the respective text segments given to participants, the digital participants had means closer to the number of thematic units in the text segment. On average for the digital brainstorming sessions, we found that the means between the text segment and the users' sticky notes were similar, and therefore the thematic units were evenly distributed throughout the sticky-notes over the nine sessions. We conclude that participants' sensemaking is better represented in the digital tool, which enabled users to create more representative and balanced sticky-notes from the case material given to them. This may be due to participants being more careful about their selection of words as they are mediated by technology, and this extra step prompted them to be more cautious when recording words within the digital system, to ensure others would be able to understand them.

7   Conclusions and Future Work

Our findings displayed that the digital tool was better in allowing participants to create more consistent thematic units across sessions, and thematic units which were more representative of the text segment given to the participants. Considering the intent of the brainstorming exercise was to encourage users to faithfully reflect the contents of the text segment given to them in their sticky notes, we found that the digital tool enabled users to create more representative thematic units than the analog tool. While the precise reason for this isn't known, we posit that this may be due to users' sense of presence in the cognitive immersive room, and their increased awareness of the technology. Their concept of a shared space that they need to interact with, as well as create data that they will share with others interactively, might influence how they chose to record data that they will share later. Further research into understanding the effect of immersive environments on users' produced data is necessary to draw more firm conclusions.

We used sensemaking theories in conjunction with user-centered design for emerging technology, which can be applied to many fields. In order to develop the digital brainstorming tool that is enabled in CISL, we used Pirolli and Card's [3] sensemaking theory in conjunction with the structured analytic techniques that are already in use by intelligence analysis, specifically the brainstorming exercise. These structured analytic techniques were sourced from intelligence analysis textbooks, which were also used to display the mental models and mindsets analysts exhibit during analysis [6]. Research concerning how users engage with immersive environments will also give insight into communication research, and how human reaction and behaviors can be captured and analyzed for remote collaboration. We are also interested in investigating how other types of technologies might influence the sensemaking process, such as collaborations with artificial intelligence, and how machine learning might lessen cognitive load during the sensemaking process.

References

1. Divekar, Rahul R., et al.: CIRA: an architecture for building configurable immersive smart-rooms. In: Arai, K., Kapoor, S., Bhatia, R. (eds.) IntelliSys 2018. AISC, vol. 869, pp. 76–95. Springer, Cham (2019). https://doi.org/10.1007/978-3-030-01057-7_7
2. Riva, G., Davide, F., Ijsselsteijn, W.A., Grigorovici, D. (eds.) Persuasive effects of presence in immersive virtual environments (2003)

3. Pirolli, P., Card, S.: Sensemaking processes of intelligence analysts and possible leverage points as identified through cognitive task analysis. In: Proceedings of the 2005 International Conference on Intelligence Analysis, McLean, Virginia (2005)
4. Ericsson, K.A., Nandagopal, K., Roring, R.W.: Toward a science of exceptional achievement: attaining superior performance through deliberate practice. Ann. N. Y. Acad. Sci. **1172**, 199–217 (2009). https://doi.org/10.1196/annals.1393.001
5. Chase, W.G., Simon, H.A.: The mind's eye in chess. In: Visual Information Processing, pp. 215–281. Elsevier (1973)
6. Beebe, S.M., Pherson, R.H.: Cases in Intelligence Analysis: Structured Analytic Techniques in Action. CQ Press (2014)
7. Rietzschel, E.F., Nijstad, B.A., Stroebe, W.: Productivity is not enough: a comparison of interactive and nominal brainstorming groups on idea generation and selection. J. Exp. Soc. Psychol. **42**, 244–251 (2006)
8. Briggs, R.O., Kolfschoten, G., Vreede, G.-J. de, Albrecht, C., Dean, D.R., Lukosch, S.: A seven-layer model of collaboration: Separation of concerns for designers of collaboration systems. In: ICIS 2009 Proceedings, p. 26, Phoenix (2009)
9. Reinig, B.A., Briggs, R.O., Nunamaker, J.F.: On the measurement of ideation quality. J. Manage. Inf. Syst. **23**, 143–161 (2007). https://doi.org/10.2753/MIS0742-1222230407
10. Fjermestad, J., Hiltz, S.R.: A descriptive evaluation of group support systems case and field studies. J. Manage. Inf. Syst. **17**, 115–159 (2001)
11. Pinsonneault, A., Barki, H., Gallupe, R.B., Hoppen, N.: Electronic brainstorming: The illusion of productivity. Inf. Syst. Res. **10**, 110–133 (1999)
12. White, M.D., Marsh, E.E.: Content analysis: a flexible methodology. Libr. Trends. **55**, 22–45 (2006)

On-Demand Lectures that Enable Students to Feel the Sense of a Classroom with Students Who Learn Together

Ryoya Fujii[1,2(✉)], Hayato Hirose[1,2], Saizo Aoyagi[3], and Michiya Yamamoto[1,2] ⓘ

[1] School of Science and Technology, Kwansei Gakuin University, Sanda, Hyogo, Japan
{fwj36188,michiya.yamamoto}@kwansei.ac.jp
[2] Graduate School of Science and Technology, Kwansei Gakuin University, Sanda, Hyogo, Japan
[3] Faculty of Information Networking for Innovation and Design, Toyo University, Kita-ku, Tokyo, Japan

Abstract. In this age of the corona disaster, face-to-face lectures have been decreasing and on-demand lectures have been increasing. However, the role of students beside oneself in real classroom interaction is important. In this study, we expand this to on-demand lectures. To understand the interaction of students in on-demand classes, we created lecture videos and a CG classroom based on a real classroom. In addition, we introduced a nodding mechanism for student characters. On the basis of this, we conducted classes using 2D video, 360-degree video, and VR using an HMD, and we clarified the differences between 360-degree video and VR. We also performed questions about movie types and character behaviors. The results of the evaluation of movie types indicate that using VR in on-demand classes can make students feel a relationship in the classroom. The results of the evaluation of character behaviors show that rash action is useful in some cases.

Keywords: On-demand lectures · VR lectures · 360-degree videos

1 Introduction

From the spring of 2020, Japanese universities started online classes with many trials and errors to prevent the spread of COVID-19, and the spreading ratio of the online classes exceeded 95% in May 2020 [1]. Online on-demand classes whose recorded contents students can watch freely have advantages in that students can study whenever and wherever they like and partake in the classes widely. But students also want face-to-face classes, and some of them have made a claim to be paid back lecture fees lest they leave a school [2, 3].

What do on-demand classes lack? We regard relationships between students as the first problem. For example, the completion rate of the online studying service Udacity [4] is less than 5 – 10%. One of the reasons for this low rate is the decline of students' motivation to study by themselves [5]. In addition, active learning is encouraged for students to stay active by relating to each other [6, 7]; however, on-demand classes do

© Springer Nature Switzerland AG 2021
S. Yamamoto and H. Mori (Eds.): HCII 2021, LNCS 12765, pp. 268–282, 2021.
https://doi.org/10.1007/978-3-030-78321-1_21

not provide the opportunity to gather in a classroom, which may lead to the decline of students' identity.

Another problem of on-demand classes is the expression of a classroom. It is known that the design and mood of the classroom affect students by their attitudes, the effectiveness of learning [8], and participation [9]. On-demand classes may have poor effects because of their big differences from face-to-face lectures. In real education, as one example of real-time lectures, N High School [10] offers a system by which students can gather in a virtual classroom and take lectures together using an HMD (head-mounted display). We can see the importance of a classroom as a place and the relationships between students through this case.

However, this is just one special example because the school is good at information technology. There seem to be many educational institutions that cannot introduce VR easily. To solve this problem, this study aims at the achievement of on-demand classes by which students can feel they are taking lectures in a real classroom, with a method that most educational institutions can introduce.

We have studied the importance of the body's effect in communication. In particular, we have developed systems by which CG characters support communication in a learning space with embodied communication, such as nodding [11, 12] or hand raising [13, 14]. In all this research, we have used a CG classroom as a place where learners' bodies are located with characters. This study applies a 3DCG classroom and the movement of CG characters in on-demand classes. With this method, which is just a subjective view, we intend for students to be able feel the classroom as a place and the relationships between students. However, our preceding studies [11–14] were based on face-to-face lectures. Therefore, there is a problem of how to stream the contents to students. Figure 1 shows the methods of on-demand classes and face-to-face lectures from the standpoint of whether there are relationships between students and an expression of the classroom or not.

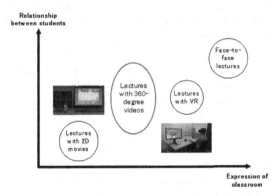

Fig. 1. The methods of on-demand lectures in classroom expression and students' involvement.

At first, with face-to face lectures, teachers can easily proceed in lessons with student-to-student and student-to-teacher interactive communication. Regarding this, teachers can conduct on-demand lectures with 2D movies easily, but it is difficult to provide a relationship between students, and there can be one-way classes. We consider that virtual lectures wherein students use an HMD for attendance are at the second position. We call this style VR lectures. With this method, we can express a classroom and other students, but we also need hardware and skills for making contents and running systems.

So we focus on 360-degree videos. This method is a kind of movie that is played in a square-like TV, and participants can look around the movie by dragging their mouse. It is the same as VR lectures in that we can generate a classroom or place CG characters in a 3D virtual space, but we play the movie not with an HMD but with PC (personal computer) Web browsers. We cannot move the eye points or operate subjects in the movies, but we can watch them with a popular service such as YouTube.

With these characteristics, on-demand lectures using 360-degree videos are considered to be a realistic way to make students feel as if they are in a real classroom with other students. In this study, we show that on-demand lectures using 360-degree videos can be practically implemented by comparing them with VR lectures in real classes, and we show to what extent they can make students feel involved with each other.

In the field of educational psychology, human relationships have been shown to be important for motivation to learn and academic achievement [15], and 360-degree on-demand classes which we focus in this study are considered to be effective. However, although there are many examples of VR lectures and 360-degree lectures that realize relationships between classroom and students at the research level [16], there are few practical examples.

2 Preliminary Experiment

As a preliminary attempt [17] leading up to this study, an online lecture using on-demand video streaming was conducted in July 2020. In this preliminary study, we found that it was difficult to hear the lecturer's voice and to see the slides when a video was simply taken of the class. Therefore, in this experiment, we decided to record the components of the lecture separately and reconstruct them in a way that would not destroy the image of the classroom and that would improve the audibility and readability of the materials. The components of the lecture are video of the class from the back of the classroom, audio recorded directly from a microphone, and slides recorded from a PC screen. By combining the recorded slides and noise-free audio with the video, we were able to create a high-quality video of the lecture. Such is called a synthesized video in what follows. An example is shown in Fig. 2 below.

The first condition was to stream synthesized videos, and the second condition was to create and stream 360-degree videos to compare them with the videos in the first condition. In the 360-degree videos, a synthesized video was placed in a CG classroom.

We then conducted online classes at a university using on-demand video streaming and conducted a questionnaire to evaluate the two conditions. Students were divided into two groups, one taking a class with 360-degree videos and the other a class with synthesized videos. After taking the class, students in both groups watched both synthesized and 360-degree videos and answered the evaluation questionnaire. Part of the results of the questionnaire item "I would like to use it again in the future," rated on a 5-point scale, and the results of Wilcoxon's signed rank test are shown on the left side

Fig. 2. A scene of a synthesized video.

of Fig. 3. While there was no significant difference between the two conditions in the results of the class mini-quiz, the synthesized video was rated higher by both groups. This may be due to the poor quality of the 360-degree videos, which made it difficult to see the slides.

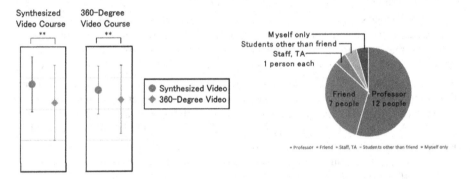

Fig. 3. Results of the preliminary experiment.

In addition, in order to investigate the students' interaction with others in a normal face-to-face class, the following question was asked: "With whom do you feel most involved in the class?" Results of this question are shown on the right side of Fig. 3. That not only professors but also friends accounted for 1/3 of the participants indicated that one's relationship with friends may be important in learning situations. However, the number of subjects was only 22, so the reliability of the results is low.

3 Method

3.1 Outline of the Lectures

On the basis of the preliminary experiment, in this study, an on-demand class using 360-degree videos was practiced at a university in January 2021. In this experiment, in addition to (2) 360-degree videos, two other conditions were prepared for the same content: (1) synthesized videos and (3) VR. In the synthesized videos condition, there are no student characters and no expression of the classroom, so it is a control condition similar to a normal on-demand class. On the other hand, the VR condition is expected to be superior to the 360-degree videos in terms of the immersion of the student characters' placement and the expression of the classroom as a place.

In this study, we conducted a questionnaire survey of students in on-demand classes about the classroom and their impressions of others. The purpose of this study is to use the results of the survey as evidence for the importance of the relationships between students and to provide supporting evidence for the usefulness of 360-degree video classes.

3.2 How to Conduct the Classes

The class consisted of two parts: a lecture and questionnaire surveys. There were three groups of students who took the lectures in different ways.

The first group is students with even student numbers among all the students of "Computer Graphics" and "Introduction to Human System Interaction" at Kwansei Gakuin University, which the author Yamamoto is in charge of. They took the class with the synthesized videos at home. The second group is students who took the 360-degree video classes on demand. This group is comprised of students with odd student numbers among all the students of "Computer Graphics" and "Introduction to Human System Interaction." The third group is students who took the VR class in the laboratory. In this study, we focused on VR classes as a method for on-demand instruction. Since it is difficult to operate the experiment and control the surrounding environment, we invited students who wished to take the class to come to our laboratory to attend it.

The two on-demand groups were asked to access different URLs to view the videos but were not told that the videos were different. For the 360-degree video group, when the videos were too arduous to play, the students were told to access the URL of the synthesized video group.

In the second half of the class, we administered three questionnaires: "Questions about the classroom and student interaction," "Evaluation of the video type," and "Evaluation of the student characters' behavior."

3.3 How to Make the CG Classroom and Student Characters

In order to implement the 360-degree video and VR classes, we created a CG model of a classroom. The CG model was created using Blender 2.79b and based on a blueprint of a classroom, measured data for details, and camera images taken. Figure 4 shows an example of a class scene using the CG classroom. The sizes of desks and chairs were measured, and classroom 402 of the School of Science and Engineering at Kwansei Gakuin University, where the class was to be held, was made into a CG model (Fig. 5 left). We also created 30 CG student characters using Pixiv's VRoid Studio (Fig. 5 right).

These CG models and student characters were loaded and placed with Unity Technologies' Unity 2019.4.3f1, along with the synthesized videos created in the preliminary experiment. The synthesized videos were placed in front of the classroom, and the

Fig. 4. An example scene of CG classroom.

Fig. 5. CG classroom we made and an example of a student character.

size and color tone of the surrounding 3D models were fine-tuned to make the virtual classroom feel seamless.

In addition, we made the student characters nod in response to the audio of the videos. Figure 6 shows an example of a nodding motion. We created a variety of nodding motions, with different numbers, speeds, and angles of nods. In addition, because it was considered unnatural for the students to be too aggressive, we also created "raggedy" motions for the characters, such as shaking their heads to the side and looking away, to convey a raggedy impression. An example of such motion is shown in Fig. 7.

Fig. 6. Nodding behavior.

Fig. 7. Raggedy behavior.

3.4 How to Make 360-Degree Videos

The 360-degree videos were recorded using UnityRecorder (Ver. 2.2.0), an asset of Unity, in a 3D virtual environment consisting of the CG classroom and student characters described in Sect. 3.3. Unity Recorder can export 360-degree videos and full-dome images.

In our preliminary experiment, we exported 360-degree videos at 4 K resolution. However, on the PC used in this study (CPU: AMD Ryzen 5 1600, video chip: NVIDIA Geforce GTX 1070, memory: 16 GB), the frame rate dropped significantly from about 100 fps when Unity Recorder was not used to about 13 fps when it was used. Therefore, we recorded the video at 0.2 times the speed of the original video and later tried to combine the audio of the original video with the exported video using Adobe Premiere Pro CC 2020.

However, this method caused a problem in which the playback speed changed in the middle of the video and the audio did not match the visuals. In order to avoid this problem, we changed to a method of cutting out each frame of the full-dome image and using ffmpeg to concatenate the frames into a movie. An example of an exported full-dome image is shown in Fig. 8.

Fig. 8. A scene of the exported video.

In addition, changing the playback speed of the video caused the nodding of the student characters, which was originally synchronized with the sound in the video, to be out of sync with the sound. We solved this problem by controlling the duration of the nodding in the original video to match the change in playback speed.

3.5   On-Demand Lectures and How to Answer the Questionnaires

To make the videos available to the students, we created a "channel" for our laboratory using the video streaming service YouTube and uploaded the videos to it. First, we uploaded the synthesized video produced in the preliminary experiment for the synthesized video course and made it available as a normal video (1080p HD).

We also uploaded the 360-degree videos created by the method described in Sect. 3.4. These videos can be viewed as general 360-degree videos by using YouTube's functions to rotate the viewpoint by dragging the screen and to zoom in on an area of interest by manipulating the mouse. An example of the streaming scene is shown in Fig. 9. If the video quality setting is left at "Automatic," there are cases where the video cannot be viewed in 4 K quality due to differences in the environment, so it is necessary to set the quality to "2160 s 4 K" when taking the course.

Fig. 9. A screenshot of a streamed 360-degree video.

In addition, Microsoft Forms, an online response form creation service, was used for answering mini-quizzes in lectures and questionnaires for various surveys. In the conventional face-to-face class, the research introduction was given for 60-70 min, followed by a mini-report. For the on-demand version of the class, we divided the research introduction video into topics such as "What is a human interface?" and "Embodied media technology." The content was the same for both courses. The lectures consisted of five sections (videos), and the mini-quizzes for "Computer Graphics" and "Introduction to Human System Interaction" consisted of 15 questions with a maximum score of 17 points. In order to be fair, the VR class in the laboratory was given before the on-demand class.

3.6 VR Lectures and How to Answer the Questionnaires

During the preliminary experiment, students were not allowed to come to the laboratory because the university was under lockdown. On the other hand, in this experiment, a small number of students who wanted to attend the class were allowed to do so in the authors' laboratory using a VR headset (Oculus Quest2, Oculus VR).

The scene is shown in Fig. 10. The CG classroom and student characters generated in Sect. 3.3 were placed in the 3D virtual environment of Unity, and the PC and VR headset were connected using Oculus Link so that the group could watch higher-quality videos than they could by watching the 360-degree videos on demand through YouTube.

For the purpose of conducting the class without taking off the VR headset, we made it possible to answer questionnaires and mini-quizzes in the 3D virtual environment of Unity, as shown in Fig. 11. We prepared a panel with questions and checkboxes on the

Fig. 10. An example scene of the VR lecture.

right side of the video and made it possible to answer questions by sending out a line from the tip of the VR headset controller and pressing the right index finger trigger when the line overlapped with the checkbox. If a question was missed in the video, the user could rewind the video for 5 s by pressing the right middle finger grip. In addition, the video could be paused by pushing the analog stick inward. We did not use the A and B buttons because we thought that they would be difficult to understand for students using the controller for the first time. For some of the questions that were answered without using the VR headset, we used Forms in the same way as the on-demand participants did.

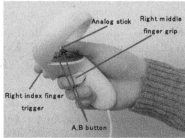

Fig. 11. How to answer the questionnaires with a VR headset.

4   Results

4.1 Lessons and Mini-quizzes

The VR course was conducted from January 12 to 15, 2021, and the on-demand courses were conducted from January 17 to 22, 2021. In the VR course, there were 12 students each for "Computer Graphics" and "Introduction to Human System Interaction." There were 22 odd and 22 even students taking the on-demand courses for "Computer Graphics" and 93 odd and 81 even students taking such courses for "Introduction to Human System Interaction." The number of students in the VR class was very small compared to in the other conditions because the class period coincided with the declaration of a state of emergency in January 2021 in Hyogo Prefecture, where the university is located.

The results of the mini-quizzes are shown in Fig. 12. The results were tested with the Mann-Whitney U-test. Out of a total of 17 points, the average score for the VR course was 13.2 points, and that for the on-demand courses was 14 points.

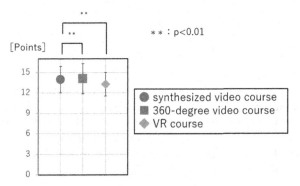

Fig. 12. Results of mini-quizzes.

The average of the ratio of viewing time to the length of the video is shown in Table 1. On average, the viewing time was longer than the playing time of the video in the VR course. On the other hand, the videos in the on-demand courses were played for 70% to 80% of the total video time on average.

Table 1. Average percentages of viewing time to playing time of the videos.

Synthesized videos	360-degree videos	VR
74.53	82.83	118.12

As data that may be useful for this interpretation, Fig. 13 shows an example of the viewer retention rate for a video on the topic of "cognitive systems engineering" for the on-demand courses. The viewer retention rate is the number of times each position in the video has been viewed divided by the number of people who have played the part. The highlighted areas in Fig. 13 show the correct answers to the questions in the mini-quizzes that accompany this video. As shown in the example of this video, the viewer retention rate was relatively high when correct answers to the mini-quizzes were given, while said rate was low in other areas.

We consider that the students who took the on-demand courses were able to watch the videos efficiently because they only watched the parts of the videos where the correct

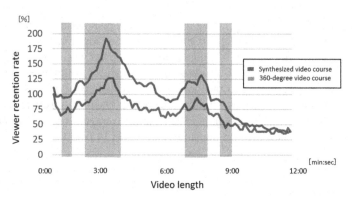

Fig. 13. An example of viewer retention rate.

answers to the mini-quizzes are explained after they checked the mini-quizzes. However, from an educational point of view, this is not a desirable behavior because we want students to watch such videos in their entirety. On the other hand, in the case of VR, there was no seek function to select the position of the videos to be played. Therefore, it was difficult for the students to watch parts of the videos as described above, which may have resulted in low scores on the mini-quizzes.

4.2 Questions About the Classroom and the Relationship Between Students

We conducted a questionnaire survey titled "What is the university classroom like for you?" First we asked, "What is the activity that gives you the strongest impression of the classroom?" We asked the students to sort four items (class, chatting, self-study, and other activities) in order of their strongest impression. Next, we asked, "Who do you think you interact with during the activity?" We asked them to sort five items (professor, staff member or TA, friend, student other than a friend, and only themselves) in order of their impressions. The results shown below are the sum of the results for each course, as they are considered to be independent of the course.

The left side of Fig. 14 shows the number of items selected as the most impressive activities. The item selected by the largest number of students was "class," which was selected by 160 students. The second was "chatting" at 56 students, followed by "self-study" at 18 students and "other" at 5 students.

Next, the students who chose "class" as the first place in the previous questionnaire had strong relationships is shown on the right side of Fig. 14. The item selected by the largest number of students was "professor," with 86 students. The runner-up to that item was "friends" (47 students), followed by "myself only" (20 students), "staff, TA" (12 students), and "students other than friends" (2 students).

The result of the most memorable activity being the class and the most memorable person being a friend may seem contradictory, but this is not the case. In face-to-face classes, there were many large classes, so students often asked questions to their friends rather than to the professor, or they were more impressed by private conversations with their friends than by their professor.

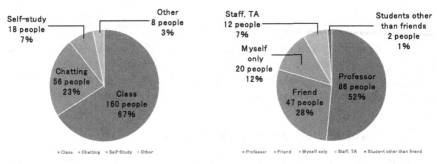

Fig. 14. The most impressive activities and persons in classes.

4.3 Evaluation of Movie Types

After the first part of the class, all groups of participants watched and compared (1) a synthesized video using the same method that was used in the preliminary experiment, (2) a 360-degree video created by the method described in Sect. 3.4, and (3) VR using the method described in Sect. 3.6. The 360-degree video was shown after the synthesized video in the on-demand courses, and in the VR class, the students watched the synthesized video, 360-degree video, and VR in that order and were asked to evaluate them. The students were asked to respond on a seven-point scale to eight items related to their satisfaction with the class and whether they felt the presence of the teacher and other students. The results are shown in Fig. 15. The results were tested with Wilcoxon's signed rank test and the Bonferroni method. In the results for the on-demand courses, the 360-degree video was rated higher on items 3 and 5, and the same trend was observed for the VR course. In addition, some items, such as item 2, were rated higher in the VR course than in the other two types of courses.

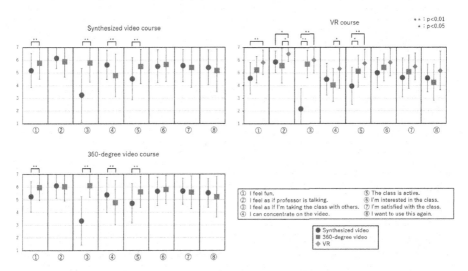

Fig. 15. Results of movie type evaluation.

Compared to in the preliminary experiment, we consider the nodding of the characters to have been effective, since there was no longer a significant difference in the item that the participants wanted to use in the future. The higher ratings of the 360-degree video and VR for item 3, "I feel as if I'm taking the class with others," and item 5, "The class is positive," can be interpreted as indicating that the students were able to feel a certain kind of involvement in the classroom.

4.4 Evaluation of the Student Characters' Behavior

Next, we evaluated the behavior of the students' CG characters. In this evaluation, the students watched 360-degree videos (e.g., 360_only) for the on-demand course and VR (e.g., VR_only) for the VR course. In this step, the students were asked to watch and evaluate the following videos in order: student characters performing no action (360_only)(VR_only), student characters performing a nodding action (360_nod)(VR_nod), and student characters performing a raggedy action such as shaking their heads (360_darake)(VR_darake). The questionnaire consisted of 8 items with 7 levels.

The results are shown in Fig. 16, tested by Wilcoxon's signed rank test and the Bonferroni method.

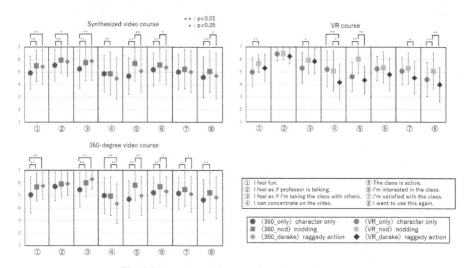

Fig. 16. Results of character behavior evaluation.

Within the on-demand courses, the results were similar; on some items, such as items 2 and 3, the raggedy action was rated as highly as nodding, and on others, such as items 4 and 5, nodding was rated more highly.

That the evaluations of items 3 and 5 differed depending on the actions of the student characters suggests that embodied communication between the students in the 360-degree video and VR was one of the factors that made students feel involved.

5   Conclusion

In this study, we conducted an on-demand class for about 200 students using 360-degree videos and VR with student characters nodding in a 3DCG classroom. Through this study, we showed that an on-demand class using 360-degree videos is feasible and can make students feel a sense of involvement with each other comparable to that felt in a VR class. In addition, a questionnaire survey conducted during the class showed that there is a certain importance of student interaction in a classroom space.

At this stage, there are still technical issues to be resolved, such as how to create videos. However, we would like to introduce 360-degree video streaming as a form of on-demand classes in the future so that students can feel a relationship to others.

Acknowledgement. This research was partially supported by JSPS KAKENHI 20H04096, 20K20121, etc.

References

1. Digital Knowledge Co., Ltd.: Survey report on the urgent introduction of online classes in Universities. https://www.digital-knowledge.co.jp/archives/22823/. Accessed 12 Mar 2021 (in Japanese)
2. Tokyo Shimbun web: <New Corona> : University rejects petition for tuition reimbursement; Nihon University student 'frustrated that nothing has changed'. https://www.tokyo-np.co.jp/article/65489. Accessed 12 Mar 2021 (in Japanese)
3. Newsweek Japan: student sues university over poor quality online classes, demands tuition back. https://www.newsweekjapan.jp/stories/woman/2020/07/post-423.php. Accessed 12 Mar 2021 (in Japanese)
4. Udacity, Inc.: Online tech courses and nanodegree programs. https://www.udacity.com/. Accessed 12 Mar 2021
5. Okada, S., Ando, T.: The effect of interaction with others on learning motivation in adult learners. Technol. Manage. Res. **16**, 17–23 (2017). (in Japanese)
6. Ministry of Education, Culture, Sports, Science and Technology-Japan: The Concept of the New Courses of Study: From Discussions at the Central Council for Education to Revision and Implementation. https://www.mext.go.jp/a_menu/shotou/new-cs/__icsFiles/afield file/2017/09/28/1396716_1.pdf. Accessed 12 Mar 2021 (in Japanese)
7. Digital Knowledge Co., Ltd.: Survey Report on the Implementation of Active Learning at Universities and Vocational Colleges. https://www.digital-knowledge.co.jp/archives/11920/. Accessed 12 Mar 2021 (in Japanese)
8. Wollin, D.D., Montagne, M.: College classroom environment: effects of sterility versus amiability on student and teacher performance. Environ. Behav. **13**(6), 707–716 (1981)
9. Wong, C.Y., Sommer, R., Cook, E.J.: The soft classroom 17 years later. J. Environ. Psychol. **12**(4), 336–343 (1992)
10. School Corporation Kadokawa Dwango Gakuen: N High School. https://nnn.ed.jp/. Accessed 12 Mar 2021 (in Japanese)
11. Watanabe, T., Yamamoto, M.: An embodied entrainment system with InterActors superimposed on images. In: Proceedings of the 11th International Conference on Human-Computer Interaction (HCII 2005), vol. 4, p. 2045 (2005)

12. Yamamoto, M., Watanabe, T.: Development of an edutainment system with InterActors of a teacher and a student in which a user plays a double role of them. In: Proceedings of the 17th IEEE International Symposium on Robot and Human Interactive Communication (RO-MAN 2008), pp. 659–664 (2008)

13. Aoyagi, S., Kawabe, R., Yamamoto, M., Watanabe, T.: Hand-raising robot for promoting active participation in classrooms. In: Yamamoto, S. (ed.) HIMI 2015. LNCS, vol. 9173, pp. 275–284. Springer, Cham (2015). https://doi.org/10.1007/978-3-319-20618-9_27

14. Aoyagi, S., Yamamoto, M., Fukumori, S.: Analysis of hand raising actions for group interaction enhancement. In: Yamamoto, S. (ed.) HIMI 2016. LNCS, vol. 9734, pp. 321–328. Springer, Cham (2016). https://doi.org/10.1007/978-3-319-40349-6_30

15. Guay, F., Boivin, M., Hodges, E.V.E.: Predicting change in academic achievement: a model of peer experiences and self-system processes. J. Educ. Psychol. **91**(1), 105–115 (1999)

16. Freina, L., Ott, M.: A literature review on immersive virtual reality in education: state of the art and perspectives. eLearning Softw. Educ. **1**, 133–141 (2015)

17. Horose, H., Aoyagi, S., Yamamoto, M.: A study on streaming 360 degree videos of classrooms for online classes. In: Human Interface Cyber Colloquium (HIC2 2020) Thesis Collection, pp. 445–456 (2020). (in Japanese)

Research on Perceptual Cues of Interactive Narrative in Virtual Reality

Entang He[1], Jing Lin[2], Zhejun Liu[1(✉)], and Yize Zhang[1]

[1] Tongji University, Shanghai, People's Republic of China
{1933635,wingeddreamer,1752565}@tongji.edu.cn
[2] Shanghai Academy of Spaceflight Technology, 3888 Yuanjiang Road, Shanghai 201109,
People's Republic of China

Abstract. From words, images, films to games, the development of narrative is inseparable from the evolution of media. With the advent of virtual reality, the traditional paradigms of storytelling need to be expanded. VR can simulate the real world vividly to an unprecedented extent, in which users can choose their own perspectives freely during their exploration. However, in addition to great immersion and presence, this powerful new technology also brings new challenges to interactive storytellers who are familiar with the traditional "frame-bound" media. To keep users focused on the storyline efficiently and naturally in VR, new ways of perceptual cues need to be explored and verified. Through a comparative experiment, we tested how well the existing perceptual cues could meet our expectations using "Secondary Mission", an original interactive narrative VR game, as a case for study. Based on our findings, design strategies for planning perceptual cues in VR interactive narratives were proposed to help in improving users' experience.

Keywords: Interactive narrative · Virtual reality · Perceptual cues

1 Introduction

Every time, the emergence of a new medium leads to a new form of narrative and new possibilities of content presentation, which is proven by its history of growth from oral narrative, literary narrative to video narrative [1]. In the age of digital technology, interactive applications and games sprang up rapidly, and the concept of the interactive narrative thus began to form.

Most of the studies on interactive narrative are mainly about the mechanisms of story generation by artificial intelligence, but few people care about user experience. While virtual reality technology appeared as an emerging medium with significant potential to enrich experiences in fields ranging from films to games [2], its high level of presence added greatly to interactive narrative and created a more realistic virtual world for users.

However, compared with finding interactable objects on a 2D screen of limited size, virtual reality technology brings higher demands to properly guided audiovisual experience. The freedom of users to look around in a 360° virtual world and intentionally

© Springer Nature Switzerland AG 2021
S. Yamamoto and H. Mori (Eds.): HCII 2021, LNCS 12765, pp. 283–296, 2021.
https://doi.org/10.1007/978-3-030-78321-1_22

choose their own perspective make it more difficult for users to interact with elements predetermined by VR experience designers [3]. The key question here is how to guide users efficiently to convey a narrative as expected. Furthermore, although there are numerous ways to guide users' attention in an interactive application, it's not easy to create an immersive narrative experience for VR story "listeners". A simple halt with a "click any key to continue" hint is definitely an inferior approach that may spoil the fluency of storytelling.

Perceptual cues used to be a hot research topic in the field of traditional film and video games. They have been proven to be particularly useful in reinforcing the existing focuses of attention or creating new ones to continuously guide the audience to focus on the storyline in a natural way [4]. Perceptual cues attracted more interested researchers when VR films have become popular because this advance of technology also brought about challenges to linear film narration. Filmmakers can no longer rely on cinematography to show the audience the intended plot points when they aspire to create a coherent narrative [5]. The strategies copied from the traditional film and game industries may be applied in VR film production, but they do not take interactivity or the unique characteristics of VR presentation into consideration and therefore can hardly be used directly in VR interactive narrative.

To guarantee that the most important information critical to the storyline reaches the users successfully, interactivity in narrative may be a good "upgrade" to linear VR films and video games. Therefore, how to find a proper way of using perceptual cues to guide users' attention both efficiently and naturally in interactive VR narratives has become an important issue to explore.

2   Related Work

2.1   Interactive Narrative

External media, from spoken language, written words to recorded images, plays a central role in various types of narrative, including dramas, novels and movies, which expand our capacity to share our individual experiences and thoughts [6, 7]. Nowadays, with the emergence of digital media, people become increasingly eager to participate in the stories themselves, and the concept of the interactive narrative thus began to form.

Interactive narrative is one type of digital entertainment in which users' actions can significantly alter the direction or outcome of the narrative experience delivered by a software, so the goal of an interactive narrative system is to immerse users in a virtual world such that they believe to be an integral part of an unfolding story [7]. The most common form of interactive narrative involves the user playing the leading role in an unfolding story in the way of role-playing games.

Interactive narrative has become studied for more than 30 years. A crucial goal of the research in this field is to create an enjoyable experience. It means that the conclusive evaluation criteria to judge the quality of an interactive narrative mainly come from end users. Existing researches mainly cope with the advances in technology, especially in the area of AI (artificial intelligence), artistic achievements and improvements in theoretical understanding, but few cares about user experience (abbr. UX) [8]. Nowadays, with the spread of VR technology to which UX so critical, it's a good timing to explore this issue from a new perspective that places emphasis on UX.

2.2 Virtual Reality

Virtual reality (VR) is a scientific method and technology created to understand, simulate, and better adapt and use the nature [1]. As a new medium, VR can create an immersive user experience with a broad potential for various applications such as films and games. With the advancement and popularity of consumer-grade VR devices such as the Oculus Rift, HTC Vive and PlayStation VR, VR has become a new medium for ordinary people's entertainment, much more accessible than ever before.

Immersion is a very important feature for both interactive narrative and VR, the goals of them are both to immerse users in a virtual world delivered by computer systems. For interactive narrative, immersion is mainly influenced by the structure and content of it, including setting, place, time and plot [9]. But the immersion in VR comes in a different way.

VR can simulate the real world of human beings very vividly. When people search for an object in a real environment, they usually perceive rich and coherent multi-sensory signals, including visions, sounds, smells, tastes, and touches. Unlike traditional media, many virtual reality applications try to emulate this richness in object-searching with multi-sensory signals, typically by combining visual and audio signals, which can create a more immersive experience [10]. Therefore, interactive narrative in VR could provide a brand-new experience of total immersion into various stories from imaginary worlds.

But the transformation of medium also poses challenge for interactive narrative. In VR, people can look at any place in the 360° virtual environment surrounding them in a totally free manner, which makes them feel like being part of that virtual world and creates an impressive experience. However, this freedom of inspection also makes it more difficult for the designers to keep users adhering to the intended storyline reliably. Therefore, it is crucial to find answers to the question of how to guide users' attention effective for the purpose of successful story-telling while keeping them freely immersed in virtual worlds at the same time.

2.3 Perceptual Cues

Most common types of perceptual cues could be further classified as audio and visual cues [11, 12]. They have been widely used in traditional films and video games. But for VR, only some of them are possibly useful in guiding users' attention [4]. Therefore, we carefully picked potentially valuable perceptual cues mainly from two sources: 1) (modified) design guidelines for VR films as a linear medium; 2) design guidelines for video games that may apply to the VR scenario.

From the first source, we investigated current research on perceptual cues for VR films. In the traditional film industry, audiences experience the story through motion pictures shown on a screen with limited size. What they see and hear is pretty predictable. In VR, however, an audience's perception continuously changes with his/her head orientation, and this autonomous movement makes it difficult to predict what he/she sees and hears in a 360° virtual environment. This freedom brings challenges to filmmakers trying to present a narrative with a specific sequence of plots [13]. Therefore, visual and audio cues are particularly crucial for VR filmmakers to guide the audience's attention to make a narrative experience possible. Moreover, VR films turn the audience from a spectator into a part of the story, which coincides with the concept of interactive narrative [7], so the research on perceptual cues in VR films has a great value as references for this study.

Similarly, there are many studies on perceptual cues in video games. A great many of the games on the market today are first-person narrative games and ask players to play roles in the virtual worlds. To prevent players from getting lost in a massive gaming world or missing an important item, perceptual cues in video games are very useful. They can help the game designers think about where to place the item, grab attention, and further decrease players' frustration and increase their engagement [14]. If used properly, they can also help us create an enjoyable experience in the form of VR interactive narrative. But sometimes, video games rely a lot on GUI based information to guide players to make judgments and take actions, and too many abrupt GUI elements can be very distracting and interrupting for users experiencing stories.

The following is a brief summary of current researches on visual and audio cues that a VR head-mounted display (HMD) can provide.

The visual modality has a higher information bandwidth than the other modalities, and it is usually used as a cue for the target-searching tasks in VR [15]. Visual cues include color, size, shape, text, material, motion, lighting, flicker, depth-of-field, blur, visual element, visual focus, alignment, face, gesture, and gaze of characters. Color has been accepted as an effective feature for a long time, and people can search for a target efficiently as long as the target and other objects' colors are not too similar [16]. Size includes features in different dimensions such as length, area and volume. Shape cue is usually used in games, and it is shown as an added virtual object that is not a part of the game world, e.g. an arrow on the ground as a direction guide. Text is a special type of shape cues that is associated with language, and it is generally used as subtitles in films and video games. Material is rarely, if not never, used as a guiding feature for visual search because it is inefficient to guide people to pay attention to one type of material against other materials as distractors [17]. Motion, including linear motion and random motion, is a very effective feature in guiding attention in the complicated scene [18]. Lighting can be an important tool in guiding the user's gaze because one will instinctively look at the position of comfortable brightness because there is usually more information [19]. Flicker be seen as the motion of lighting, and it is a general dynamic change feature [20]. Changing depth-of-field and creating partial blur can immerse people in a virtual environment by reducing the effective visual area [9]. Multiple visual elements, if presented with a hierarchy of importance, can organize the viewer's attention purposefully [21]. As in traditional frame-bound medium, creating a specific visual focus in both space and time could guide the viewer's attention effortlessly [21]. Aligning the plot points or perspectives of the scene in the same direction can effectively guide the viewer's attention towards the intended orientation during scene transitions [9, 21]. Finally, human faces are quite easy to find among other objects, but it is hard to say how they can work as a guiding feature. A review by Frischen argued that facial expressions of emotion influence the process of visual searching [22], and Pillai found that the character directly facing the viewer creates and sustains interest [21]. The character's gaze can also guide the viewer's attention to a particular position [9] or lead them into another scene [19].

Although audio is less efficient in conveying information than images, it can increase the presence of the virtual environment [4] and guide users from one area of the scene to another effectively [23]. Audio cues include spatial sound effects, non-spatial sound effects and conversations. Spatial sound effects can also refer to those diegetic sound that exists in the narrative and 3D environment, which could be the most effective feature to guide one's attention when the target object is out of his/her sight [9]. Non-spatial sound effects can refer to those non-diegetic sounds from outside the narrative, such as background music, that is usually used to render the story's atmosphere and affect people's feelings [19].

Current research on perceptual cues is limited. Most studies only divide them into two categories according to modality, namely visual and audio cues. It might work in some cases, but is not likely to explain most of the questions regarding user experience. To understand better how perceptual cues could affect user experience in VR interaction narratives, a new framework needs to be established for the purpose of design and evaluation.

3   Our Framework

3.1   Efficiency, Naturalness and Robustness

We assumed that there are three factors to consider in designing the way of guiding users' attention in VR interactive narrative.

The first factor is *efficiency* required by the medium virtual reality itself. The freedom of inspection in VR as mentioned above has changed the traditional method of narrative and made it very important to guide users' attention efficiently so as to keep them focused on the storyline. How to grab users' attention successfully without consuming too much time and effort from them is the key to guarantee a more coherent narrative. This is one of the most important goals of perception cues as far as efficiency is concerned.

The second factor is *naturalness* required by the narrative. An interactive narrative in VR aims to immerse users in a virtual world so vivid that they begin to believe to be an integral part of a story [7]. However, an abrupt and unreasonable hint in a VR experience is likely to kick a user out of an unfolding story. To make a user feel that the virtual environment conforms with his/her understanding of the real world to ensure an immersive narrative experience is an important requirement of perceptual cue design regarding naturalness.

Usually, there is a trade-off between efficiency and naturalness. To balance between these two factors and make full use of them, we need to consider how to integrate the perceptual cues with the narrative of the virtual environment but also make them stand out from the environment visually at the same time [24], which is undoubtedly challenging.

The third factor is *robustness*. This word comes from computer science and means the ability of a system to resist changes and maintain its initial stability. As for VR interactive narrative, the initial stability of *efficiency* and *naturalness* are commonly confronted with the changes brought about by interactivity.

In a VR experience, users can change not only head orientation but also body location in the virtual environment. They not only see different objects in the same location but also observe the same object from different locations in different directions. This change of perspectives may highly likely invalidate a perceptual cue, so they should be considered in a 3D environment rather than from a stationary point of view. Secondly, even if a user is told to search for a target, he/she cannot recognize the basic feature of it clearly if it's far away. Therefore, attention guidance in VR should happen in a continuous manner, which means guiding a user from one location to another with perceptual cues continuously until they reach the target destination, helping to create a coherent narrative experience.

The studies on perceptual cues from the film and game industries are mostly dealing with efficiency and naturalness.

In VR films, for example, filmmakers need to guide the audience to focus on the storyline, but this cannot be done by telling them what to look at in a direct but clumsy way. Since audience passively accepts stories from the films, VR filmmakers usually follow the bottom-up attention theory and use some visual or audio properties of some aspects in the scene that attract more attention than others [25]. These properties, including movement, color and brightness, usually attract attention subconsciously [26]. Therefore, it is not only possible but also necessary for the perceptual cues in VR films to be natural.

Unlike VR films, video games usually tell a player more directly what to search for, and the player needs to actively explore the story by interacting with the objects in the environment. As a result, video games usually follow the top-down attention theory that biases people towards choosing the object related to their goal when they are observing a scene [25]. Studies have shown that the top-down method controls a player's attention more than the bottom-up method if the video game is highly goal-oriented [14]. Therefore, the design of perceptual cues in games are usually more efficient.

Interactive narrative in VR is more like the experience between VR films and video games/VR games. Compared with VR films, it offers more interactivity for the user to explore the story actively, while compared with video games, it places a higher demand on the naturalness of narrative.

In conclusion, efficiency and naturalness are the first two factors to be considered in VR interactive narrative, and they decide how the perceptual cues can guide users to make judgments and take actions at a specific moment. Robustness is the third factor to be considered. It decides how perceptual cues should be planned continuously across the entire 3D environment and provide users with a coherent narrative experience.

3.2 Reclassification

In this study, we try to propose our design strategies about the perceptual cues of VR interactive narrative deriving from the design guidelines for VR films and video games. To make the analysis clearer and more focused, we screened out all those visual and audio cues not applicable to VR, then further categorized them as contrast cues, dynamic cues, character cues and scene-based cues.

Contrast cues are the most significant type of perceptual cues, which refer to those features of different forms. Contrast cues mainly include color, lighting, size, shape, text, material and non-spatial sound. Dynamic cues refer to those changing features like motion, flicker and spatial sound. Character cues, which are features only belong to characters in the virtual world, include the gestures, faces, conversations and gaze of characters. Scene-based cues refer to features taking place in structured scenes, including depth-of-field, blur, visual element, visual focus and alignment. Although they may not be guiding features in their own right, they can modulate the effectiveness of other features [8]. For example, playing with the depth-of-field or creating partial blur can change the apparent size and guide people's attention with it.

This classification is mainly based on how these perceptual cues function instead of through which sensory channel they work, and works more effectively from the perspective of a designer.

4 The Experiment

4.1 Design the VR Interactive Narrative

"Secondary Mission" is an original interactive narrative VR game developed by ourselves. It tells a sci-fi story about a survivor of a space catastrophe trying to rescue the data storage device and collect other floating objects from scattered debris after his space station was shattered in an accident. There is another surviving astronaut to keep in touch with during the process. Since the environment of this game, space and space station, confronts almost everyone with an unfamiliar environment, perceptual cues become especially important in this situation, making it a very good candidate as the test material. Immersed in the unfamiliar dark space, a player desperately needs perceptual cues to help them accomplish the task.

First, it's necessary to determine which perceptual cues should be selected to include in the experiment. For all types of cues in these four categories (contrast cues, dynamic cues, character cues and scene-based cues), we estimated their efficiency [25] and naturalness using heuristic evaluation. Then we chose five cues that are believed to have better performance in both efficiency and naturalness, including linear motion, lighting, color, visual focus and spatial sound. We also chose another five cues that may perform weakly in one or both aspects, including random motion, shape, material, text and non-spatial sound. Moreover, since other researchers believe that the combination of cues performs better than a single one in guiding attention [9] in a VR interactive narrative, we combined the color and spatial sound cues, and also the material and non-spatial sound cues.

We used two versions of the game in a comparative experiment to verify if these chosen perceptual cues meet our expectations. (See Fig. 1. and Fig. 2.) We labelled them version A (using linear motion, lighting, color and spatial sound, visual focus) and version B (using random motion, shape, material and non-spatial sound, text).

In this experiment, totally four objects could be found, namely three personal belongings and a data storage device. Each object corresponded to one type of cue. The first object was a medical kit placed near the start location, and it was used to compare linear motion cue and random motion cue. The second object was a desktop planter to compare lighting cue and shape cue, with which we wanted to continuously guide the user with a series of cues. In version A, we used several glowing space capsules on the path as a

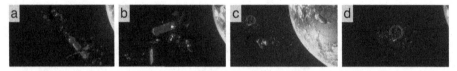

Fig. 1. Version A: (a) linear motion, (b) lighting, (c) color and spatial sound, (d) visual focus.

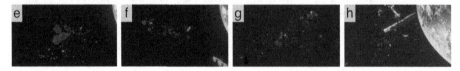

Fig. 2. Version B: (e) random motion, (f) shape, (g) material and non-spatial sound, (h) text.

guide; in the other version, we placed several arrows close to the same group of space capsules which were dark this time. The third object was a phonograph hidden inside a large cluster of debris and other trivial objects, which made this task very difficult. Therefore, we use the combination of visual and audio cues as mentioned above, namely "color and spatial sound" in version A and "material and non-spatial sound" in version B. In version A the phonograph had a red tint, making it different from the surrounding white debris, and sent out a spatial sound locatable through the headset. In version B, the phonograph was white but had a very rough surfacing, different from the metallic debris around it. It was also hearable, but the sound was 2D, which means it did not change according to the position of the player. The final object was the data storage device used to compare the text cue and the visual focus cue. In version A, the device was placed at the center of the ring part of the space station, making it naturally the intended focus of vision. In version B, it was placed in a location that was not special, but there was a 3D text close to it reading "data core".

Participants must pilot the spacecraft to find the data storage device in order to complete the mission. Finding the other three personal belongings was optional (Fig. 3.).

Fig. 3. The cockpit of spacecraft.

Once the player found an object, a conversation with the surviving astronaut would be triggered to offer the player more information about the story.

To avoid *learning effect* and keep the irrelevant variables under control, we built slightly different scenes for two versions. The locations of the four objects in each scene are nudged by small distances, but the participants' start location, the order of the four objects, and the distance between them are same or similar. (see Fig. 4 and Fig. 5).

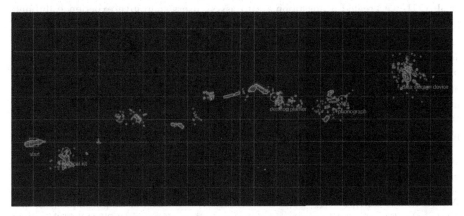

Fig. 4. The locations of the start point and four objects in version A.

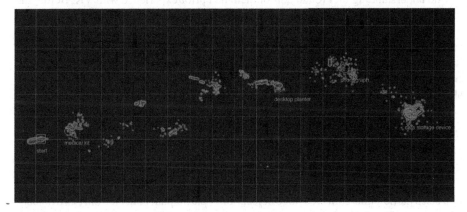

Fig. 5. The locations of the start point and four objects in version B.

4.2   Measures

We measured the completion time of each task and the accuracy of them (how many objects they found) as the indicator of *efficiency* [26]. As for naturalness, it is quite difficult to measure it objectively. Existing studies mostly talk about how to make cues more natural [24] rather than how to measure their naturalness. By definition, a perceptual cue is natural if its existence in the scene conforms to the real-world logic. But this kind of measurement does not take the difference among users into consideration. For example, some people may seem a glowing arrow on the ground to be unnatural, while others may feel the opposite. Therefore, we decided to use an interview to figure out the naturalness of perceptual cues. By using words like "abrupt," "natural," "reasonable", we made sure the participants understood the question correctly and enlighten them to think about the question "Do you think this guiding method is natural for you? And why?" more reasonably.

Based on the dimensional model of the user experience proposed by Christian Roth et al. [27], we created a questionnaire to assess users' narrative experience through five dimensions: Curiosity ("I was curious about what will happen next during the experience."), Suspense ("I was worried about whether I can find the target object."), Aesthetic Pleasantness ("I think the graphics I saw during the experience gave me great pleasure."), Self-enhancement ("I felt a sense of accomplishment after the experience.") and Optimal Task Engagement ("I felt completely detached from the real world during the experience."). There were some other questions asking about preference ("Which VR interactive narrative experience do you prefer? And why?") and basic demographic information.

4.3 Participants and Procedure

A total of 10 participants (7 males and 3 females) were invited as participants in this pilot study. They were all university students with a median age of 23. All of them had a basic understanding of VR and half of them had prior experience with VR. Before the experiment, they were briefly informed of the background of the story and how to pilot the spacecraft. Participants were asked to test the two versions in a random order. Each session lasted about 10 min, followed by a questionnaire. We also recorded how long it took each participant to find the objects and the number of objects found. After the two sessions were finished, the participants were asked to give their subject opinions in short interviews.

5   Results

From the result of the number of objects collected by the participants, the cues of version A are more effective than version B. With version A, 6 participants collected all the four objects, 3 participants collected three objects and 1 participant collected only two objects. With version B, only 3 participants collected all the four objects, 3 participants collected three objects, and 4 participants collected two objects. In terms of the efficiency of each optional object being collected, more than 7 participants collected the first two objects, and it showed that motion, lighting and shape are all very effective cues, especially all the participants collected the object guided by the lighting. For the third object, 7 participants collected it with version A, but only 3 participants collected it with version B, which showed that the non-spatial sound is inefficient in searching for objects.

From the result of the task's completion time, the first and second objects showed the most meaningful difference. The mean value suggested that the linear motion (average time = 45.63 s) was more efficient than the random motion (average time = 56.00 s), and the lighting (average time = 207.8 s) was more efficient than the shape (average time $= 237.8$ s). As for the third and fourth objects, because it was difficult to recognize precisely the moment a participant began his/her search, no conclusion could be drawn reliably.

The narrative experience questionnaire gave the following result (Table 1).

Table 1. Results from the narrative experience questionnaires.

	Curiosity avg.	Suspense avg.	Aesthetic pleasantness avg.	Self-enhancement avg.	Optimal task engagement avg.
Version A	3.6	2.9	3.7	3.5	3.9
Version B	3.4	3.3	3.9	3	3.6
A > B	n = 4	n = 4	n = 0	n = 4	n = 3
A = B	n = 4	n = 1	n = 8	n = 4	n = 7
A < B	n = 2	n = 5	n = 2	n = 2	n = 0

As the result demonstrates, version A performed better in *Curiosity*, *Self-enhancement* and *Optimal Task Engagement*, but slightly worse in *Suspense* and *Aesthetic Pleasantness*.

According to the answer to the question "Which VR interactive narrative experience do you prefer? And why?", 7 out of 10 participants preferred version A. Unexpectedly, the participants who had much experience with VR preferred version B because they preferred the gamified experience brought by the GUI like arrows and text. Some of them claimed that clear guidance like arrows and text would make them feel more secure in the space environment and they did not want to spend too much effort searching for objects. But there were also other participants complained that the arrows were very abrupt and dragged them out of the experience.

In the short interview, we asked participants about their subjective opinions about the efficiency and naturalness of each perceptual cue. Most participants thought that linear motion and visual focus cues were more natural but less efficient, while random motion and text cues were the opposite. All the participants thought that lighting and spatial sound cues were very natural, color cue was very efficient, and spatial sound cue was natural and efficient at the same time. Material cue was complicated, most participants thought that the efficiency of it as a guide depended on the context.

6 Discussion and Design Suggestions

According to the experiment results, we can positively confirm the value of well-designed perceptual cues may improve the user experience of VR interactive narrative. They can help users to track down key clues in the story with less effort and make it less likely for them to miss items or plots of great importance.

From the sophisticated fields like film or game creation, we may find a large collection of perceptual cues at our disposal. However, not all these cues perform similarly as far as VR interactive narrative is concerned. Sometimes, the requirement of efficiency and naturalness even conflict with each other. For instance, spatial sound was believed by many to be very natural, but to spot precisely the location of sound is not very easy for most people and thus making it not that efficient. On the other hand, a text with an arrow might be very efficient, but this kind of "incompatible" cue may easily and strongly remind the user of the artificial nature of the environment. To understand the advantages and disadvantages of each type of cues and to use and tweak them wisely might be more important than easily conclude which one is superior.

Our experiment also suggested efficient and natural perceptual cues performed better in *curiosity, self-enhancement* and *optimal task engagement*. We believe that it was due to the fact that more efficient and natural perceptual cues made the game process smoother and kept the participant more focused on the story itself instead of being distracted or interrupted by the being cued. As for *aesthetic pleasantness*, the difference was minor: 8 out of 10 participants gave the same score and 2 thought version B was better. It might suggest that perceptual cues did not have a direct impact on aesthetic value. Version B also outperformed A in *suspense*, but we don't think the result is very meaningful because this story was too short to incubate suspense.

Finally, an unexpected result showed that experience VR users preferred efficient but unnatural cues more than the others. A reasonable guess is that a player's mental definition of the experience plays an important role in their appraisal of the perceptual cues. When the experience is more like a storytelling film, naturalness is the most important factor that helps to maintain the presence. However, when the experience is perceived as a game, as those experience VR gamers did in this experiment, efficiency tends to overwhelm naturalness in importance. Since the test material we used in this research was in the transitional area on the spectrum, the participants were likely to understand it differently.

Based on our findings, we propose the following design suggestions for planning perceptual cues in VR interactive narrative, which might be useful for those who want to create an efficient and natural narrative experience.

(1) Lighting cue is an implicit one that guides users' attention naturally, and it remains very efficient regardless of its distance to the user in a 3D environment.
(2) Spatial sound cues immerse users in the virtual environment better by emphasizing the atmosphere, which can be very helpful for narratives. When combined with visual cues, it can be even more efficient and natural.
(3) Shape and text cues are very clear and efficient. They can effectively avoid users from feeling lost, especially in an unfamiliar environment. However, they can also be very unnatural and jeopardize users' feeling of presence easily.
(4) Material cues are very uncertain features that need sufficient test before being used in a VR narrative experience.
(5) Visual focus cues in structured scenes are also very implicit and clever. They are usually efficient but can easily fail to work when viewed from a perspective different than the intended one, and therefore making sure via experience design that your players always encounter the cues as expected.
(6) Many scene-based cues such as blur and visual element common to film and games might not be applicable to VR interactive narrative.

7 Conclusion

In this study, we explored the possibilities of perceptual cues to improve user experience in VR interactive narrative and proposed our design suggestions.

We selected perceptual cues from the fields of VR films and games, analyzed them using three coefficients, namely "efficiency", "naturalness" and "robustness", and classified them into four categories, namely "contrast cues", "dynamic cues", "character cues" and "scene-based cues". We hope this effort provides a new perspective of understanding how perceptual cues might be functional in the scenario of VR interactive narrative.

The experiment we carried out yielded results that met our expectations to a certain extent. Most of the participants believed that they had a more immersive experience and spent less time accomplishing the task when the perceptual cues met the requirement of efficiency and naturalness. However, there are still many limitations in this study. Firstly, the test material we used was too short to incubate suspense or include character cues, which limited the scope of this research. Secondly, this pilot study only involved 10 participants and the results need further verification with a larger sample size. Anyway, we believe more future researches are needed in the field of interactive narrative in VR to meet with the rapid advancement and blooming popularity of this technology.

References

1. Murray, J.H.: Research into interactive digital narrative: a kaleidoscopic view. In: Rouse, R., Koenitz, H., Haahr, M. (eds.) ICIDS 2018. LNCS, vol. 11318, pp. 3–17. Springer, Cham (2018). https://doi.org/10.1007/978-3-030-04028-4_1
2. Desurvire, H., Kreminski, M.: Are game design and user research guidelines specific to virtual reality effective in creating a more optimal player experience? Yes, VR PLAY. In: Marcus, A., Wang, W. (eds.) DUXU 2018. LNCS, vol. 10918, pp. 40–59. Springer, Cham (2018). https://doi.org/10.1007/978-3-319-91797-9_4
3. Ko, D.-U., Ryu, H., Kim, J.: Making new narrative structures with actor's eye-contact in cinematic virtual reality (CVR). In: Rouse, R., Koenitz, H., Haahr, M. (eds.) ICIDS 2018. LNCS, vol. 11318, pp. 343–347. Springer, Cham (2018). https://doi.org/10.1007/978-3-030-04028-4_38
4. Brillhart, J.: The language of VR (2016)
5. Nielsen, L.T., et al.: Missing the point: an exploration of how to guide users' attention during cinematic virtual reality. In: Proceedings of the 22nd ACM Conference on Virtual Reality Software and Technology, pp. 229–232, November 2016
6. Zhao, Q.: A survey on virtual reality. Sci. China Ser. F Inf. Sci. **52**(3), 348–400 (2009)
7. Riedl, M.O., Bulitko, V.: Interactive narrative: an intelligent systems approach. AI Mag. **34**(1), 67 (2013)
8. Koenitz, H.: Five theses for interactive digital narrative. In: Mitchell, A., Fernández-Vara, C., Thue, D. (eds.) ICIDS 2014. LNCS, vol. 8832, pp. 134–139. Springer, Cham (2014). https://doi.org/10.1007/978-3-319-12337-0_13
9. Pillai, J.S., Verma, M.: Grammar of VR storytelling: analysis of perceptual cues in VR cinema. In: European Conference on Visual Media Production, pp. 1–10 (2019)
10. Chen, T., Wu, Y.S., Zhu, K.: Investigating different modalities of directional cues for multi-task visual-searching scenario in virtual reality. In: Proceedings of the 24th ACM Symposium on Virtual Reality Software and Technology, pp. 1–5, November 2018

11. Lescop, L.: Narrative grammar in 360. In: 2017 IEEE International Symposium on Mixed and Augmented Reality (ISMAR-Adjunct), pp. 254–257. IEEE, October 2017

12. Marples, D.: The influence of intrinsic perceptual cues on navigation and route selection in virtual environments (Doctoral dissertation, University of Huddersfield) (2017)

13. Xu, Y., et al.: Gaze prediction in dynamic 360 immersive videos. In: Proceedings of the IEEE Conference on Computer Vision and Pattern Recognition, pp. 5333–5342 (2018)

14. El-Nasr, M.S., Yan, S.: Visual attention in 3D video games. In: Proceedings of the 2006 ACM SIGCHI International Conference on Advances in Computer Entertainment Technology, pp. 22–es, June 2016

15. Way, T.P., Barner, K.E.: Automatic visual to tactile translation. I. human factors, access methods and image manipulation. IEEE Trans. Rehabil. Eng. **5**(1), 81–94 (1997)

16. Nagy, A.L., Sanchez, R.R.: Critical color differences determined with a visual search task. JOSA A **7**(7), 1209–1217 (1990)

17. Wolfe, J.M., Myers, L.: Fur in the midst of the waters: visual search for material type is inefficient. J. Vis. **10**(9), 8 (2010)

18. Chun, M.M., Wolfe, J.M.: Visual attention. Blackwell handbook of perception, 272310 (2001)

19. Li, Y.M., Joo, J.W.: The change of animation narrative structure caused by virtual reality technology. J. Digit. Contents Soc. **20**(3), 459–467 (2019)

20. Kunar, M.A., Watson, D.G.: Visual search in a multi-element asynchronous dynamic (MAD) world. J. Exp. Psychol. Hum. Percept. Perform. **37**(4), 1017 (2011)

21. Pillai, J.S., Ismail, A., Charles, H.P.: Grammar of VR storytelling: visual cues. In: Proceedings of the Virtual Reality International Conference-Laval Virtual 2017, pp. 1–4, March 2017

22. Frischen, A., Eastwood, J.D., Smilek, D.: Visual search for faces with emotional expressions. Psychol. Bull. **134**(5), 662 (2008)

23. Cockburn, A., Brewster, S.: Multimodal feedback for the acquisition of small targets. Ergonomics **48**(9), 1129–1150 (2005)

24. Dillman, K.R., Mok, T.T.H., Tang, A., Oehlberg, L., Mitchell, A.: A visual interaction cue framework from video game environments for augmented reality. In: Proceedings of the 2018 CHI Conference on Human Factors in Computing Systems, pp. 1–12, April 2018

25. Wolfe, J.M., Horowitz, T.S.: Five factors that guide attention in visual search. Nat. Hum. Behav. **1**(3), 1–8 (2017)

26. Chen, T., Wu, Y.S., Zhu, K.: Investigating different modalities of directional cues for multi-task visual-searching scenario in virtual reality. In: Proceedings of the 24th ACM Symposium on Virtual Reality Software and Technology, pp. 1–5, November 2018

27. Roth, C., Vorderer, P., Klimmt, C.: The motivational appeal of interactive storytelling: towards a dimensional model of the user experience. In: Iurgel, I.A., Zagalo, N., Petta, P. (eds.) ICIDS 2009. LNCS, vol. 5915, pp. 38–43. Springer, Heidelberg (2009). https://doi.org/10.1007/978-3-642-10643-9_7

Avatar Twin Using Shadow Avatar in Avatar-Mediated Communication

Yutaka Ishii[1]([⊠]), Satoshi Kurokawa[2], and Tomio Watanabe[1]

[1] Okayama Prefectural University, Kuboki 111, Soja, Okayama, Japan
{ishii,watanabe}@cse.oka-pu.ac.jp
[2] Graduate School of Okayama, Prefectural University, Kuboki 111, Soja, Okayama, Japan
satoshi@hint.cse.oka-pu.ac.jp

Abstract. Shadow is an effective projection for presence of real objects, and has important roles in three-dimensional effects and reality of virtual space. Therefore avatar's shadow would be able to have interactive effects with the avatar's self in a virtual space. In our previous research, we reported that an auto-generated interactive avatar's motion which is different from the talker's own motion not always supports human communication. However, by using an avatar-shadow, talkers could communicate effectively without inconsistency between their own motions and their avatar's motions in a virtual space, in which the avatar-shadow's motions are expressed by combining the talker's own motions with auto-generated interactive motions. Accordingly we propose an embodied entrainment system called "Avatar Twin" using shadow avatar. This system is constructed by talkers' shadow avatars based on their own motions, and their avatars based on auto-generated entrained motions, in addition to the previous avatar-shadows. In previous studies, an avatar was a subject as a substitute for the user, and the main purpose was to investigate the effects of shadow support in communication. In this paper, we propose the master-slave inversed approach by swapping between measured and entrainment motions in shadow avatar communication system, and develop the system prototype.

Keywords: Shadow avatar · Embodied communication · Entrainment

1 Introduction

With the development of information technology, including the influence of coronavirus, avatar technology that exists as oneself in a remote space is rapidly developing. Takeuchi et al. propose a telework called "avatar work" using avatar robot that enables people with disabilities to engage in physical works such as customer service [1]. In avatar work, disabled people can remotely engage in physical work by operating a proposed robot with a mouse or gaze input. In addition, voice interaction agents have been developed, and communication interfaces that allow people to experience face-to-face interaction in a remote space have been proposed [2, 3]. Therefore, various studies have been conducted to clarify the relationship between avatars and user operations in practical situations. In

© Springer Nature Switzerland AG 2021
S. Yamamoto and H. Mori (Eds.): HCII 2021, LNCS 12765, pp. 297–305, 2021.
https://doi.org/10.1007/978-3-030-78321-1_23

particular, Narumi proposes "Ghost Engineering" for enabling us to change our cognitive functions as we hope by modifying our body perception and recognition [4].

On the other hand, shadows help to create the visual sense of a talker's identity, in real situations [5]. In virtual reality, they play an important part in the perceived three-dimensionality of computer-generated (CG) characters and objects, as well as simulating other real visual effects. Shadows are particularly useful in avatar-mediated communication. A shadow cannot exist independently of the avatar that is representing the talker. For example, one talker can react to another's avatar based on the behavior of the avatar's shadow rather than that of the avatar itself. This could encourage the entrainment of rhythms embodied in the speaker's voice to improve the ease of communication.

In this study, we introduce avatars and shadows into an embodied virtual communication system (EVCOS) for human interaction analysis by synthesis. In our previous system, an avatar's motions were based on the talker's own motions, and the avatar-shadow's motions are constructed on the basis of the talker's own motions and also on auto-generated nodding motions by using InterRobot Technology (iRT) [6]. This system can enhance the effectiveness of communication by avatar-shadow's nodding. This is because the avatar-shadow's nodding encourages the talker's embodied entrainment, and the talker's motions do not contradict the avatar's motions by using the avatar-shadow.In addition, we have developed an embodied avatar-shadow system using color to express talkers' feelings based on an interaction-activated communication model, and have confirmed the effectiveness of the system by the sensory evaluation of a communication experiment [7]. In an embodied interaction, both speech and nonverbal behavior such as nodding and body movements are rhythmically related and mutually synchronized between the talkers.

In this paper, we propose an embodied entrainment system called "Avatar Twin" using shadow avatar. This system is constructed by talkers' shadow avatars based on their own measured motions, and their avatars themselves based on auto-generated entrained motions. In previous studies, an avatar was a subject as a substitute for the user, and the main purpose of these studies was to investigate the effects of shadow support in communication. Accordingly, we propose the master-slave inversed approach by swapping between measured and entrainment motions in shadow avatar communication system, and develop the system prototype.

2 Avatar Twin Using Shadow Avatar

2.1 Concept

Figure 1 shows the concept of Avatar Twin using shadow avatar. In real space, a shadow cannot be independent of the object that creates it. As mentioned above, a slave shadow is used helpful for visual effects to impress the embodiment of master object. We propose a communication system that uses shadow avatars as masters by replacing the function that reproduces the original body movements of avatars. The slave avatar gives awareness of role-play communication by its appearance and promotes rhythm synchronization by the automatically generated entrainment motions. The master shadow avatar directly reflects the user's own actions. It can be edited into a characteristic expression by being limited to the motion presentation of one color.

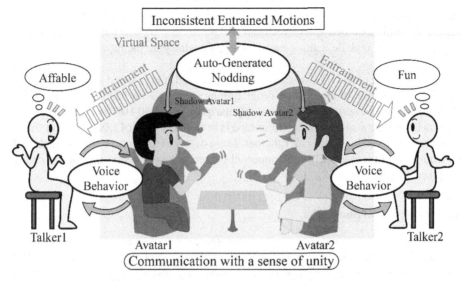

Fig. 1. Concept of Avatar Twin using shadow avatar.

When an avatar character performs nodding responses based on iRT explained in the Sect. 2.4, it activates embodied entrainment between talkers via virtual space. The nodding of talkers occurs thereby the nodding responses of avatar. The avatars and the shadow avatars that include the embodied entrainment create a communication environment that is more united. In general, avatars in virtual space can freely change their appearance characteristics, and there are many scenes in which communication is performed by impersonating the character. However, on the contrary, its appearance characteristics may be a constraint on role play. Therefore, by changing the shadow avatar without changing the appearance of the avatar, it is possible to make the illusion that the avatar has changed. In addition, it is possible to influence the communication environment by adding characteristic changes that cannot be expressed by avatars to the shadow avatar as the subject.

2.2   Embodied Virtual Communication System (EVCOS)

VirtualActor (VA) is an interactive avatar that represents the talker's upper body motions and voice on the basis of his or her verbal and nonverbal information in a virtual face-to-face communication environment [9]. Two remote talkers can communicate through their respective VAs. The motions of head, both arms, and body for each VA are accurately represented on the basis of the positions and angles measured by four magnetic sensors (Polhemus FASTRAK) placed on the top of talker's head, both wrists, and back. The VAs are seated on opposite chairs in the room in the virtual space section. Talkers can communicate as they observe the interaction of upper body motions and the voice, through VAs in the same space. The creation of VAs focuses on the embodied rhythms of communication. Talkers can confirm a correspondence between each talker and each VA by the behaviors of both VAs. In addition, in EVCOS, the analysis-by-synthesis for interaction in communication is performed by processing the behavior of the VAs under various conditions of their spatial relations and positions [10]. The behavior processing of the VAs could include actions such as cutting or delaying the motions and voice of the VAs.

The rendering of the 3D characters can be performed at high speed by using DirectX, which is a collection of application programming interfaces for handling tasks related to multimedia. The system is sampling angular position of 30 Hz from four magnetic sensors to operate avatars. Each user's voice is sampled at 16 bit 11,025 Hz, and transmitted and received by the gigabit Ethernet. These data are recorded on the hard disk, and transmitted and received between connected two computers directly in order to create the place of communication.

2.3 Rendering of Shadow

A shadow avatar is generated with a semi-transparent gray texture on polygons of VA. This texture is expressed like a real shadow which is projected on a floor and a wall. The shadow avatar is rendered on the wall, so that it is easily visible to talkers. Moreover, the shadow avatar is also rendered on the floor to represent the part extending to the wall. The rendering of the shadow avatar on the floor is 2-dimensional. The shadow avatar on the wall is rendered in the "billboard" style, which is one of types of rendering. The parts of polygons, which have a depth value that is smaller than the threshold of depth buffer test, are drawn. If the drawn polygons are behind the other ones, they overlap in 2-dimension. As a result, the colored part of the overlapping polygons is deeper than the other parts, and the entire shadow avatar becomes uneven in color. We apply a stencil buffer test to the shadow avatar, to mask the color of overlapped polygons. The stencil buffer test is a rendering method that configures in pixels drawing and part of overlapped polygons trying not to draw. This way makes the shadow avatar get an even color, by eliminating the unevenness due to overlapping polygons. The change of the color and improved evenness after the stencil buffer test are shown in Fig. 2. Then Fig. 3 shows the slave avatar-shadow motion by superimposing the nodding response on the VA's motions, on the basis of the speaker's speech.

Without stencil buffer test.   With stencil buffer test.

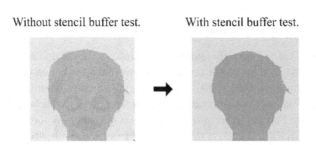

Fig. 2. Enhancement of shadow avatar after the stencil buffer test.

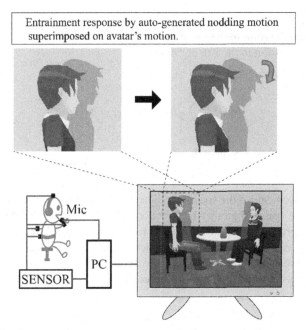

Fig. 3. Embodied entrainment motion in the avatar-shadow system.

2.4 Interaction Model for Auto-generated Entrained Motion

A listener's interaction model of the slave avatar includes a nodding reaction model [8] that estimates the nodding timing from a speech ON-OFF pattern and a body reaction model linked to the nodding reaction model. A hierarchy model consisting of two stages, macro and micro (Fig. 4), predicts the timing of nodding. The macro stage estimates whether a nodding response exists or not in a duration unit that consists of a talkspurt episode $T(i)$ and the following silence episode $S(i)$ with a hangover value of 4/30 s. The estimator $M_u(i)$ is a moving-average (MA) model, expressed as the weighted sum of unit speech activity $R(i)$ in (1) and (2). When $M_u(i)$ exceeds the threshold value, the nodding $M(i)$ is also an MA model, estimated as the weighted sum of the binary speech signal $V(i)$ in (3). The body movements are related to the speech input at a timing over the body threshold. The body threshold is set lower than that of the nodding prediction of the MA model, that is expressed as the weighted sum of the binary speech signal to nodding. The mouth motion is realized by a switching operation synchronized with the burst-pause of speech. In other words, when the InterActor works as a listener for generating body movements, the relationship between nodding and other movements is dependent on the threshold values of the nodding estimation.

$$M_u(i) = \sum_{j=1}^{J} a(j)R(i-j) + u(i) \tag{1}$$

$$R(i) = \frac{T(i)}{T(i) + S(i)} \tag{2}$$

$a(j)$: linear prediction coefficient.
$T(i)$: talkspurt duration in the i-th duration unit.
$S(i)$: silence duration in the i-th duration unit.
$u(i)$: noise

$$M(i) = \sum_{k=1}^{K} b(j) V(i-j) + w(i) \tag{3}$$

$b(j)$: linear prediction coefficient.
$V(i)$: voice.
$w(i)$: noise.

The body movements of the speaker are also related to the speech input by operating both the neck and one of the other body actions at a timing over the threshold, that is the speaker's interaction model estimates as its own MA model of the burst-pause of speech to the entire body motion. Because speech and arm movements are related at a relatively high threshold value, one of the arm actions in the preset multiple patterns is selected for operation when the power of speech is over the threshold.

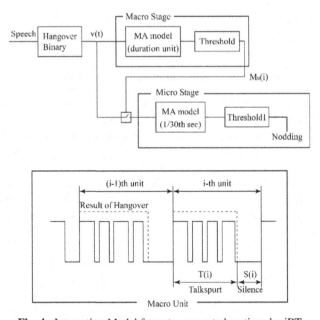

Fig. 4. Interaction Model for auto-generated motions by iRT.

2.5 Character and Room Models

A 3D modeling software Metasequoia is used for modeling the characters and mapping textures in the embodied virtual communication system. Skeletal animation technology, which configures virtual joints and bones, is used so that the CG characters have skeletal structures similar to a human body. The weight of each apex of a CG character's surface affected by virtual bones is set to express real and smooth movements and changes in the appearance of clothes and skin. We have prepared human avatars for daily scene (Fig. 5, 6) and nurse and patient avatars for nursing communication education to provide nursing students with a virtual experience [11].

Fig. 5. Avatar characters in the embodied virtual communication system.

Fig. 6. Virtual cafe room as a daily scene.

3   Development of the System Prototype

Figure 7 shows the system prototype developed in this study. In the shadow avatar system, the avatar operates as a voice-driven embodied entrainment character, and the shadow avatar links the movements of the user measured by magnetic sensor. In addition, the shadow avatar is separated from the avatar and is located outside the avatar. In this research, the system has a function to change the size of the shadow avatar according to the amount of speech of the user as a communication feature to be expressed using the shadow Fig. 7 (b). Each speaker's utterance would be controlled by changing the size of the shadow avatar of the conversation partner according to the speaker's utterance volume.

(a) Initial state.   (b) Larger character state of speaker.

Fig. 7. Example of communication scene using the prototype.

4 Conclusion

In this paper, we propose an embodied entrainment system called "Avatar Twin" using shadow avatar, and develop the system prototype. This system is constructed by talkers' shadow avatars based on their own measured motions, and their avatars themselves based on auto-generated entrained motions. In the future, the system evaluation will be conducted by role-play communication using various avatars.

Acknowledgments. This work was supported by JSPS KAKENHI Grant Number 19K12067.

References

1. Takeuchi, K., Yamazaki, Y., Yoshifuji, K.: Avatar work: telework for disabled people unable to go outside by using avatar robots "OriHime-D" and its verification. In: Companion of the 2020 ACM/IEEE International Conference on Human-Robot Interaction (HRI 2020 Companion) (2020). https://doi.org/10.1145/3371382.3380737
2. Kawamoto, S., et al.: Design of software toolkit for anthropomorphic spoken dialog agent software with customization-oriented features. IPSJ J. **43**(7), 2249–2263 (2002). (in Japanese)
3. Lee, A., Oura, K., Tokuda, K.: An open-source toolkit realizing attractive voice interaction systems. MMDAgent. IEICE Technical Report, NLC2011–51, SP2011–96, vol. 111, no. 364, pp. 159–164 (2011). (in Japanese)
4. Narumi, T.: Ghost engineering: designing our cognitive functions by modifying our body. Cogn. Sci. **26**(1), 14–29 (2019). (in Japanese)
5. Miwa, Y., Itai, S., Watanabe T., Nishi, H., Shadow awareness enhancing theater space through the mutual projection of images on a connective slit-screen. J. Int. Soc. Arts Sci. Technol. (SIGGRAPH 2011 Art Paper) **44**(4), 325–333 (2011)
6. Esaki, K., Inoue, S., Watanabe, T., Ishii, Y.: An embodied entrainment avatar-shadow system to support avatar mediated communication. In: Proceedings of the 24th IEEE International Symposium on Robot and Human Interactive Communication (RO-MAN2015), pp. 419–424, 31 August 2015
7. Ishii, Y., Watanabe, T., Sejima, Y.: Development of an embodied avatar system using avatar-shadow's color expressions with an interaction-activated communication model. In: Proceedings of the 4th International Conference on Human-Agent Interaction (HAI 2016) pp. 337–340, 4 October 2016

8. Watanabe, T.: Human-Entrained Embodied Interaction and Communication Technology, pp. 161–177. Emotional Engineering, Springer (2011). https://doi.org/10.1007/978-1-84996-423-4_9
9. Watanabe, T., Ogikubo, M., Ishii, Y.: Visualization of respiration in the embodied virtual communication system and its evaluation. Int. J. Hum. Comput. Interact. **17**, 89–102 (2004)
10. Watanabe, T., Okubo, M., Ishii, Y., Nakabayashi, K.: An embodied virtual communication system with human virtual actor and abstract virtual wave. J. Hum. Interface Soc. **2**(2), 1–10 (2000)
11. Yamamoto, M., Takabayashi, N., Watanabe, T. Ishii, Y.: A nursing communication education support system with the function of reflection. In: Proceedings of the 2015 IEEE/SICE International Symposium on System Integration (SII2015), pp. 912–917, 11 December 11 2015

Effects of Interpupillary Distance and Visual Avatar's Shape on the Perception of the Avatar's Shape and the Sense of Ownership

Tokio Oka[✉], Takumi Goto, Nobuhito Kimura, Sho Sakurai, Takuya Nojima, and Koichi Hirota

The University of Electro-Communications, Tokyo 182-8585, Japan
{Tokio_Oka,goto.takumi,no.kimura,sho,hirota}@vogue.is.uec.ac.jp,
tnojima@nojilab.org

Abstract. Owing to the spread of virtual reality (VR) content, avatars that can be used have become more diverse. Several studies have been conducted on the effects of avatars on users. The investigation on the effects of avatar shape on the perception of object size and the sense of ownership of the avatar has been limited; however the effects on the perception of the avatar's shape have not been investigated. Studies focusing on visual effects have reported that the interpupillary distance (IPD) influences object-size perception and distance perception. It has also been reported that lowering the height of the camera and shortening the IPD can make the surrounding space seem larger. This suggests that the IPD may affect the shape perception and the sense of ownership of avatars. In this study, we focused on changing both the shape of the avatar and the IPD. We prepared avatars with a shape similar to that of the participant's body and avatars with extremely long/short limbs. Then, we examined the changes in the shape perception and the sense of ownership of the avatar by manipulating the prepared avatar based on one of the three IPDs (short, normal, and long). The results confirmed that the length of arms and legs and IPD may affect the perception of length and thickness of arms and legs and the sense of ownership of avatars.

Keywords: Avatar · Interpupillary distance (IPD) · Sense of ownership · Virtual reality (VR)

1 Introduction

Owing to the spread of head-mounted displays (HMDs), contents and services that enable communication using avatars in virtual reality (VR) environments are becoming increasingly common. The avatar is a user's alternative body in a VR environment, and its appearance can be diverse. However, if the types of avatars available in VR content are limited by the content administrator, then the shape of the user's body and the avatar may not necessarily resemble each other.

In this regard, the effects of the appearance and shape of the avatar on the sense of ownership have been studied. The sense of ownership is the perception that "this

© Springer Nature Switzerland AG 2021
S. Yamamoto and H. Mori (Eds.): HCII 2021, LNCS 12765, pp. 306–321, 2021.
https://doi.org/10.1007/978-3-030-78321-1_24

is self-body" and is mainly obtained by the temporal and spatial coincidence of visual and somatic sensations [1, 2]. Some of these studies have reported on the effects of differences between the shape of the user's real body and "user's alternative body" on the occurrence of the sense of ownership.

For example, Kirteni et al. have reported that a sense of ownership can be generated even in arms that are approximately three times longer than the original arm, by extending the arm step-by-step with haptic stimuli in a VR environment [3]. According to Lin et al., when operating a VR hand that moves in response to the movement of the user's hand, a sense of ownership of the VR hand occurs even if the VR hand is larger than the user's real hand [4]. Wittkopf et al. created a user's alternative right hand by presenting a mirror image of his/her left hand on the right side of the body with the right hand of the user hidden instead of the avatar's. Their study showed that, when the mirror image of the left hand was enlarged or reduced, the sense of ownership of the alternative right hand decreased [5].

Although there is a difference in whether the "user's alternative body" is an avatar or a mirror image, there are contradictory findings regarding the effects of the shape of the "alternative body" on the sense of ownership. In addition, many of these studies limit shape changes to the arms and hands, and only a few of them focus on the entire body. Most of these studies have verified the effects of the shape of some body parts on the perception of the shape or the sense of ownership of body parts. Few studies have addressed shape perception and the sense of ownership of the entire body.

However, the effects of the shape of the "alternative body" on the sense of ownership and on the perception of object size or distance around the "alternative body" are being studied. Van et al. presented an image of a person touching a doll's leg from the first-person perspective using an HMD and showed that the user's leg in the same posture as the doll was touched simultaneously. Then, they showed that the sense of ownership occurred on the legs of the visually presented doll despite the size of the doll being different from the actual size of the user. In addition, they found that when a sense of ownership occurred, the perception of the size and distance of surrounding objects in the images changed depending on the size of the doll [6]. It has also been reported that when a VR hand with a realistic appearance is moved in response to the movement of the user's hand in the VR space, the perception of the size of the objects around the VR hand occurs owing to the size of the VR hand [7]. In contrast, there is a report stating that the size of the VR hand does not affect the sense of ownership.

This report suggests that the perception of the size of surrounding objects may have changed based on the perceived size of one's own body, in the same way that the perception of the size of objects changes because of the influence of the size of surrounding objects [8]. In this regard, we predicted that the shape of the avatar could affect not only the surrounding object but also the perception of the avatar's size. In particular, the shape of some body parts of the avatar may affect the perception of the shape or size of other body parts.

The aforementioned studies focus on the shape of the "alternative body." However, one of the other factors that can affect the perception of the shape of the avatar and the sense of ownership of the avatar is the interpupillary distance (IPD) on the HMD, for viewing the VR space from the first-person perspective.

IPD is the distance between the eyes. As the IPD on the HMD changes, the deviation of the retinal image seen by the left and right eyes also changes. The HMD displays images taken from two positions on the software or in reality, and changing the distance between these two positions is synonymous with changing the user's IPD.

Regarding the relation between the IPD and shape perception, Nishida et al. proposed a system that allows the user to experience the visual field of children by manipulating the viewpoint perceived through an HMD. In this system, two cameras capture the image of the real world and present the image to the user through an HMD. These cameras are installed at the height of the user's waist, and the distance between the cameras is smaller than the distance from the user's pupils. It has been reported that this system design enables the user to see the real world with a lower line of sight and a narrower IPD and to make the surrounding space feel wider than it really is [9]. In addition, Kim et al. reported that the visual perception of the size of objects and distance varies because of changes in the IPD [10].

These findings suggest that changing the IPD may also change the perception of the avatar's shape. Therefore, it is possible that changes in the IPD would indirectly affect the sense of ownership.

In these studies, changing the shape of the avatar and changing the visual effect have been considered independently, and it has not been investigated how the shape perception and the sense of ownership change when the shape of the avatar and the visual effect are changed simultaneously. In particular, the effect of the IPD on the perception of avatar's shape has not been investigated and may be different from the effect of size perception on objects.

Therefore, in this study, we investigate the effects of the arm, leg, and IPD length on the shape perception of the avatar. Furthermore, we investigate the effect of each length on the sense of ownership.

2 Experiment

The purpose of this experiment was to investigate the effects of the arm, leg, and IPD length on the shape perception of the avatar and the sense of ownership. We conducted experiments to investigate the effects of the avatar's arms and legs, which are alternative bodies for the user in the VR space, and the IPD on the HMD on the perception of the avatar's shape and the sense of ownership of the avatar.

2.1 Design of the Experiment

The perception of the shape of a visual object changes relatively because of the influence of the shape of the avatar that is visually perceived at the same time [6, 7]. This phenomenon would occur in the avatar, that is, in the body of the user in the VR space. It has been reported that the morphological similarity between an object and the real body of the user affects the occurrence of the sense of ownership of the object [1]. The IPD on the HMD has been reported to affect the size perception and distance perception of objects [10], which may also affect the shape perception of the avatar. Therefore, we predicted that the IPD on the HMD and the shape of the avatar would affect the sense of ownership of the avatar.

- H1: The shorter the legs, the longer the arms are perceived.
- H2: The shorter the arms, the longer the legs are perceived.
- H3: The shorter the IPD in the software, the longer the arm is perceived.
- H4: The shorter the IPD in the software, the longer the leg is perceived.
- H5: The shorter the IPD in the software, the thicker the arm is perceived.
- H6: The shorter the IPD in the software, the thicker the leg is perceived.
- H7: If the arm length of the avatar is different from that of the real body, the sense of ownership is reduced.
- H8: If the leg length of the avatar is different from that of the real body, the sense of ownership is reduced.
- H9: The closer the ratio of the IPD in the software to the IPD in the real body is to the ratio of the avatar's arms and legs to the real body, the higher the sense of ownership of avatars.

To verify these hypotheses, we created a VR system that allows the user to move freely in the VR space using the prepared avatar while changing the IPD on the HMD in various ways.

This system presents the VR space from the viewpoint of the avatar through an HMD (HTC VIVE PRO, HTC Vive). The VR space was built using Unity (Version 2019.4.18f). The user's activity range in the VR space was 2 m × 3 m. The avatar was created using MakeHuman (Version 1.2.0) and Blender (Version 2.8.0). This VR system tracks the motion of the user's head, hips, arms, legs, elbows, and knees using the HMD, two controllers (VALVE INDEX and Valve Corporation), and six trackers (HTC VIVE Tracker and HTC Vive).

For this experiment, we prepared nine types of avatars with different lengths of arms and legs, as shown in Fig. 1. We first constructed an avatar with the shape of the standard shown in the center of Fig. 1. This avatar had a realistic male appearance, with a height of 17 cm, an arm length of 73 cm, and a leg length of 75 cm. Based on this avatar with the shape of the standard, we set the following three conditions for the length of the arms and legs of avatars: 0.5 times, 1 time, and 2 times. Combining these conditions, we created nine types of avatars. The thickness of the arms and legs of the avatars was constant for all the conditions. During the experiment, the head of the avatar was hidden to prevent it from entering the field of view. In addition, the standard IPD was set to 66 cm. This IPD was also changed to three conditions: 0.5 times, 1 time, and 2 times.

In this experiment, 27 conditions were prepared as independent variables according to the combination of conditions on the avatar's arm length, leg length, and IPD. However, the lengths of the arms and legs and the IPD were of relative sizes on Unity.

To evaluate the effect of the IPD on the body schema ("body schema" refers to the perception of where each part of the body is located in spatial coordinates), we designed a task to measure the misalignment between the positions of the fingertips of the avatar' s limbs and the positions that participants believe are their fingertips. For the measurement, we placed a sphere and a cube at the reference point after the participant moved in the VR space with an avatar in each condition. At this time, we hid everything, including the avatar and floor, except the sphere and cube. Then, the participant moved the sphere and the cube to the positions that the participant thought were the fingertip and the toe tip using the controller, respectively, and the positions were measured. After

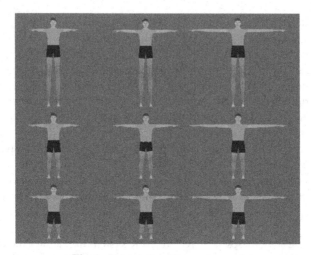

Fig. 1. Schematic of the avatar used

the measurement, the distance from the reference point to the position of the tips of the limbs indicated by the sphere and the cube was compared with the distance of the tips of the avatar's limbs operated by the participant immediately before this task. Because the perception of the size and distance of the sphere and cube is affected by the IPD, the perceptual positions of the tips of the limbs are expected to change under the influence of the IPD [10].

2.2 Participants

Thirteen Japanese undergraduate and graduate students participated in the experiment. The age of the participants ranged from 21 to 24 years. The participant's height ranged from 162.0 to 176.5 cm (169.1 cm $\pm$ 1.4 SD), arm length ranged from 64.0 to 74.0 cm (70.7 cm $\pm$ 3.0 SD), and IPD ranged from 5.8 to 7.2 cm (65.4 cm $\pm$ 3.5 SD). We obtained informed consent from all the participants before the experiment.

2.3 Procedure

First, we invited participants to our laboratory and explained the purpose of this experiment as follows: investigation on the effect of the shape of avatars on the behavior of the user of the avatar in VR space. We then measured the height, arm length, and IPD of the participants. We also asked the participants to attach devices to track the motion of their bodies, as shown in Fig. 2(a). After the attachment, we checked whether the avatar in each of the conditions was moving according to the movements of the participants. In cases where the participants felt discomfort or unnaturalness for the avatar's shoulder height or foot orientation, we adjusted them in Unity so that they felt comfortable.

Next, to help the participants to be familiarized with moving in the VR space using avatars, we asked them to perform the task of touching the three-dimensional objects appearing in the VR space with their hands and feet as quickly as possible (hereinafter "touch task") (Fig. 3). This task started when they touched a sphere and a cube presented in the participants' field of view with their hands and feet, respectively. For the first 20 s from the start, they were required to touch a sphere that appeared at random positions with their hands. For the next 20 s, they were required to touch a cube that appeared at random positions with their feet. The sphere and cube disappeared once touched, and a new sphere or cube appeared in randomly different positions. The diameter of the sphere was 10 cm, and the length of one side of the cube was 10 cm. However, the shapes of these objects differed depending on the IPD on the HMD. The participants performed these tasks three times. After these tasks were completed, the floor of the avatar and the VR space were hidden.

After completing the touch task, we asked the participants to perform a task to evaluate whether the body schema changed (hereinafter "body schema task"). In this task, they were required to move a sphere displayed at the position corresponding to the right shoulder joint of the avatar used in the "touch task" to the perceptual positions of the fingertips of their own body. Similarly, they were required to move a cube displayed at the position corresponding to the waist height in front of the avatar to the perceptual positions of the toes of their actual body. At this time, the participants assumed the posture shown in Fig. 2(b) and moved the sphere and cube by controlling the stick of the controllers. We calculated the difference between the position of the avatar's right fingertip and the perceived position of the participant's right fingertip: the distance was obtained by subtracting the length of the avatar's arm used in the touch task from the distance between the avatar's right shoulder joint and the sphere, which the participants moved. Similarly, we calculated the difference between the position of the avatar's toe and the perceptual position of the participant's toe: the distance was obtained by subtracting the length of the avatar's legs used in the touch task from the distance between the waist height of the avatar and the position of the cube, which was moved by the participants. These calculated distances were used to compare the body scheme in each condition.

After completing these two tasks, the participants were asked to remove the HMD and controllers. Then, they completed the following questionnaire using a 7-point Likert scale. We instructed the participants to provide comments in a free-text format on the avatar or experiment.

- Q1: The avatar was perceived as my body… (1 = do not perceive, 7 = perceive).
- Q2: Avatar was moved by my will… (1 = I do not feel it, 7 = I feel it).
- Q3: Avatar's arms were comparable to my real arms… (1 = short, 7 = long).
- Q4: Avatar's legs were comparable to my real legs… (1 = short, 7 = long).
- Q5: Avatar's arms were comparable to my real arms… (1 = thin, 7 = thick).
- Q6: Avatar's legs were comparable to my real legs… (1 = thin, 7 = thick).

One trial consisted of the aforementioned tracking setting, "touch task," "body schema task," and questionnaire. This trial was performed under all the conditions. After all trials were completed, participants answered in a free-text format on how often they usually wear HMDs, how often they use avatars in VR space, and their overall impression of this experiment.

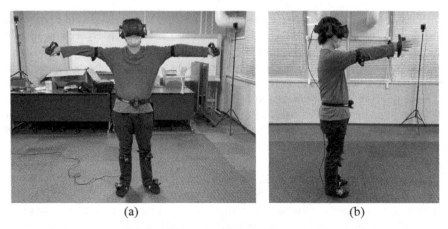

(a)   (b)

Fig. 2. Postures assumed by the participant

Fig. 3. Touch task

3   Results

3.1   Evaluating the Body Schema

To evaluate the differences in the body schema calculated in the "body schema task," we conducted a three-factor repeated-measures analysis of variance (ANOVA) for each condition on the length of the avatar's arms (fARM), length of the avatar's legs (fLEG), and IPD on the HMD (fIPD). The conditions common to these factors are as follows: 0.5 times (short condition), 1 time (normal condition), and 2 times (long condition).

The results showed a significant main effect of fIPD on the misalignment between the avatar's position and the position that the participants believe (fingertip: $F(2,24) = 7.89$ and $p = .002$; toe: $F(2,24) = 8.39$ and $p = .002$). This ANOVA test also showed a two-way interaction between fARM and fIPD ($F(4,48) = 12.79$; $p < .001$) and fLEG and fIPD ($F(4, 48) = 2.72$; $p = .040$) for fingertip misalignment. For the toe misalignment, there was a significant two-way interaction between fLEG and fIPD ($F(4, 48) = 3.26$; $p = .019$). Figures 4, 5 and 6 show the results of multiple comparisons using the Holm method for the evaluation values in each condition on the factors that have a simple main effect on these two-way interactions.

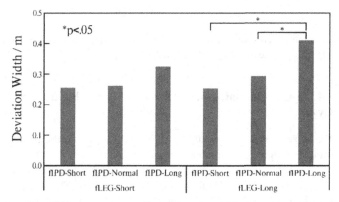

Fig. 4. Results of multiple comparisons of conditions in which there was a simple main effect in the two-way interaction between fLEG and fIPD for fingertip misalignment

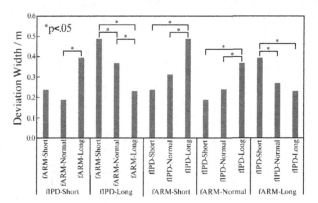

Fig. 5. Results of multiple comparisons of conditions in which there was a simple main effect in the two-way interaction between fARM and fIPD for fingertip misalignment

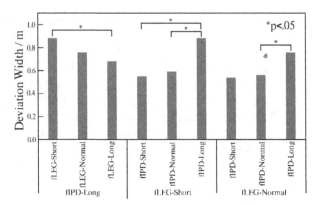

Fig. 6. Results of multiple comparisons of conditions in which there was a simple main effect in the two-way interaction between fLEG and fIPD for toe misalignment

3.2   Results of the Questionnaire on the Sense of Ownership and the Sense of Agency

We used the three-factor repeated-measures ANOVA to evaluate the results of Q1 and Q2. This test detected significant main effects of fARM (Q1: $F(2, 24) = 8.26$ and $p = .002$; Q2: $F(2, 24) = 4.43$ and $p = .023$) and fIPD (Q1: $F(2, 24) = 12.63$ and $p < .001$; Q2: $F(2, 24) = 12.14$ and $p < .001$) for the evaluation values. There was a significant two-way interaction between fARM and fIPD in Q1 ($F(4, 48) = 3.03$; $p = .026$).

Figure 7 shows the results of multiple comparisons using the Holm method for the evaluation values for Q1 in each condition on fARM and fIPD, indicating a simple main effect on these two-way interactions. Figure 8 shows the results of multiple comparisons for the evaluation values for Q2 under each condition of fARM and fIPD, for which the main effect was observed.

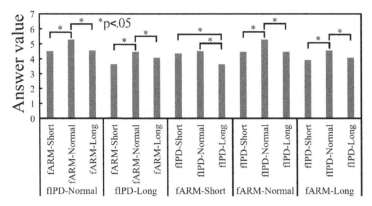

Fig. 7. Results of multiple comparisons of conditions in which there was a simple main effect in the two-way interaction between fARM and fIPD for Q1

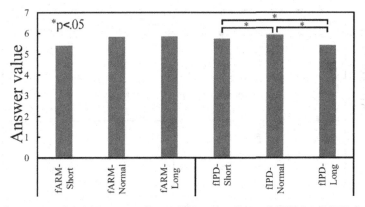

Fig. 8. Results of multiple comparisons of the main effects of fARM and fIPD for Q2

3.3 Length Perception for Arms and Legs

We used the three-factor repeated-measures ANOVA to evaluate the results of Q3, Q4, Q5, and Q6. This test detected significant main effects of fLEG for Q3 ($F(2, 24) = 47.58$; $p < 0.001$), Q4 ($F(2, 24) = 57.40$; $p < 0.001$), and Q6 ($F(2, 24) = 96.50$; $p < 0.001$), fARM for Q3 ($F(2, 24) = 270.56$; $p < 0.001$) and Q5 ($F(2, 24) = 56.94$; $p < 0.001$), and fIPD for Q3 ($F(2, 24) = 63.38$; $p < 0.001$), Q4 ($F(2, 24) = 4.83$; $p = 0.017$), Q5 ($F(2, 24) = 42.74$; $p < 0.001$), and Q6 ($F(2, 24) = 80.09$; $p < 0.001$) for the evaluation values. Table 1 shows the significant interactions in the test.

Table 1. Interactions for Q3–Q6

Number	Conditions in which the interaction was significant
Q3	fLEG and fARM ($F(4, 48) = 6.00$; $p = 0.001$)
	fARM and fIPD ($F(4, 48) = 17.74$; $p < 0.001$)
Q4	fLEG and fARM ($F(4, 48) = 3.64$; $p = 0.012$)
	fLEG and fIPD ($F(4, 48) = 3.28$; $p = 0.019$)
Q5	Three-way interaction ($F(8, 96) = 2.12$; $p = 0.041$)
Q6	fLEG and fARM ($F(4, 48) = 3.98$; $p = 0.007$)

The result of the test for Q5, in which there was a three-way interaction, showed that the simple two-way interaction of fARM and fIPD in Leg-Normal was significant ($F(4, 48) = 4.36$; $p = 0.004$).

Figures 9, 10, 11, 12, 13, 14, 15, 16, 17 and 18 show the results of multiple comparisons using the Holm method for the evaluation values in each condition on the factors that have a simple main effect on these two-way interactions or the simple two-way interaction.

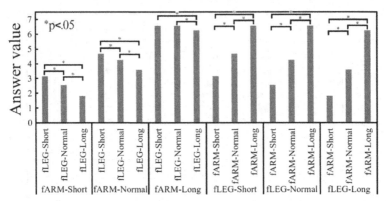

Fig. 9. Results of multiple comparisons of conditions in which there was a simple main effect in the two-way interaction between fLEG and fARM for Q3

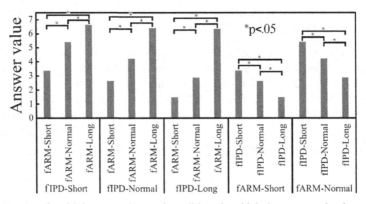

Fig. 10. Results of multiple comparisons of conditions in which there was a simple main effect in the two-way interaction between fARM and fIPD for Q3

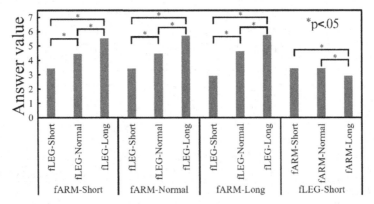

Fig. 11. Results of multiple comparisons of conditions in which there was a simple main effect in the two-way interaction between fLEG and fARM for Q4

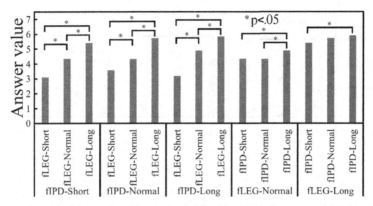

Fig. 12. Results of multiple comparisons of conditions in which there was a simple main effect in the two-way interaction between fLEG and fIPD for Q4

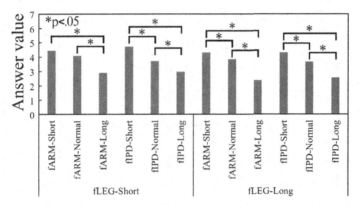

Fig. 13. Results of multiple comparisons of conditions in which there was a simple main effect of fLEG-Short or fLEG-Long in the three-way interaction between fLEG, fARM, and fIPD for Q5

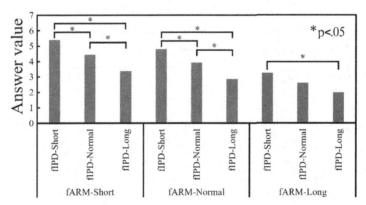

Fig. 14. Results of multiple comparisons of conditions in which there was a simple main effect of fARM in the three-way interaction between fLEG, fARM, and fIPD for Q5

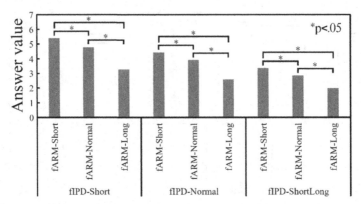

Fig. 15. Results of multiple comparisons of conditions in which there was a simple main effect of fIPD in the three-way interaction between fLEG, fARM, and fIPD for Q5

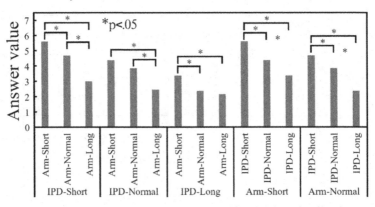

Fig. 16. Results of multiple comparisons of conditions with a simple main effect in Arm and IPD in Leg-Normal, where there was a simple interaction in the three-way interaction between fLEG, fARM, and fIPD for Q5.

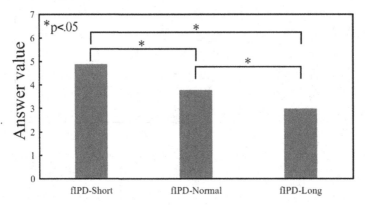

Fig. 17. Results of multiple comparisons of main effects of fIPD for Q6

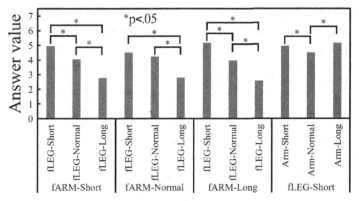

Fig. 18. Results of multiple comparisons of conditions in which there was a simple main effect in the two-way interaction between fLEG and fARM for Q6

4   Discussion

The results of the analysis of Q1 showed that the sense of ownership in fIPD-Normal was significantly higher in the fARM-Normal level than in the other two levels. This supports hypothesis H7. However, hypothesis H8 was not supported because there was no significant difference between the levels of fLEG. It was also confirmed that fIPD-Normal was significantly higher than the other levels in all fARM conditions. Therefore, hypothesis H9 was rejected. In the evaluation of the body schema, when the size of the IPD was set to be inversely proportional to the arm length, it was confirmed that the misalignment between the fingertip position of the avatar and the fingertip position perceived by the participant increased. There was no fIPD condition in which the misalignment between the avatar's position and the participant's perceived position was significantly lower than in fIPD-Normal.

The results of the evaluation values for Q2 showed that changing the IPD from its normal length decreases the sense of agency. However, because the mean of the values at each level was more than five, the sense of agency was assumed to be generated to some extent when the IPD was set to 2 or 0.5 times.

The results of the evaluation values for Q3 showed that the shorter the leg of the avatar, the longer the arm was perceived, and thus, H1 was supported. Except in the fARM-Long condition, the shorter the IPD on the HMD, the longer the arm was perceived. This supports a part of H3. If the avatar's arm is very long, it appears that the perception of arm length may be less susceptible to the IPD. In contrast, in the fARM-Long condition, the average evaluation value for Q3 exceeded six, regardless of the condition for fIPD. It may be possible to clarify the threshold of the length of the avatar's arm where the perception of the arm length changes under the influence of the IPD through numerical evaluations of the perception of the arm length.

The results of the evaluation values for Q4 showed that only in the fLEG-Short condition, the shorter the arm of the avatar, the longer the leg was perceived. This scarcely supports H2. Furthermore, there was a significant difference in the evaluated values only in the fLEG-Normal and fLEG-Long conditions on fIPD. This scarcely supports H4. By contrast, there was a significant difference between the evaluation values for all fLEG conditions on fARM and fIPD. This result indicates that the perception of the leg length is highly dependent on the avatar's leg length, with little effect on the arm length of the avatar or the IPD. In other words, the perception of the arm length and leg length likely differs according to the influencing factor.

The results of the evaluation values for Q5 showed that the shorter the IPD, the thicker the arm was perceived. This supports H5. Moreover, the shorter the arm of the avatar, the thicker the arm was perceived. This shows that the perception of the arm thickness is strongly dependent on the arm length. The results of Q3 showed that the IPD may have little effect on the perception of the arm length when the arm is long. This suggests that the non-significant difference between the levels of fIPD in the fARM-Long condition may be caused by the fact that the IPD did not affect the perception of the arm length.

The results of the evaluation values for Q6 showed that the shorter the IPD, the thicker the leg was perceived, and thus, H6 is supported. In addition, the shorter the leg of the avatar, the thicker the leg was perceived. This shows that the leg length also affects the perception of the leg thickness.

Based on these results, we confirmed that the effects of the IPD on the perception of the arm length and the leg length were different. Unless the head was tilted sideways, the change in the retinal image caused by the change in the IPD was limited to the lateral direction, and this may be the reason why the effect of the IPD was different on the arms and legs. Although the arms could move in various directions in front of the face in this experiment, there was little movement of the legs other than placing the legs forward, and we believe that the retinal image did not shift much in the leg-length direction.

The sense of ownership was decreased by changing the IPD from its normal length. The results of Q5 and Q6 suggest that the IPD affects the perception of the arm and leg thickness. Therefore, this indicates that changing the normal IPD cannot reduce the misalignment between the avatar and the position of the body on the senses but can affect the perception of thickness; this causes increased discomfort, which may have caused the sense of ownership to decrease.

5   Conclusion and Future Work

The purpose of this study was to investigate the effects of the lengths of an avatar's arm and leg and the IPD on the HMD on the shape perception of the avatar and the sense of ownership. The experiments confirmed that the perception of the avatar's arm length was affected by the leg length and the size of the IPD on the HMD, whereas the perception of the avatar's leg length was scarcely affected by the size of the IPD on the HMD and avatar's arm length. We also confirmed that the perception of the avatar's arm thickness and leg thickness was affected by the respective length and the HMD's IPD size. In addition, the avatar's arm length and the HMD's IPD size affected the sense of ownership.

This study suggests that unless there is a special reason, it is desirable to match the shape of the avatar and the HMD's IPD with the real body to have a comfortable VR experience. However, we did not directly change the thickness of the avatar's arms or legs in our experiments. There is a possibility that this thickness and the IPD mutually affect the perception of the avatar's shape. This may require further research. Moreover, the effect of the IPD on the perception of the length of the avatar's arms was also different from the effect of the IPD on the perception of the length of the avatar's legs. In this regard, the effect of the IPD on the perception of the object's size and distance may differ vertically and horizontally. It is possible that the effect of the IPD on the perception of the avatar's size and length of various body parts differs between the vertical and horizontal directions when the object is stationary. In the future, we plan to verify these possibilities predicted based on the analysis of this experiment.

Acknowledgments. This work was supported by JSPS KAKENHI Grant Numbers JP19H04230, JP 19H05661.

References

1. Argelaguet, F., Hoyet, L., Trico, M., Lécuyer, A.: The role of interaction in virtual embodiment: effects of the virtual hand representation. Proc. IEEE VR' **16**, 3–10 (2016)
2. Botvinick, M., Cohen, J.: Rubber hands 'feel' touch that eyes see. Nature **391**, 756 (1998)
3. Kilteni, K., Normand, J.-M., Sanchez-Vives, M.V., Slater, M.: Extending body space in immersive virtual reality: a very long arm illusion. PLoS ONE **7**(7), e40867 (2012)
4. Lin, L., Normovle, A., Adkins, A., Sun, Y..: The effect of hand size and interaction modality on the virtual hand illusion. In: 2019 IEEE Conference on Virtual Reality and 3D User Interfaces (VR), Osaka, Japan (2019)
5. Wittkopf, P.G., Lloyd, D.M., Johnson, M.I.: Changing the size of a mirror-reflected hand moderates the experience of embodiment but not proprioceptive drift: A repeated measures study on healthy human participants. Exp. Brain Res. **235**(6), 1933–1944 (2017)
6. van der Hoort, B., Guterstam, A., Ehrsson, H.H.: Being barbie: the size of one's own body determines the perceived size of the world. PLoS ONE **6**(5), e20195 (2011)
7. Ogawa, N., Narumi, T., Hirose, M.: Virtual hand realism affects object size perception in body-based scaling. In: 2019 IEEE Conference on Virtual Reality and 3D User Interfaces (VR), Osaka, Japan (2019)
8. Franz, V.H., Scharnowski, F., Gegenfurtner, K.R.: Illusion effects on grasping are temporally constant not dynamic. J. Exp. Psychol. Hum. Percept. Perform. **31**(6) (2005)
9. Nishida, J., Matsuda, S., Oki, M., Takatori, H., Sato, K., Suzuki, K.: Egocentric smaller-person experience through a change in visual perspective. In: Extended Abstracts of the 2019 CHI Conference on Human Factors in Computing Systems (2019)
10. Kim, N.-G.: independence of size and distance in binocular vision. Front. Psychol. **9**, 988 (2018)

Impact of Long-Term Use of an Avatar to IVBO in the Social VR

Akimi Oyanagi[1](✉), Takuji Narumi[2](✉), Kazuma Aoyama[1,2](✉), Kenichiro Ito[1](✉), Tomohiro Amemiya[1,2](✉), and Michitaka Hirose[3](✉)

[1] Virtual Reality Educational Research Center, The University of Tokyo, Tokyo 113-8656, Japan
{oyanagi,aoyama,ito,amemiya}@vr.u-tokyo.ac.jp
[2] Graduate School of Information Science and Technology,
The University of Tokyo, Tokyo, Japan
narumi@cyber.tu-tokyo.ac.jp
[3] Research Center for Advanced Science and Technology,
The University of Tokyo, Tokyo, Japan
hirose@cyber.tu-tokyo.ac.jp

Abstract. The illusion of virtual body ownership (IVBO) gives us the feeling of perceiving a virtual body as our own body. It affects our behavior and the sense of presence. The recent emergence of lower-cost VR devices and social VR environments have created the ability for consumers to possess and manipulate avatars in an immersive virtual environment. In other words, IVBO can be induced even in consumer environments. However, apart from laboratory experiments, few studies have investigated IVBO in social VR, because it is necessary to have the same participants go to the laboratory and participate in experiments every day. Thus, we investigated how long-term use of an avatar impacts IVBO. Our experiment was conducted on social VR platform VRChat. Participants are instructed to pick up 30 objects during at the start of experiment. Then, they played VRChat up to 2 h. This task was lasted 30 day per user. Our results shown IVBO could be enhanced day by day. Specifically, IVBO can be significantly enhanced at day 14 regardless of the type of an avatar. However, this effect was not collated with avatar-identification.

Keywords: Virtual body ownership · Avatars · Social virtual reality

1 Introduction

In a virtual environment, users manipulate a virtual body as a digital representation instead of their own physical body. This digital representation that identifies a user is referred to as an avatar. The user's body positions are reflected by the avatar to manipulate it in an immersive virtual environment, using virtual reality (VR) devices, as if they were moving their own physical body. This allows them to intuitively interact with the virtual environment via an avatar. Interestingly, previous works have reported the illusion of virtual body ownership (IVBO), where users recognize a virtual body as their own body [1], and the proteus effect in which an avatar affects the user's attitude and behavior

© Springer Nature Switzerland AG 2021
S. Yamamoto and H. Mori (Eds.): HCII 2021, LNCS 12765, pp. 322–336, 2021.
https://doi.org/10.1007/978-3-030-78321-1_25

[2]. Hence, in recent years, the study of human interaction in the VR environment has focused on the effect of avatar use.

Generally, laboratory experiments use sophisticated tracking devices such as MVN XSens (https://www.xsens.com/) and OptiTrack (https://www.optitrack.jp/) that are able to track many of the user's joint positions (both hands, both legs, knees, elbows, waist, chest, head, shoulder, and so on) to manipulate an avatar. However, the recent emergence of lower-cost VR devices and social VR environments, such as VRChat (https://vrc hat.com/), and Altspace (https://altvr.com/), have created the ability for consumers to possess and manipulate avatars in an immersive virtual environment. This means that even consumer apparatus can create an environment that can elicit IVBO and the proteus effect. Indeed, Steed et al. reported that the existence of an avatar enhances the sense of presence and embodiment even in consumer equipment such as low-cost VR devices using smartphones [3]. Furthermore, Eubanks et al. reported that IVBO can be induced even by inverse kinematics (IK) [4], which is adapted in a social VR as the common manner of manipulation of an avatar [5].

There are many differences in terms of treatment of an avatar between laboratory experiments and consumer environments. In laboratory experiments, participants manipulate a specified avatar prepared by experimenter using sophisticated apparatus for a limited time (about 1–2 h). Conversely, social VR users manipulate one or more unique avatars customized by them using the IK method (a poor tracking method in comparison with sophisticated apparatus) and consumer devices as their own digital representation in the social VR community for long periods of time. Furthermore, users carry out special communication in communities where they give positive feedback, such as praising and stroking other avatars. This means that they regard an avatar as their own alter-ego in social VR [6].

However, apart from laboratory experiments, few studies have investigated IVBO in social VR, because it is necessary to have the same participants go to the laboratory and participate in experiments every day to investigate these factors. This is impractical for a number of reasons, and has led to many practical findings that cannot be obtained in laboratory experiments. In particular, finding factors that enhance IVBO in a social VR leads to the effective use of an avatar and appropriate risk management.

IVBO is an illusion that is constructed with a sense of ownership (SBO) and a sense of agency (SoA) [7]. These factors can be elicited by presenting synchronous visuomotor information [8]. The bottom-up factor is that which elicits a sense of ownership by presenting sensory information. Generally, IVBO can be elicited from avatars that differ from the user's physical body. There is a possibility that IVBO can be elicited even in the animation-character-type avatar preferred by social VR users.

On the other hand, consistency between pre-existing body representation and avatar appearance (e.g., appearance and pose) is referred to as the top-down factor. For instance, previous studies have reported that IVBO can be enhanced by an avatar that resembles the user's physical body [9] and wears the same clothing [10]. It is assumed that the top-down factors are important to enhance IVBO in a social VR because these platforms use low-accuracy devices in comparison with a laboratory experiment (i.e., users can only use visual cues).

We have carried out an experiment on a social VR and reported that social VR users can significantly enhance IVBO when using their own avatar (i.e., an avatar that users use in daily life as their digital representation in the social VR) in comparison with an avatar created by researchers [11]. In this experiment, the authors recruited participants who had used social VR content over one month. However, it is not clear what the timing is for enhancing IVBO in the long-term use of an avatar.

We defined avatar identification as the degree to which the players regard their avatar to be himself or herself. It is not clear how avatar identification with an avatar impacts IVBO. Identification is the most characteristic feature of online games. Previous studies examining MMORPGs have reported that users feel an identity with an avatar [12–14]. Thus, it is assumed that the similar effect of personalization to an avatar [9] is evoked when users have strong identification to an avatar.

2   Related Work

2.1   Illusion of Virtual Body Ownership

Initial work has reported the illusion of body ownership through the classical rubber hand illusion. Botvinick and Cohen have reported that, when the experimenter simultaneously strokes the participants' hands while hidden from their vision and an artificial hand, the participants perceive the artificial hand as part of their own body [15]. This illusion can be induced when stimulation is synchronously represented both temporally and spatially. Ijsselsteijn et al. [16] and Slater et al. [17] confirmed that this illusion can be induced using virtual reality. IVBO mainly consists of two factors: (1) the sense of body ownership (SBO) is a sensation that describes "this is part of my body" and (2) the sense of agency is a sensation that describes "this action is driven by my intention" [7, 18].

Related work on IVBO has confirmed two dimensional factors that elicit or enhance this illusion [19]: (1) bottom-up factors are related to multisensory integration (e.g., synchronous visuo-tactile stimulation and visuo-motor inputs), and (2) top-down factors are related to consistency between observed virtual body parts and pre-existing body representation (e.g., posture [20], corporeal [21], skin texture [22], and anthropomorphism [1]). This means that these factors involve conceptual interpretations, such as memories and experiences. Many researchers use bottom-up factors as strong triggers for IVBO. Consequently, previous studies have reported that IVBO can be elicited on various avatars, even those apparently differing from the user's physical body (age [23], opposite gender [24], skin texture [25], robot [1], and animal [26]). Apart from these factors, in recent years, some studies have pointed out that there are individual differences in terms of the sense of body ownership and have investigated how internal factors (e.g., personality traits [27, 28], sensory suggestibility [29], and age [30]) impacted on it.

It is assumed that the effect of IVBO into an avatar becomes a characteristic top-down factor in social VR. Waltemate et al. used 3D scan technology to treat avatars that resemble the user's physical body, and reported that the 3D scanned avatar significantly enhanced the intensity of IVBO [9]. Although personalization commonly works over the user's physical body, it also works over an avatar in online games [31, 32]. We defined avatar identification as the degree to which the players regard their avatar to be himself or herself. Hence, it is assumed that the effect of avatar identification for IVBO can be elicited over the user's avatar in a social VR. Our previous work has reported that IVBO can be enhanced using a user's own avatar, as opposed to an avatar prepared by an experimenter [16]. This means that if the effect is strengthened by consistency between the user's body image and an avatar, it is assumed that the effect of avatar identification is stronger as the relationship between an avatar and users becomes deeper. We hypothesize that avatar-identification will increase by continuing to use an avatar, thus IVBO can be enhanced. Furthermore, it is assumed that the user who already played social VR (i.e., they already possess their avatar) has stronger avatar-identification than the user who are not social VR player.

2.2 Avatar-Identification

An avatar depicts the user's physical representation in online games, and users experience activities and adventure in the digital world by manipulating and customizing an avatar. They understand and attach themselves to an avatar by imparting it with a unique behavior, intention, style, and sense of personality [31]. Users feel distressed when their avatars are harmed by other participants' malicious actions, such as attacks [32]. Hence, an avatar plays a central role in a virtual world where there are communications and dynamics for self-representation. A few studies have examined self-representation in social VR [6]. Freeman et al. conducted a survey on self-expression among users of different social VR platforms, and reported that different platforms regard self-expression differently [6]. The characteristic property of social VR is embodiment into an avatar, which occurs through the reflection of body movements. This leads to a feeling of intimacy with an avatar and motivates customization. Interestingly, the direction of customization varies for each platform. Whereas in AltSpace and Rec Room, the avatars are likely to resemble their own physical body, users are likely to have an avatar that apparently differs from their physical body, such as a bird, in a VRChat. Customized avatars are seen as an extension of the self, even if they are different from the user's physical body. Previous studies have already investigated how personalization to a 3D-scanned avatar impacts IVBO [9]. However, it is not clear how identification to an avatar in social VR impacts IVBO.

Some social VR platforms (e.g., VRChat) can be used without the use of a VR device (called the desktop mode because it takes place on a PC, similar to non-VR games). If identification to avatars affects IVBO, users who are already using social VR in desktop mode are expected to see the effect of improved IVBO at an earlier stage than those who are not. Therefore, in this study, we investigate the change in IVBO for these two types of users.

2.3 Research Question and Hypothesis

We investigated the following research questions:

RQ1. We investigated long-term changes in IVBO.

RQ2. We identified the timing when significantly enhancing IVBO.

RQ3. We investigated correlation among IVBO, identity.

H1. Avatar identification with avatars increases over time, thereby improving IVBO

H2. Users who are already playing VRChat they already have a high avatar-identification with avatars, so the effect of avatar identification on IVBO is higher from the start compared to VRChat beginners.

3 Experiment

We conducted an experiment on the social VR platform, VRChat. We recruited two types of users: those who already played VRChat (referred to as VRChat players) and those who were new to VRChat (referred to as non-VRChat players). Both types of users had never used any VR equipment before (VRChat has a mode that can be played with only a desktop PC), and the main participants were those who had just purchased VR equipment during the recruitment period. We also approved participants who have a few VR experiences (once or twice). We recruited users via social media (Twitter) and explained the details of the experiment using video chat tools. In addition, we also recruited VRChat users who have a humanoid-avatar. We defined a humanoid avatar as an avatar that has a human morphology and does not extremely deviate from normal body size.

Upon approving informed consent, the participants were asked to sign a consent form and send our organization. A total of 14 participants were gathered. There were eight non-VRChat users and seven VRChat users. The experiment was conducted over a period of one month, during which time the IVBO was evaluated. In the end, one participant dropped out of the study and the data of 13 participants were used for the analysis. Since this experiment was conducted with online users, the equipment for the immersive devices was owned by the users. When asked in advance what equipment they were using, all 13 users were using Oculus Quest. Three of them were playing VRChat with Oculus Link.

3.1 Procedure

The experiment lasted for one month, and participants were instructed to play for no more than 1-2 h each day. After completing the embodiment task to elicit IVBO (described below), participants completed a questionnaire and played VRChat for 1–2 h a day. Since the content of play cannot be completely controlled, we left the play style up to the users in this experiment. In addition, the reason why we determined these procedures is there are no users who cooperate with limited VR experience for one month's experiment. In addition, playing VRchat works as constructing identity into an avatar.

The embodiment task consisted of picking up 30 box-shaped objects placed in the VRChat space in order. A mirror was placed in front of the user, and the user was able to see the synchronization of visual-motor sensations from the first-person perspective and in the mirror (see Fig. 1). Based on existing researches, this body synchronicity prompted the generation of IVBO [1, 9, 11, 13]. After acquiring 30 objects, users were asked to move their bodies freely with a mirror in front of them for 3 min. This was not only to encourage the generation of the IVBO, but also to have the user recognize the avatar that he or she was wearing. Immediately after the completion of this embodiment task, we paused and asked the participants to answer a questionnaire. After that, we restarted VRChat and had the participants play for 1–2 h. These are summarized in Fig. 2.

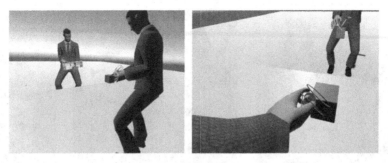

Fig. 1. The VR environment and participant's point of view.

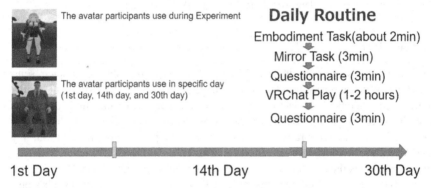

Fig. 2. Summary of experimental procedure.

During the experiment, the non-VRChat users were asked to continue using the avatars shown left side of the picture in Fig. 3. This avatar was selected because it was the default avatar available in VRChat and was the preferred animated character-type avatar in VRChat. Avatars of the opposite sex to the user were presented (male avatars are shown on the left side of the first avatar on Fig. 3; female avatars are shown on the left side of the second avatar on Fig. 3). The VRChat users were allowed to use their usual avatars in desktop mode for the experiment (We show an example on the right side of Fig. 3). On the first, second week, and last day, they were asked to use a more realistic same-sex avatar prepared by the experimenter shown in center side of the picture in Fig. 3. This allowed us to measure the change in IVBO over time and compare it to use of the same sex avatars to investigate the change in IVBO with continuous use, regardless of the type of avatar.

Fig. 3. Avatars that participants use: Left) The avatar non-VR users use, Central) the avatar participants use on specific days, Right) and the avatar VRChat users use.

3.2 Procedure

The main methods of measuring VBO were subjective rating, self-acceptance, sensory drift (the shift in the position of the self's arm as perceived by the participant), and skin conductance response [33]. However, it has been reported that the subjective sense of ownership and the implicit sense of ownership (e.g., proprioceptive drift) do not always coincide [34]. In fact, the brain areas where they are important are different [35, 36]. In social VR platforms, in order to bridge the differences in physical characteristics (such as height) between the user and avatar, many of them change the interocular distance (the distance between the stereo cameras in the VR space) according to the avatar's height. As a result, the self-acceptance sensory congruence between the user and the avatar is roughly the same for all avatars. It depends on the size of the avatar's limbs, but does not deviate from the user's sense of self-acceptance unless the morphology of the body is too deformed. Therefore, IVBO in social VR seems to be highly dependent on top-down factors or internal factors of the user (personality traits, strength of self-concept, and so on). Many studies have reported that top-down and internal factors are involved in the subjective self and do not significantly affect the implicit self [27, 28]. Therefore, even in this online experiment, where it is difficult to prepare the equipment, only a subjective rating is considered sufficient.

Considering the fatigue that comes from answering the questions every day, we referred to traditional questionnaires and adapted 4 items. We referred IVBO-scale to measure the sense of ownership and the sense of agency [7]. Additionally, we also adapted the item g1 of IPQ scale to measure the sense of presence [37]. In addition, we created an original questionnaire consisting of 14 items to measure avatar-identification. Participants were instructed to answer identification questionnaires once every one week. In the research of CSCW, although previous work has reported -Scale to measure avatar identification [38], we created this questionnaire because there are a lot of differences regarding functions, achievement, and interaction in VRChat. We summarize these questionnaires on Table 1.

Table 1. Questionnaire

Label	Item		
		Daily questionnaire	
My body	Q1	To what extent did you feel as if an avatar is your own body?	1 Not at all 7 very much
Mirror	Q2	To what extent did you feel as if the body you saw in the mirror is your own body?	1 Not at all 7 Very much
Agency	Q3	To what extent did you feel an avatar was controlled with your motor intention?	1 Not at all 7 Very much
Presence	Q4	In the computer-generated world I had a sense of "being there"?	1 Not at all 7 Very much
		Weekly questionnaire	
	Q1	To what extent does an avatar's appearance establish as yourself in VRChat?	1 Fully disagree 7 Fully agree
	Q2	Avatars are only a visual representation for others to recognize themselves	1 Fully disagree 7 Fully agree
	Q3	To what extent do you hold an image of your real body when interacting with other users?	1 Fully disagree 7 Fully agree
	Q4	When you use a different avatar, how much do you feel that it is outwardly inconsistent with your usual avatar?	1 Fully disagree 7 Fully agree
	Q5	Evaluate the relationship between self and an avatar when using an unusual avatar, based on Fig. 4	From1 to 9
	Q6	To what extent do you identify yourself as a VRChat user?	1 Not at all 7 Very much
	Q7	To what extent do you superimpose yourself and your avatar while playing VRChat?	1 Not at all 7 Very much
	Q8	To what extent do you perceive yourself as part of the VRChat community?	1 Not at all 7 Very much
	Q9	To what extent are you happy that you are part of the VRChat community?	1 Not at all 7 Very much
	Q10	There are moments when I reflect back to myself the setting and personality I envision for Avatar	1 Rarely 7 Very often.
	Q11	While playing VRChat, there are moments when you behave yourself according to your avatar's appearance	1 Rarely 7 Very often.
	Q12	While you are living in real life, there are moments when you behave like the avatars you use in VRChat	1 Rarely 7 Very often.
	Q13	Avatar is close to my ideal image of myself	1 Fully disagree 7 Fully agree
	Q14	My ideal image came closer to Avatar	1 Not at all 7 Very much

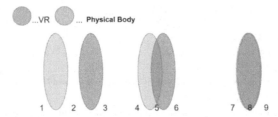

Fig. 4. Figure for item of Q5. Participants answered the evaluation of Q5 based on this.

4 Results

Parts of some users forgot to fill out daily questionnaire for a part of days. It is difficult to analyze if we remove these user's data. Therefore, we supplemented these missing values by using the median values for the 5 days before and after. We used the average value of BO1 and BO2 of Table 1 to analyze IVBO. We compared the data taken in the first week with the subsequent data, since it is difficult to acquire avatar-identity on day 1 for non-VRChat users. Since there were many missing identity changes on the last day, we did not use them. We used the data from the first, second and third week. In addition, some users filled out the questionnaire a few days late. We used those data, because the sample for the analysis would be extremely small when we excluded them. First, we analyzed for correlation between the intensity of IVBO and the number of days by single regression analysis. We analyzed the combined data of VRChat users and non-VRChat users. We also analyzed separated VRChat user and Non-VRChat users. The significance level was set at a p-value of 0.5. We show our results in Fig. 5(left). As a result, there were significant positive correlations for all analyses (All Users: r = 0.2773, p < 0.0000, Non-VRChat Users: r = 0.4691, p < 0.0000, VRChat Users: r = 0.1880, p = 0.013). The correlation coefficient of Non-VRChat users was higher than VRChat users. It is assumed that the changes of VRChat user's IVBO were not dramatically increased, because they strongly felt IVBO at the moment of day 1. Next, we analyzed between day 1 and next day using a Wilcoxon signed-rank test to investigated what timing of the effect of improving IVBO has appeared. If there is no significant difference between them, the comparison was repeated with subsequent days (e.g., day 1 vs. day 3). If the significant difference was found, we used its day as a day of reference for a comparison (e.g., day 3 vs. day 4). We show our results in Fig. 5(right). As a result, there was a significant difference between day 1 and day 10 (V = 7, p = 0.0390). Then, we compared between day 10 and later days. There was a significant trend between day 10 and day 25 (V = 1.5, p = 0.0937). However, there were no significant differences after that, and p-values of after day 25 were not lower that day 25. That means, the effect of enhancing IVBO by long-term use saturate after day 10.

Therefore, we investigated whether there is a significant difference in IVBO depending on the type of user (i.e., Non-VRChat users vs. VRChat users). A mixed two-way ANOVA was conducted with two different types of avatars as between-subjects factors and the number of days (day 1, day 14, and last day) as within-subjects factors, since the Shapiro-Wilk test showed normality in all groups. As a result, there was a main effect

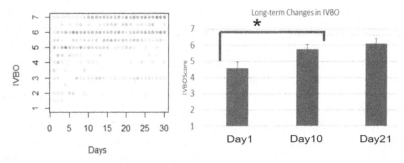

Fig. 5. Results of experiment; left) Correlation between the intensity of IVBO and days, right) Comparison regarding IVBO among days. Error bar shows a standard error.

between the number of days (F = 5.440, p = 0.0120). We examined whether avatar-identity changed with the number of days, or whether it differed among two-types of users (Non-VRChat User vs. VRChat User). We conducted a two-way mixed ANOVA, with user type as a between-subjects factor and number of days as a within-subjects factor. We show our results in Fig. 6. The results showed that there was no main effect between either user type or number of days (Avatar: F = 0.207, p = 0.6591, Day: F = 0.094, p = 0.9110). A multiple regression analysis was conducted with IVBO score as the dependent variable, identity score as the independent variable, and days as the covariate. The results of the analysis showed that the correlation coefficients between them were not significant (identity score (p = 0.298), days (p = 0.422)). That means, this result indicates that the IVBO was improved day by day, regardless of the type of user. This result also implies avatar-identification is not correlated with IVBO. It is assumed that inconsistency between their proprioception and the avatar's body cause the low evaluation, because participants who scored low mentioned to that effect.

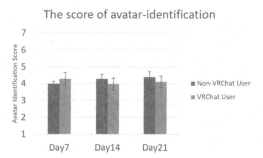

Fig. 6. Results of avatar-identification. Comparison regarding the intensity of avatar-identification among day 7, day 14, and day 21. Error bar shows a standard error.

Next, regardless of an avatar, we investigate is it possible to enhance IVBO. Since the Shapiro-Wilk test did not show normality in parts of groups, a mixed two-way ANOVA applied by the aligned rank transform (ART) was conducted with two different types of avatar (Avatar used in daily life vs. Avatar used in specific days) and days (day 1, day 14, and day 30) as within-subjects factors. When a significant difference is found, tukey's method multiple comparisons was conducted. We show our results in Fig. 7(right). As a result, there were significant differences between avatars (F = 27.2130, P < 0.0000). There was significant trend interaction between the avatar factor and days factor, although there was no significant interaction between them. Multiple comparisons showed that there was no significant difference between avatars on the first day, but there was a significant difference between at the time of comparison with day 14 and the last day (avatar daily used on day 14 and same gender avatars on day 1 (p = 0.0177), avatar daily use on day 30 and same gender avatars on day 1 (p = 0.0099), avatar daily use on day 14 and same gender avatar on day 14 (p = 0.0022), avatar daily use on day 14 and same gender avatar on day 30 (p = 0.0035), avatar daily use last day and same gender avatar on day 14 (p = 0.0011), avatar daily use on day 30 and same gender avatar on day 30 (p = 0.0018)). The avatars used every day were not real avatars of the same sex, as used in the laboratory, but cartoon avatars of the opposite sex. In other words, continuous avatar use improves IVBO regardless of the avatar's appearance.

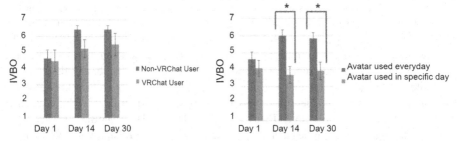

Fig. 7. Results of experiment; left) Comparison regarding IVBO between Non-VRChat users and VRChat users. Error bar shows a standard error, right) Comparison regarding the intensity of IVBO between day 1 and another day

Likewise, we conducted analysis regarding the sense of agency and the sense of presence using the two-way ANOVA applied with ART with two different types of avatar (avatar factor: Avatar used in daily life vs. Avatar used in specific days) and the number of days (days factor: day 1, day 14, and last day) as within-subjects factors. There was a significant difference regarding the sense of agency between the number of days (F = 4.3184, p = 0.0169). This result implies the sense of agency can be improved day by day regardless of avatar types. On the other hand, there was no significant difference regarding the sense of presence.

5   Discussion

Our results show correlation between the number of days and the intensity of IVBO. Furthermore, there was a significant difference between day 1 and day 10. After that, there was a significant trend day 10 and day 25, although there were no significant differences between day 10 and later days. That means, IVBO can be enhanced day by day. After that, this effect saturates on the day 10, because there was no significant difference between day 10 and later. There was also a significant difference between avatars except for day 1 (i.e., week 2 and the last day). The avatars used every day were not real avatars of the same sex, as used in the laboratory, but cartoon avatars of the opposite sex. Therefore, the improvement effect of IVBO can be effective regardless of the avatar's appearance. We concluded for RQ1 and RQ2 as below; 1) regardless of the type of avatars, IVBO can be enhanced day by day, 2) this effect can be significantly elicited at day 10. In addition, this effect saturates on the day 14. It is assumed that IVBO can be improved by top-down factors, because participants played over 1 h on day 1 (i.e., it is satisfied by visuo-motor synchronization the bottom-up factor at the moment of day 1). However, there was no significant difference regarding avatar-identification that is assumed to become the top-down factor between day 1 and the last day. In addition, there was no significant correlation between the number of days and the value of IVBO. These results imply the effect of improving IVBO was not caused by avatar- identification. In this experiment, Non-VRChat participants were instructed to manipulate the specified avatar. It is assumed that the reason why participants evaluate low avatar-identification is they did not customize an avatar. Previous work has reported that avatar-based customization can influence avatar identification [38].

In VRChat, there is a limit to the number of polygons of avatars that can be displayed so as not to overload the machine due to the performance of Oculus Quest. Thus, it is assumed that VRChat users feel strange from what they expected when compared to desktop mode, which has no limit. In addition, although we recruited VRChat users who use human avatars with limbs, their avatars are handmade avatars and the quality varies from person to person. It is assumed that it causes inconsistency between their avatars and their proprioception. In the fact, these were seen in participant's comments. Some users mentioned that they felt uncomfortable, as if the avatar of the same gender reflected in the mirror (i.e., the avatar used on a particular day) was not themselves. In other words, avatars used on a daily basis have become established as body images, and this has led to an improvement in IVBO. According to this, it is assumed that IVBO can be enhanced by continuing to use for long time, even in the case of users who have strong avatar-identification. Taken together, we concluded for RQ3, H1, H2 as below; 1) Our results showed avatar-identification is not collated with IVBO, 2) and it only be collated with the number of days. 3) Regardless of the type of users, IVBO can be enhanced day by day. However, our experiment could not completely prove this.

6 Limitation and Future Work

Our results showed avatar-identification is not correlated with IVBO. However, we have not collected enough data to prove this. For example, regardless of avatar-identification, it is assumed that participants simply got used to seeing the avatar during the experiment of latter half. It is also assumed that IVBO improves in conjunction with an improved sense of agency as a result of learning how to exercise an avatar. In addition, we did not collect a lot of VRChat users who has already strong avatar-identification (i.e., some of the VRChat users collected in this paper did not have strong avatar-identification). If avatar-identification are related with IVBO, participants having strong it evaluates low IVBO to an avatar prepared by experimenter even they use an avatar long-term.

Properties of the avatars (e.g., shape, appearance, height and length of limbs) to be recruited cannot be completely controlled. However, we should recruit participants who have an avatar consistent with the avatar's body position and the proprioception, which play a major role in bottom-up factors. In the future work, we plan to address these issues to further investigate the effects of avatar-identification on IVBO.

7 Conclusion

We investigated how long-term use of an avatar impacts on the IVBO. Previous our work has reported that user's avatar (i.e., users identify an avatar as a digital representation in social VR) can significantly enhance IVBO in comparison with an avatar prepared by researchers. In this paper, we conducted a long-term experiment in social VR platform VRChat to investigate changes in IVBO. In addition, we also investigated the correlation between changes in IVBO and avatar-identification. Our results showed the correlation between the number of days and IVBO. That means, the intensity of IVBO can be enhanced day by day. Eventually, the results showed what could be called an improvement effect in the second week. On the other hand, avatar-identification could not be correlated with the number of days. Hence, it is assumed that the effect of improving IVBO cannot be elicited by avatar-identification. However, the fact that the IVBO improves after about two weeks of use of any avatar indicates the possibility of more immersive learning by creating a mechanism that allows avatars used in social VR platforms to be used in training VR applications.

Acknowledgments. This work was partially supported by JST Moonshot R&D (Grant Number JPMJMS2013). This work was also supported by Council for Science, Technology and Innovation, "Cross-ministerial Strategic Innovation Promotion Program (SIP), Big-data and AI- enabled Cyberspace Technologies". (funding agency: NEDO)

References

1. Lugrin, J.-L., Latt, J., Latoschik, M.E.: Anthropomorphism and illusion of virtual body ownership. In: Imura, M., Figueroa, P., Mohler, B. (eds.) In Proceedings of the 25th International Conference on Artificial Reality and Telexistence and 20th Eurographics Symposium on Virtual Environments, pp. 1–8, (2015)
2. Yee, N., Bailenson, J.: the proteus effect: the effect of transformed self representation on behavior. Hum. Commun. Res. 33(3), 271–290 (2007)
3. Steed, A., Frlston, S., Lopez, M.M., Drummond, J., Pan, Y., Swapp, D.: An 'In the Wild' experiment on presence and embodiment using consumer virtual reality equipment. IEEE Trans. Vis. Comput. Graph. 22(4), 1406–1414 (2016)
4. Buss, S.R.: Introduction to inverse kinematics with jacobian transpose, pseudoinverse and damped least squares methods (2004)
5. Coleman, E.J., Moore, A.G., Fishwick, P.A., McMahan, R.P.: The Effects of Body Tracking Fidelity on Embodiment of an Inverse-Kinematic Avatar for Male Participants. In: 2020 IEEE International Symposium on Mixed and Augmented Reality, pp. 54–63. Porto de Galinhas (2020)

6. Freeman, G., Malony, D.: Body, Avatar, and me: the presentation and perception of self in social virtual reality. Proc. ACM Hum.-Comput. Interact. **4**, 239 (2020)
7. Roth, D., Lugrin, J.-L., Latoschik, M,E., Huber, S.: Alpha IVBO – construction of a scale to measure the illusion of virtual body ownership. In: Proceedings of CHI Conference Extended Abstracts on Human Factors in Computing Systems, pp. 2875–2883. ACM, Denver (2017)
8. Spanlang, B.: How to build an embodiment lab: Achieving body representation illusions in virtual reality. Front. Robot. AI **27**, 1–22 (2014)
9. Waltemate, T., Gall, D., Roth, D., Botsch, M., Latoschik, M,E.: The impact of avatar personalization and immersion on virtual body ownership, presence, and emotional response. IEEE Trans. Vis. Comput. Graph. **24**(4), 1643–1652 (2018)
10. Dongsik, J., Kangsoo, K.: The impact of avatar-owner visual similarity on body ownership in immersive virtual reality. ACM VRST Conference '17, pp. 77–78. ACM, Gothenburg (2017)
11. Oyanagi, A., Narumi, T., Ohmura, R.: An Avatar that is used daily in the social VR contents enhances the sense of embodiment. Trans. Virtual Reality Soc. Japan **25**(1), 50–59 (2020)
12. Ducheneaut, N., Ming-Hui, W., Yee, N., Wadley, G.: Body and mind: a study of avatar personalization in three virtual worlds. In Proceedings of the SIGCHI Conference on Human Factors in Computing Systems, pp. 1151–1160. ACM, Boston (2009)
13. Schroeder, R.: The Social Life of Avatars: Presence and Interaction in Shared Virtual Environments. Springer Science & Business Media (2012)
14. Milik, O.: Persona in MMO Games: Constructing an identity through complex player/character relationships. Personal Stud. **3**(2), 66–78 (2017)
15. Botvinick, M., Cohen, J.: Rubber hands 'feel' touch that eyes see. Nature **391**, 756 (1998)
16. IJsselsteijn, W.A., de Kort, Y.A.W., Haans, A.: Is this my hand I see before me? The rubber hand illusion in reality, virtual reality, and mixed reality. Presence: Teleoper. Virtual Environ. **15**(4), 455–464 (2006)
17. Slater, M., Perez-Marcos, D., Ehrsson, H., Sanchez-Vives, M.V.: Towards a digital body: the virtual arm illusion. Front. Hum. Neurosci. **2**(6) (2008)
18. Tsakiris, M., Prabhu, G., Haggard, P.: Having a body versus moving your body: how agency structures body-ownership. Conscious. Cogn. **15**(2), 423–432 (2006)
19. Tsakiris, M.: My body in the brain: a neurocognitive model of body-ownership. Neuropsychologia **48**(3), 703–712 (2010)
20. de la Peña, N.: Immersive journalism: immersive virtual reality for the first-person experience of news. Presence: Teleoper. Virtual Environ. **19**(4), 291–301 (2010)
21. Tsakiris, M., Carpenter, L., James, D.: Hands only illusion: multisensory integration elicits a sense of ownership for body parts but not for non-corporeal objects. Exp. Brain Res. **204**(3), 343–352 (2010)
22. Haans, A., IJsselsteijn, W.A., de Kort, Y.A.W.: The effect of similarities in skin texture and hand shape on perceived ownership of a fake limb. Body Image **5**(4), 389–394 (2008)
23. Banakou, D., Groten, R., Slater, M.: Illusory ownership of a virtual child body causes overestimation of object sizes and implicit attitude changes. Proc. Natl. Acad. Sci. **110**(31), 12846–12851 (2013)
24. Slater, M., Spanlang, B., Sanchez-Vives, M.V., Blanke, O.: First person experience of body transfer in virtual reality. PLoS ONE **5**(5), (2010)
25. Peck, T.C., Seinfeld, S., Aglioti, S.M., Slater, M.: Putting yourself in the skin of a black avatar reduces implicit racial bias. Conscious. Cogn. **22**(3), 779–787 (2013)
26. Krekhov, A., Cmentowski, S., Kruger, J.: Vr animals: surreal body ownership in virtual reality games. In: Proceedings of the 2018 Annual Symposium on Computer-Human Interaction in Play Companion Extended Abstracts, pp. 503–511. ACM, New York (2018)
27. Krol, S.A., Thériault, R., Olson, J.A., Raz, A., Bartz, J.A.: Self-concept clarity and the bodily self: malleability across modalities. Soc. Pers. Soc. Psychol. **46**(5), 808–820 (2020)

28. Dewez, D., Fribourg, R., Argelaguet, F., Hoyet, L., Mestre, D., Slater, M., Lécuyer, A.: Influence of personality traits and body awareness on the sense of embodiment in virtual reality. In: 2019 IEEE International Symposium on Mixed and Augmented Reality (ISMAR), pp. 123–134. Beijing, China (2019)
29. Marotta, A., Tinazzi, M., Cavedini, C., Zampini, M., Fiorio, M.: Individual differences in the rubber hand illusion are related to sensory suggestibility. PLoS ONE **11**, (2016)
30. Marotta, A., Zampini, M., Tinazzi, M., Fiorio, M.: Age-related changes in the sense of body ownership: new insights from the rubber hand illusion. PloS one **13**(11) (2018)
31. Wolfendale, J.: My avatar, my self: Virtual harm and attachment. Ethics Inform. Technol. **9**(2), 111–119 (2007)
32. Inkpen, K.M., Sedlins, M.: Me and my avatar: exploring users' comfort with avatars for workplace communication. In: Proceedings of the ACM 2011 conference on Computer supported cooperative work, pp. 383–386. ACM, New York (2011)
33. Armel, K.C., Ramachandran, V.S.: Projecting sensations to external objects: evidence from skin conductance response. Proc. R. Soc. Lond. B Biol. Sci. **270**, 1499–1506 (2003)
34. Rohde, M., Di Luca, M., Ernst, M.O.: The rubber hand illusion: feeling of ownership and proprioceptive drift do not go hand in hand. PLoS ONE **6**(6), (2011)
35. Ehrsson, H.H., Spence, C., Passingham, R.E.: That's my hand! Activity in premotor cortex reflects feeling of ownership of a limb. Science **305**, 875–877 (2004)
36. Kammers, M.P., Verhagen, L., Dijkerman, H.C., Hogendoorn, H., de Vignemont, F., Schutter, D.J.: Is this hand for real? Attenuation of the rubber hand illusion by transcranial magnetic stimulation over the inferior parietal lobule. J. Cogn. Neurosci. **21**(7), 1311–1320 (2008)
37. Viaud-Delmo, I. (n.d.). Igroup Presence Questionnaire (IPQ) item download. http://www.igr oup.org/pq/ipq/IPQinstructionsFr.doc
38. Turkay, S., Kinzer, C.K.: The effects of avatar-based customization on player identification. Int. J. Gaming Comput.-Med. Simul. **6**(1), 1–25 (2014)

Multi-modal Data Exploration in a Mixed Reality Environment Using Coordinated Multiple Views

Disha Sardana[1]([✉]) [iD], Sampanna Yashwant Kahu[1] [iD], Denis Gračanin[1] [iD], and Krešimir Matković[2] [iD]

[1] Virginia Tech, Blacksburg, VA, USA
{dishas9,sampanna,gracanin}@vt.edu
[2] VRVis Research Center, Vienna, Austria
matkovic@vrvis.at

Abstract. Immersive analytics is an emerging field of data exploration and analysis in immersive environments. The main idea is to use visual analytics in a fully immersive 3D space. The availability of immersive Extended Reality systems has increased tremendously recently, but it is still not as widely used as conventional 2D displays. We describe an immersive analysis system for spatio-temporal data and compare how it performs in an immersive environment and on a conventional 2D display. We provide a novel view called map-plot that enables analysis and exploration of spatial time series data. We also design an embodied interaction for the map-plot view. The approach is realized based on the coordinated multiple views paradigm. In addition to the map-plot view, several standard views are available. Voice commands and spatial audio are used to support interaction and increase users' embodiment. The findings from a user study show that developed system is much more efficient in a real immersive environment than using conventional 2D displays.

Keywords: Immersive analytics · Mixed reality · Embodied interaction · Coordinated multiple views

1 Introduction

Data visualization helps scientists, domain experts, and laypersons to explore and understand complex data. In spite of recent advancements in automatic data analysis technology, such as deep learning, visualization remains a premium choice for data analysis in case of complex and ill defined problems where human imagination and intuition play a crucial role.

Traditionally, visualization is depicted on 2D screens. Recent developments and omnipresence of 3D displays motivates a research question if immersive spaces can be used for 3D visualization. Visualization in immersive environments

---

D. Sardana and S. Y. Kahu—The authors contributed equally to the paper.

© Springer Nature Switzerland AG 2021
S. Yamamoto and H. Mori (Eds.): HCII 2021, LNCS 12765, pp. 337–356, 2021.
https://doi.org/10.1007/978-3-030-78321-1_26

may lead to a better chance of recognizing interesting patterns, correlations between parameters and outliers [5]. Immersion helps in understanding data more intuitively in terms of its geometry and spatial context, and improves retention of the realized correlations in the data [12].

Immersive visualization can make it easier to find connections in data in a high-dimensional space. Users can walk through the data, get immersed completely and can increase their pattern recognition capability. Immersive analytics aims to remove the barriers between humans, data and technology for data analysis and decision making [13]. However, immersion on its own is not enough, and needs to be accompanied by embodiment. User interactions must take advantage of the human bodies as an interaction device [16].

We combine immersive analytics and interactive visual analysis by means of coordinated multiple views (CMV). We deploy a real 3D view which supports embodied interaction to better relate spatio-temporal data to its origin. We still use 2D views in immersive environment, where the use of 3D is not justified. The newly introduced spatio-temporal map-plot view outperforms conventional, 2D, multiple curves view for spatio-temporal data. Addition of spatial sound and embodied interaction enhances analysis further. Usage of spatial sound allow assigning sound sources to individual data points (objects) in the mixed reality (MR) space. The aggregated, spatialized sound, that depends on the user position and orientation in immersive, or MR space (relative to data points), provides audio cues to inform and aid visual analysis. Using the whole body to navigate and interact with the MR space provides affordances for embodied interactions. We describe user feedback on immersive analytics and demonstrate the usefulness of our approach on a complex spatio-temporal data set from the IEEE VAST Challenge 2019 [1].

2   Related Work

Studies have not only hinted at the potential of immersive environment over the desktop system to considerably enhance users' efficiency for structure and feature detection tasks in data analytics, but also suggested that users learn to perform tasks in a virtual environment fairly quickly because of the perceptive nature of the environment [5].

Bach et al. used a desktop environment, a TabletAR (tablet + tangible object) mode and an ImmersiveAR (HoloLens) environment to compare performance for tasks with varying levels of spatial perception, degrees of freedom for interaction and the spatial and cognitive proximity between the two [6]. Tasks requiring high level of manipulation (selection and cutting plane) were performed the best in Immersive tangible AR scenario, in terms of precision and task completion time, respectively. Even though the HoloLens device was new to most of the participants, ImmersiveAR scenario was as precise as Desktop, given any task. The TabletAR mode performed the worst overall, possibly due to the mismatch between the dimensionality of the perception and the interaction space.

Filho et al. analyzed multidimensional data obtained through PCA technique in immersive and desktop environments and found out that the effort to find information and navigation in 3D scatterplot in immersive spaces is lesser than in desktop based 2D or 3D conditions [26]. Prouzeau et al. focused on the design spaces for visual links routing between physical and virtual objects in immersive environments [22]. Blum et al. enable users ability to arrange various data visualizations in a large display immersive environment to study sense-making in three different settings: a very large wall display system, virtual reality (VR) headset with Sparse Peripheral Display (SPD) and non-SPD VR environment [8].

Su et al. conducted a comparative study for multiple coordinated 2D and 3D visualizations to get an assessment of using the hybrid visualizations in data analysis [25]. Their system supported linking and brushing across 2D (Large High Resolution Display) and a complete immersive HMD based 3D environment.

Recently, various researchers have developed tool-kits [11] or environments [9] that support a visualization pipeline in XR environments and can visualize large amount of data points. The developed systems allow users to visualize their data at a room-sized scale [11], interact naturally and intuitively with them in real-time [9], and have the capability to allow tangible interactions.

The main goal of engaging in data in new ways is to find out how to derive insight from big and complex data. Adding modalities other than vision is a possible approach. Currently, the multi-modal aspects in the field are under exploited [15]. Sound and touch in addition to visuals have largely been overlooked [10]. Dynamic spatial audio can play a fundamental role in immersive scenarios [18]. With eyes one can see in front, but with sound one can hear things from all around. Spatial audio enables users to more precisely connect to spatial data in a novel and engaging manner than the non-spatialized sonification techniques [7]. Certain sound attributes like loudness can be geo-mapped to the data points, making the interaction less time consuming [17], and leading to a better spatial perception of the data. Our ears are excellent signal-to-noise filters allowing events to be recognized above the background noise [7]. Studies [17,20] suggested a significant improvement in search based tasks when auditory cues are used.

There are many challenges related to the development of MR CMV tools. It will be worth exploring data structures with higher complexity, e.g. curves, complete cubes, complex shapes, etc. Many studies have utilized point cloud visualizations for doing comparative studies, since it incorporates visualizations like 3D scatterplots, space-time cubes, even biomedical images, and flow fields [6]. Very little interaction is observed in the current studies [6]. It may be intentional, to manage the effect of the interactions on user performance/efficiency. An example is limited exploration of multi-modal interactions. Sound and touch have been largely overlooked, including limited use of voice commands [24]. Multimodal interaction allows sensory distribution, reducing visual overwhelming, and improving multisensory view management system [20].

Data understanding requires a human interaction to visualize and listen to it to comprehend what is there. That brings a need for designing a system that would allow easy interaction and manipulation of the displayed data. Fully immersive VR isolates the user from the real world and enormously diminishes the affordances for embodied interactions. MR environments provide sufficient immersion while maintaining those affordances. Hence, the goal is to take advantage of embodied cognition in an MR environment, which in turn will lead to better user experiences and more effective analysis [16].

3   Approach

We propose an approach for embodied data exploration and immersive analysis of multi-dimensional heterogeneous data using coordinated multiple views (CMV) in an MR environment. Leveraging the surrounding space, we systematically position multiple coordinated 3D and 2D views in the space, allowing user interaction and linking between the displayed visualizations. The main idea of CMV is to show several views which depict various attributes of multi-dimensional data. The user can then interactively select some of the data items in one view (this is called brushing), and all items that belong to the same records will be highlighted in all other views (linking).

We use a 3D view for parallel visualization of spatio-temporal data sets. Additional views are then added, as needed. Interaction is provided by using a virtual slicing plane to allow users to navigate across the temporal information in the same space. That also helps to avoid the mismatch between the dimensionality of the perception and the interaction space. Sonification of the data points provides complementary audio cues to aid in the visual analytics process. Similarly, voice commands complement the gestures. The MR environment (including the physical space) provides affordances for embodied interactions.

As a proof of concept, we wanted to gauge the reaction of users traditionally conditioned to see visual plots of the data in 2D environments. It was also vital to ascertain if users would note the multi-modal aspects of the designed demo. We chose to present and analyze data in an MR environment using the Microsoft HoloLens device on the well known Iris dataset [3] and International Reference Ionosphere—IRI dataset (2016) [2].

Figure 1(a) shows the rendered 3D scatter plot from Iris data set and Fig. 1(b) shows the scatter plot in an MR environment. The points in the dataset are represented using databall objects (i.e., spheres of a fixed size). The R, G and B values for the color of each databall are mapped to their respective X, Y and Z coordinates. All the axis labels, axis limit labels and the plot title are programmed such that they swivel around their respective y-axis. The amount of swivel is controlled by the position of the user in the 3D space such that all the labels always face towards the user.

A spatial sound source can be placed in a 3D space so it appears to be originating from a given point in that space. The sound source is modelled as a point source and the sound can be changed by providing it any standard audio file configured to play in an infinite loop. In this implementation, we place audio sources at the center of each of the databalls. A sine-square wobble sound is

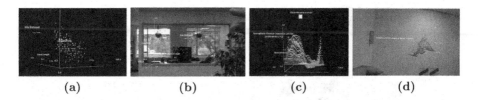

Fig. 1. Snapshots of our proof-of-concept in action. **(a)**, **(b)**: Iris dataset scatter plot. **(c)**: Electron density plot at 08:00:00 UTC shown in the Holographic Remoting player. **(d)**: Electron density plot at 09:00:00 UTC in MR.

played in an infinite loop from this sound source, mimicking the sound of a bird calling. This sound is chosen because natural sounds are found to be relaxing and have proved helpful in the perception and interpretation of the data [21].

4   Prototype

Our design and approach is informed by an extensive literature survey and user feedback from the proof of concept [22,27]. We try to leverage natural geometry of rooms by arranging the elements of our design as if they were placed on the walls and floor of the room in the virtual environment. To show the CMV, the user can issue voice commands, such as 'show brushed data', 'show bar charts' and 'show scatter plot'. The IEEE VAST challenge 2019 data [1] was used. The task of the challenge is to apply visual analytics to help a city grapple with the aftermath of an earthquake that damages their nuclear power plant. We specifically focus on the first part of this challenge which deals with visual analysis of crowd-sourced data about the different kinds of damages citizens observe during the course of the disaster (such as, sewers, roads or bridges, etc.).

4.1   Map-Plot View

In order to visualize the spatial dimension of the dataset effectively, we leverage the geographical map of the region. Each record in the dataset provided by citizens is associated with the location from which it originated. We placed the map of the city on a plane which shows the borders between its 19 sub-regions, which we term as map-plot. For each record, we plot the associated data within the border of its corresponding region on this map-plot. The intuition behind this approach is to visually associate each record with its corresponding location in the map thereby providing a more natural sense of localization to the user for the data points within the maps (Fig. 2(a)).

The dataset contains time-stamps associated with each record denoting the time at which a record was received. This makes it essential to visualize the data in a temporal fashion, maintaining the temporal relations between the records. We do this by placing two map-plots on the opposite walls of the room. The

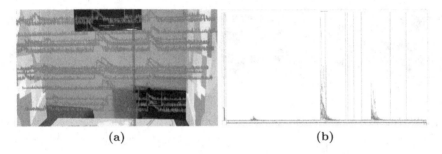

(a)   (b)

Fig. 2. 3D and 2D comparison of time-series data. (**a**): The new map-plot view. Data is linked with each neighborhood. (**b**): The conventional 2D view shows multiple time series.

maps on one of these planes is the original version and on the other is the mirror reflection around the vertical axis. This helps in aligning each neighborhood between the two maps when facing each other. Labels indicating the time-stamp represented by these planes are placed vertically above them. Further, each corresponding pair of neighborhoods on these two maps are connected with each other by a line-plot between them displaying temporal information. The spatial perspective (Fig. 2(a)) provides additional context and provides better insight into data compared to the corresponding conventional 2D view (Fig. 2(b)).

4.2   Coordinated Multiple Views (CMV)

The dataset contains multiple heterogeneous dimensions. Although these dimensions are being shown as an aggregate on the time series line plots, a more detailed and segregated view of the data is necessary. In order to achieve that, CMV of the individual data dimensions are shown. The linked CMV exhibit brushing, as shown in Fig. 3(b)). If a user selects a data sphere on the floor view, its color toggles, and the color for all elements corresponding to the same region in all other linked CMV also toggles.

We created two scatter plots in our visualization. One of them is placed on the floor map-plot as shown in Fig. 3(a). It shows a region-wise aggregate of all reported values for a dimension (i.e., shake intensity) for the time-stamp selected by the slicing plane. In another scatter plot, we plot one dimension from the dataset (i.e., power) on a map-plot by plotting a single sphere hovering over each neighborhood as shown in Fig. 3(e). The radius and the color of the sphere are determined by the number of reports received and by the average of the intensity values of all the damage reports received at a given point in time. (Fig. 3(g)).

Multiple 3D bar charts (one for each region) are aligned in a direction normal to the map-plot plane, leveraging the 3D space as compared to a 2D plot. The height and color are changed based on the intensity value of the dimension being plotted (Fig. 3(f)). The 2D bar chart shows the aggregate data for the brushed

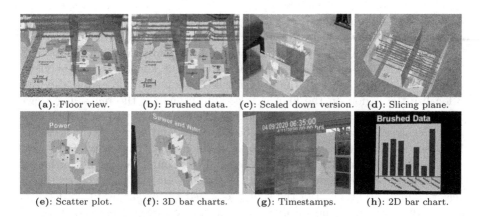

(a): Floor view.   (b): Brushed data.   (c): Scaled down version.   (d): Slicing plane.

(e): Scatter plot.   (f): 3D bar charts.   (g): Timestamps.   (h): 2D bar chart.

Fig. 3. The components of the plot container.

regions on the floor plot. The distribution of the number of reports received for each of the dimensions in the dataset is shown in the 2D bar chart (Fig. 3(h)).

4.3   Slicing Plane

The data plotted on each of the CMV are shown only for any given timestamp. However, there needs to be a mechanism to allow users to navigate along the time axis. We introduce a slicing plane to select an instant of time. The slicing plane is parallel to the two vertical map-plots to the left and the right of the time-axis and can move only along a single direction (Fig. 3(d)) to allow users a precise control in the 3D space. The place on the time-series line plots where this slicing plane intersects with them is the timestamp used to display the data on each of the CMV. Whenever users move this slicing plane along the time axis, the data on all the CMV changes to reflect the data at the timestamp where the slicing plane intersects the time axis. A label showing the timestamp the slicing plane is currently at, is placed vertically on top of it (Fig. 3(g)).

4.4   Sound Design

In scenarios with complex visual information, busy visualizations, or where the user is unable to look at multiple visuals, parallel listening can be of great advantage [14]. For example, when a user is sliding the slicing plane, the change in data in the other CMV, like on the floor, or behind the user on a wall, might not be easy to pay attention to visually. We added sound to all the CMV in the environment. It was important that the sounds are non-fatiguing to allow data exploration for longer periods of time. To sonify multiple streams of data in various CMV concurrently, we mapped the sounds intuitively to the data attributes wherever possible. The differences in these sounds allow easy distinction between various data types while parallel listening, therefore reducing the cognitive load.

We added a sound source to the biggest databall in the scatterplot placed on the floor map-plot, for any given time-stamp. The amplitude of this source is directly proportional to the radius of the databall from which it is emanating. Since the floor map-plot (along with the scatter plot and the map-plot view) could be scaled to room size, the direction of the sound would enable the user to localize the region on the map it is coming from. For the remaining views (2D bar chart/3D bar chart/scatterplot), we assigned a sound source to each of their centres. Since the CMV could be moved around anywhere in the room, the closer a user is to a view, the louder the sound is. That is why, we used phase modulation for mapping data in the CMV to sound. As the magnitude of the data increases, the tempo at which the sound is emanating increases (like a Geiger counter). The pitch of the sound attached to the slicing plane depends on the sum of all the damage values reported at the current time stamp.

5   Cross-Platform Implementation and System Performance

We deployed our prototype in two MR environments using Microsoft HoloLens 1, and Microsoft HoloLens 2 devices to perform a cross-platform system performance comparison based on the prototype's scalability and functionality. We used the Unity 3D Engine (v 2019.4.11f1 LTS) and Model View Controller framework. We also deployed our prototype using WebGL to compare between an immersive 3D environment setup (HoloLens) and a non-immersive 3D environment setup (WebGL). Navigation in WebGL is done via a first person controller character using mouse/trackpad and keyboard (W-A-S-D keys or arrow keys). Further, the interaction with the slicing plane, brushing of the floor data and dragging of the CMV is supported using the left mouse button. The sound of various data attributes in different views is also provided. We use the Web Speech API [4] to enable voice commands providing the same modes of interaction, visual and aural, in both environments.

To evaluate the performance of our prototype, we conducted an experiment to measure the memory and CPU utilization under different operating conditions and usage patterns of the prototype. For memory utilization, we measure the total space in MB used by all the applications running on the device. Specifically, we record the active list size on the device. To see the trends in the memory usage of our application, we leave all the other applications unchanged throughout the experiment's duration. We record this value for a span of 10–15 seconds for each of our readings and average them.

Table 1 shows the recorded values for Microsoft HoloLens 1 and 2 devices. While the CPU usage for both the devices is comparable, HoloLens 2's memory usage is almost double that of HoloLens 1. The CPU usage drops considerably for large room-sized plot for both HoloLens 1 and HoloLens 2. This could be because of the large size of the plot, only a part of it is visible to the user and only the objects in the current field of view of the user are rendered.

Table 1. Memory and CPU usage with respect to the size of the prototype and the level of functionality in Microsoft HoloLens 1 (HL1) and HoloLens 2 (HL2).

Factor	HL1		HL2	
	Mem. (Mb)	CPU (%)	Mem. (MB)	CPU (%)
Scalability				
Small (Laptop-sized)	1504.995	51.06	3034.341	53.18
Medium (Table-sized)	1569.712	54.96	3224.193	52.89
Large (Room-sized)	1511.883	23.92	3265.011	45.21
Functionality				
Default	1522.461	57.67	3029.027	53.69
Default + CMV	1513.578	65.05	3025.037	59.1
Default + CMV + Brushing	1513.516	59.95	3056.986	59.95
Default + CMV + Brushing + Slicing Plane movement	1521.532	59.59	3099.244	57.56

While comparing different levels of functionality in the default scale, the memory usage does not fluctuate as much since the scale of the default mode allows the user to have the entire plot in the field of view, no significant change in memory usage is observed for both HoloLens 1 and 2. Further, the entire dataset is always loaded in the memory. As a result, no on-demand or batch-wise transfer of data is being done to and from the disk. Another point to be noted is that when the CMV are not visible (i.e., they are hidden), they are still getting rendered with the visibility setting disabled. These arguments further justify the relatively lower variance in the memory usage. Lastly, against the intuition that with increased level of functionality the CPU usage should increase, we do not see a clear trend in the recorded values that could explain this intuition.

6   User Study Design

We designed a between-subjects user study to evaluate how effective it is to analyze data in an immersive 3D environment using a Microsoft HoloLens 2 device, as compared to doing the same analytics task in a non-immersive WebGL environment on a 2D desktop screen.

6.1   Tasks and Procedures

The participation in the study was individual. The entire procedure was split into two sections: in-person study and online study. The in-person study took place in a room where users were free to walk around. The online study took place over a Zoom session.

After filling out the eligibility criteria questionnaire, the users were scheduled for their respective study versions. The participation was voluntary, and it was made sure that they understood they could quit the study if they felt uncomfortable for any reason. The study began with a pre-study questionnaire to collect the users' demographics. We then gave a demonstration of the prototype to the users using a presentation. The presentation's content was the same for both the scenarios to train the users identically.

In the in-person study, the users were then given the HoloLens 2 device to get comfortable with, train for the gestures, and explore the prototype's features. Correspondingly, in the online study, the users were given a link to the WebGL application to open in a Google Chrome browser to explore the prototype's features. They were asked to share their computer screen and sound.

Table 2. Exploratory tasks.

ID	Type	Description
t1	Search	Name the region with a nuclear plant
t2	Count	How many hospitals are there in the entire city?
t3	Identification of important features	Brush the biggest data sphere at timestamp '04/08/2020 12:00:00'. Enter the name of the region linked to this data sphere
t4	Correlation, Count	Count and enter the total number of individual bars for the brushed region in the 3D bar chart view
t5	Correlation, Identification	Name the least damaged utility based on the number of reports for the brushed region in the 2D bar chart view
t6	Understanding of data	What is the timestamp when maximum number of damage reports are received, indicated by the line-plots?
t7	Trend determination	How is the sound of the slicing plane correlated with the damage reports? Positive/Negative/Zero correlation
t8	Anomaly detection	Identify the timestamp when the power outage occurs for the Broadview region (i.e., the line-plot goes flat)

The training section was designed to provide an overview of our prototype's features, including linking, brushing, voice commands, and interaction with the slicing plane. A timed questionnaire was given to the users, which showed the question, and the users were expected to answer those questions by exploring the developed prototype. In the HoloLens environment, the questionnaire appeared in a browser window placed on their right-hand side. In the WebGL environment, the users were told to use keyboard shortcuts to switch between the two tabs: the WebGL application and the questionnaire.

After completing the training section, the users moved to the task section. A series of exploratory tasks were given to them, as described in Table 2. The task section began with a search-type task followed by a counting-type of task. The counting-type tasks prevent users from disengaging once the target information is found, providing them understanding of the global context. Afterward, we had tasks related to understanding data, identifying important features, finding correlations, trend determination, and anomaly detection. All of these tasks are evaluated based on the completion time and accuracy.

We recorded users' point of view in the in-person study to observe their interaction patterns in the HoloLens environment for various tasks. Once the task section was completed, the users filled out a post-study questionnaire reflecting their experience, the challenges they faced, the most useful elements of the prototype, and the prototype's intuitiveness. The entire study was done within 60 min for either scenario.

6.2 Data Collection

We used Qualtrics as the tool for implementing questionnaires (i.e. pre-study questionnaire, training questionnaire, task questionnaire, and post-study questionnaire) to collect data during the user study. In the online study, users used their mouse/trackpad and keyboard to fill data into these questionnaires. For the in-person study, the training and task questionnaires were filled using the in-app MRTK virtual keyboard in a Microsoft Hololens 2 device. The pre-study and post-study questionnaires in the in-person study were filled using a keyboard, mouse and monitor that we had made available to the users at the venue.

To measure the accuracy and the time taken to complete each of the tasks in the training and task section, the data entered by users in the Qualtrics questionnaires was recorded. Further, the time-stamp of the first click, the time-stamp of the last click, the time-stamp at which the question was submitted and the total number of clicks were recorded. This click-related data was recorded only if the users clicked anywhere on the questionnaire web page. In the online study, the users were asked to open the questionnaires in a different browser window, and they used keyboard shortcuts to switch between the WebGL browser window and the questionnaire window. In the in-person study, we implemented a hand menu with two buttons to open an in-app web browser for displaying the training and task questionnaires respectively. We asked the users to position the web browser to their right hand side at the beginning of the training and task sections. Tap and scroll gestures of the HoloLens 2 device were used to interact with the web page and the click-related data similar to the online study was collected.

6.3 Hypothesis

The null hypothesis H_0 for the experiment is that the task completion times in an immersive MR environment (HoloLens) are greater than the time taken by a user to complete the same task in a non-immersive 3D environment (WebGL). The alternative hypothesis H_α is that the task completion times in an immersive MR environment (HoloLens) are lower than the time taken by a user to complete the same task in a non-immersive 3D environment (WebGL).

6.4 Demographics

Non-immersive 3D Environment on a 2D Computer Screen: A total of 18 users (11 female, 7 male) participated in the study. 16 of the users were in the

Table 3. Task accuracy and median task completion times (in seconds).

ID	HoloLens		WebGL	
	Accuracy	Time (seconds)	Accuracy	Time (seconds)
t1	1	11.86	1	23.84
t2	0.75	23.52	0.61	33.20
t3	0.81	35.57	0.67	51.76
t4	0.75	32.61	0.94	82.82
t5	0.94	34.24	1	56.67
t6	0.50	37.80	0.44	118.45
t7	1	28.01	0.94	38.64
t8	0.87	24.81	0.61	69.50

20–33 age group. Two users were predominantly left-handed, and one of them was ambidextrous. None of them had physical disability or mobility issues that could prevent them from using a 2D computer screen, keyboard or mouse. They all had prior experience with visual analytics, e.g., using bar charts, graphs, scatter plots. None had prior experience with WebGL applications. Only one had prior experience with analyzing data using sound.

Immersive 3D Environment Using a Microsoft Hololens 2 Device: A total of 16 users (10 male, 6 female) participated in the study. 14 of the users were in the 20–27 age group. One user was predominantly left-handed. None of them had physical disability or mobility issues that could prevent them from making hand gestures or moving within the experiment-space. All of the users had prior experience with visual analytics, e.g., using bar charts, graphs, and scatter plots. 10 users had prior experience with XR, and eight of them had prior experience with MR applications. Nine of them wore glasses. 10 users had prior experience with gesture-based interaction devices such as Leap Motion, Kinect, etc. Only two users had prior experience with analyzing data using sound.

7   User Data Analysis

Each user submitted their responses for the eight data analytics tasks using a Qualtrics questionnaire described above. One of the users in the in-person study could not finish the task section as the device battery got critically low. In total, we got 127 (16×8 - 1) responses for the HoloLens environment, and 144 (18×8) responses for the WebGL environment.

7.1   Statistical Analysis

Figure 4 shows side-by-side boxplots of users' performance in terms of task completion times, for both modes of study: HoloLens and WebGL. Task completion

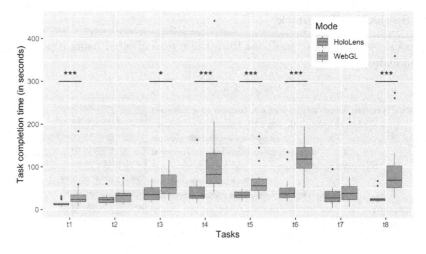

Fig. 4. User performance measured as task completion times (in seconds).

times are taken as the timestamp of the first-click on the task questionnaire. This was done to avoid any bias in terms of familiarity levels of the users with a physical keyboard versus a virtual keyboard. A task was considered as done, as soon as a user clicked on the answer textbox to start typing an answer. The users were specifically told not to click anywhere on the page until they found an answer. This was monitored in the training section as well. Table 3 shows the accuracy and median time taken to complete a task (in seconds). The accuracy is normalized since the number of users for both modes of studies are different.

Wilcoxon Rank Sum test, a non-parametric test, was used to find significant differences in the task completion times for each task. We chose this because when we tested the normality assumption using the Shapiro Wilk test [23], only task t3 satisfied this assumption. Further, we used the Levene's test [19] to test the homogeneity of variance condition for task t3 and it was not satisfied. Therefore, for all the tasks, we chose to do the Wilcoxon Rank Sum test.

We used an α level of 0.05 in all significance tests and report the significance values for $p < .05$, $p < .01$, and $p < .001$, abbreviated by ($*$), ($**$) and ($**$ $*$), respectively. The p-value from the Wilcoxon Rank Sum test for task t1 is 0.0002914 ($* * *$), task t2 is 0.05087, task t3 is 0.02518 ($*$), task t4 is 7.967e-06 ($* * *$), task t5 is 0.0002914 ($* * *$), task t6 is 2.569e-06 ($* * *$), task t7 is 0.1484 and task t8 is 8.813e-07 ($* * *$).

We conclude that there is sufficient sample evidence to reject our null hypothesis H_0 for tasks t1, t3, t4, t5, t6 and t8. There is enough evidence in the data to show that the task completion times in an immersive MR environment (HoloLens) for certain types of tasks are significantly lower than the same in a non-immersive 3D environment (WebGL). For tasks t2 and t7, there is no significant statistical difference in the task completion times between the two modes. Therefore, we fail to reject the null hypothesis H_0 for tasks t2 and t7.

7.2 Subjective Feedback

Figure 5 shows subjective feedback from the users collected using a post-study questionnaire.

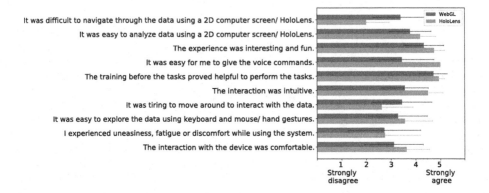

Fig. 5. Mean user responses for the post-study questionnaire.

WebGL: One of the most common feedback received from the users for the WebGL environment was about navigation issues in the environment, which included orienting the camera, moving around the space, point-and-click interactions, and aligning the CMV plots by dragging and dropping them. Three of the users mentioned that although it was hard in the beginning, once they figured out how to navigate, it got easier. Some of the users were trying to zoom in and out like they would have done it on a 2D interface instead of navigating in the 3D environment. This confusion could have happened because of the non-immersive 3D environment on a 2D screen.

Another frequent issue exclusively experienced in the WebGL environment was that the voice commands were sometimes not getting detected. This could have happened because of accent issues, the varying quality of microphones in the users' computers, and lower detection accuracy of the Web Speech API used. It was also observed that female users found it more challenging to issue voice commands as compared to the male users.

One of the training section asked them to read the timestamp from the left-hand side map, and five of them read the date wrong, due to orientation issues, i.e., 4/8/2020 instead of 4/6/2020.

Four users mentioned that the sounds were annoying to them. Multiple simultaneous sound sources distracted them while doing a task which needed only a single sound source enabled to be able to answer correctly. While some users used hide commands to mute the sounds, not all of the users used that feature.

When asked whether the 3D visualization was helpful in analyzing the data, 13 users said a definite yes to this question. One of the users said "provided a more real sense of the data. Moving the slicing plane was very informative in understanding the data as time changed." Another one said, "it enabled viewing objects from different angles." One user mentioned, "it allows multiple data points and types to be simultaneously visualized". Along similar lines, one user said, "Yes, I can simultaneously look at multiple parameters." "A lot of information was available at a time with relative ease to access them", which was one of the design goals of the developed prototype. Users were also able to identify linking between the coordinated views. One user said, "I think it was an interesting way to view the data and one might see new links between the data in this format." One user stated that it was not helpful in terms of the line plots—"Not helpful for viewing the line graphs which had a lot of detail to begin with, and viewing them in 3D made it harder. They all seemed to blend together."

Twelve users found certain aspects of the sound useful, but some of them had mixed feeling about the sounds. One of the users said, "Yes, sound acted like an added dimension to the 3D environment, enhancing the ease of access to data." Users found the sounds most useful to report the timestamp of the maximum damage reports, and they were able to relate it with the movement of the slicing plane. Another user said, "Yes, the different pitch and tone relayed the urgency of the disaster and it gave an accurate sense of the damage." However, they did not understand or find the sounds much useful for the other views. Further, they found the sounds for the continuous warning signals annoying and repetitive and were glad that they could turn those off when needed. Some users found the concurrent sounds distracting and one of the users specifically mentioned that more than two simultaneous sounds proved to be distracting to do a task. Notification sounds gave aural confirmation for the voice commands, so those were favorable. In this study, users were not trained about what the sounds meant, to gauge whether they are able to intuitively relate it to the data, and it turns out 17 out of the 18 users were able to do so. The users expressed that the sounds would be more advantageous with training about the meaning of the sounds.

HoloLens: When asked about the challenges faced by the users in the in-person study in the HoloLens environment, one of the aspects was about unfamiliarity with the device. One of the users said that the most challenging part was "Getting used to the nuances of the device. Figuring out the proper distance for touching the virtual screens and such."

Another issue was selecting the questionnaire window once a user interacted with the prototype. Users found it difficult to bring the question into focus again, and it took a few tries before they could begin typing the answers. One user said, "I think the hand gestures require you to use a certain motion in a certain range, which made it difficult to use, i.e., you have to do it the way the machine wants you to, or it doesn't recognize it." This indicated the relatively steep learning curve for learning the device's gestures.

Another user said, "The tap command was not working very well, and also the edge window kept getting deselected. the keyboard kept following me and I couldn't change that."

Interestingly, a challenging aspect for one user was "Getting used to turning my head or walking around in order to search for what information I needed because I'm very used to sitting down at a desk to analyze data on a computer."

When asked how helpful the 3D visualization was in analyzing the data, 100% of the users expressed that it was helpful. Three of the users specifically mentioned that walking around the data was helpful. One of the users said, "Yes, I can walk up to the part of data I need to view instead of awkward mouse movements for aligning it."

Many users mentioned how intuitive and better it was to visualize data in this environment as compared to viewing multiple charts on a computer screen, which would have taken multiple screens, or a lot of effort to link between them in the first place. Visualizing all the data combined with 2D representations such as maps in one place at the same time, as compared to needing many plots, especially when those plots change with respect to time, made the analysis intuitive. This emphasizes the usefulness of the design of our prototype.

One of the users specifically mentioned that the coordination between the line plots and the regions on the map plots was intuitive to have in one view. Users reported that the data was presented in an efficient way and they could easily find the information they needed. One user, whose job is data visualization, said that this might be the future of visualization.

Nine of the 16 users in the HoloLens environment said that the audio was helpful in analyzing the data. While most of the feedback was similar to the corresponding feedback received for the WeGL environment, like the notification sounds, and the aid in reporting the time stamp of maximum damage reports, one of the users also said "the audio helped provide another layer of context to the data." Another user noted that the sounds were a better indication than the visualizations for plots such as line-plots. Some users however, due to the lack of training, took longer to discover that the sound was related to the data. One user reported, "Yes, while looking at so many plots and charts, the audio helped focus on specific aspects that were required to look at the data." Another user expressed, "when it was just one audio input, I could find the maximums/minimums of data using the change of the pitch or volume." One of the users described that they could associate the sounds with a particular feature.

Limitations: One of the limitations that users indicated is that WebGL navigation could be better. Currently, the mouse/ trackpad is used for changing the first person point of view, and the mouse left click is used for brushing the data spheres and dragging the CMV and the slicing plane. This added complexity in terms of adjusting the orientation of the visualizations while interacting with the slicing plane or CMV. Further, some users felt that the presentation before the training section was overwhelming due to many visual and aural features of the prototype that were described.

For the HoloLens environment, entering answers in the questionnaire could be made simpler. The users found it challenging to type answers using a virtual keyboard and selecting the browser window after interacting with the prototype. We selected this data collection approach to have consistency in both modes of study. However, if an experiment is set up only in the HoloLens environment, think-aloud protocol or even speech dictation could be used instead of AR finger tapping. Due to the narrow field of view of the device, the users sometimes experienced that the virtual keyboard or the browser window was not fully visible. However, this was resolved by walking away from those holograms or adjusting their scale.

8   Discussion

Task t4 and t5 showed significant differences in task completion times. This could have happened because these tasks involved the use of voice commands. In the HoloLens environment, the voice commands worked very smoothly, whereas in the WebGL environment, it took multiple tries for some users to get the voice commands to work. Once possible reason for this difference in the experience of issuing voice commands could be because of the different voice recognition software/APIs used in these two environments.

Additionally for task t4, to count the number of bars in the HoloLens environment, it was pretty easy for the users to either walk towards the 3D bar chart or drag it near to them and count the individual bars. Whereas, in the WebGL environment, users had to navigate towards the 3D bar chart and orient it in a position and angle such that they could count the individual 3D bars. That navigation and orientation added significant time to this task.

The same reason can be justified for the significant differences in task completion times for tasks t6 and t8, where the users spent time orienting themselves to visualize the line plots, and to drag the slicing plane to either report the time of maximum damage reports or identify the time when the line-plot goes flat for a region, indicating a power outage. This was however very intuitive for the users to do in the HoloLens environment as shown by the less time taken.

The significant differences for the task completion times in the task t1 could be accounted for the fact that the floor of the prototype was closer to the floor of the room, and users had to just look over at the floor map to read. Whereas in the WebGL environment, users were not oriented for a top view and they had to navigate on the map to search the asked region. The field of view in the WebGL environment was limited to the computer screens of the users. In the HoloLens environment, they were immersed in the MR environment and could easily turn their head or walk and look over the floor map to search data.

In task t2, no significant results were found, since users may have already oriented themselves towards the floor map in the previous task, so no further time might have been spent for the orientation, but only to do the counting task.

Task t3 accuracy level for the WebGL environment is lower because the projection of a 3D sphere on a 2D screen viewed from a sub-optimal angle, made it harder to tell which one was the biggest sphere.

Task t7 accuracy levels are among the highest for both the environments. Even with no training about the sound, users were able to identify how the sound was correlated with the data. This indicates that sound not only enhances the experience, but also helps users in analyzing the data.

9   Conclusion

We presented an approach for embodied data exploration in an MR environment with CMV containing 2D and 3D views. Using the surrounding space, we displayed traditional 2D displays on the walls, 3D scatter plots on the floor and time-series lineplots in the 3D space. A slicing plane is used as an interaction tool to allow navigation between various timestamps of the spatio-temporal data. Linking and brushing between the CMV allowed users to get details from the displayed overview of the data.

In a traditional desktop-based environment, click and drag features of the mouse allow navigation on the 2D screen. Mouse cursor can move only in two directions, while the hand can move only a few inches. We achieved a similar interaction in the 3D environment through a slicing plane, where users could move their whole bodies freely, and at the same time constraining the movement of the slicing plane in a single direction allowing precise navigation in an immersive environment. The seamless design of the slicing plane removes any barriers between the interaction and the perception space. This is important as the interaction movements and interaction effects are in the same space, with zero spatial distance, and therefore lowering the cognitive load. Moreover, the invisible linking in our system surpasses the problem of occlusion and clutter, which is a major problem when visual links are drawn between the CMV views.

The ability to walk around the displayed data, the placement of the coordinated views in the surroundings and the real-time interaction with high spatial proximity supports embodiment rather than just visual immersion, which is the main aim of the approach we proposed. Moreover the use of sound for multiple streams of data allowed parallel exploration of the various dimensions of the data. This proved to be most helpful, especially for large scale plots, when the entire plot is not in view and sound spatialization is used to draw immediate attention towards the region of interest. The user study conducted with a total of 34 users validates our approach and the results show significantly better user performance in an immersive mixed reality environment compared to a 2D desktop screen.

The future work will focus on improving the sound design. In this study, continuous warning signals (often high-pitched) were used to attract users' immediate response to a spike in any of the plot view. Shorter duration sounds will be implemented in the future design. Furthermore, the maximum number of concurrent sounds in a mixed reality environment for immersive analytics will be studied. Mute commands will be implemented, which will only mute the audio sources, and not hide the views. In the WebGL environment, users found it awkward or unnatural to talk to a computer screen without an activating command, such as 'Hey Google!'. A similar voice activating command will be used to address the software and make the experience better for users.

Acknowledgements. VRVis is funded by BMK, BMDW, Styria, SFG, Tyrol and Vienna Business Agency in the scope of COMET - Competence Centers for Excellent Technologies (879730) which is managed by FFG.

References

1. IEEE VIS 2019 conference: VAST challenge: mini-challenge. https://vast-challenge.github.io/2019/. Accessed 12 Feb 2021
2. International Reference Ionosphere - IRI (2016). Community Coordinated Modeling Center (CCMC) | NASA. https://ccmc.gsfc.nasa.gov/modelweb/models/iri2016_vitmo.php. Accessed 12 Feb 2021
3. UCI Machine Learning Repository: Iris Data Set. https://archive.ics.uci.edu/ml/datasets/iris. Accessed 12 Feb 2021
4. Web Speech API - Web APIs — MDN. https://developer.mozilla.org/en-US/docs/Web/API/Web_Speech_API. Accessed 12 Feb 2021
5. Ams, L., Cook, D., Cruz-Neira, C.: The benefits of statistical visualization in an immersive environment. In: Proceedings of the IEEE Virtual Reality 1999 Conference, pp. 88–95 (1999)
6. Bach, B., Sicat, R., Beyer, J., Cordeil, M., Pfister, H.: The hologram in my hand: how effective is interactive exploration of 3D visualizations in immersive tangible augmented reality? IEEE Trans. Visual Comput. Graphics **24**(1), 457–467 (2018)
7. Barrett, N., Mair, K.: Sonification for geoscience: listening to faults from the inside. In: EGU General Assembly Conference Abstracts, vol. 16 (2014)
8. Blum, S., Cetin, G., Stuerzlinger, W.: Immersive analytics sensemaking on different platforms. In: Proceedings of the 27th International Conference in Central Europe on Computer Graphics, Visualization and Computer Vision (WSCG 2019), pp. 69–80 (2019)
9. Cavallo, M., Dolakia, M., Havlena, M., Ocheltree, K., Podlaseck, M.: Immersive insights: a hybrid analytics system for collaborative exploratory data analysis. In: Proceedings of the 25th ACM Symposium on Virtual Reality Software and Technology, pp. 1–12. ACM (2019)
10. Chandler, T., et al.: Immersive analytics. In: Proceedings of the 2015 Big Data Visual Analytics (BDVA 2015), pp. 1–8. IEEE (2015)
11. Cordeil, M., et al.: IATK: an immersive analytics toolkit. In: Proceedings of the 2019 IEEE Conference on Virtual Reality and 3D User Interfaces (VR 2019), pp. 200–209. IEEE (2019)
12. Donalek, C., et al.: Immersive and collaborative data visualization using virtual reality platforms. In: Proceedings of the 2014 IEEE International Conference on Big Data (Big Data 2014), pp. 609–614. IEEE (2014)
13. Dwyer, T., et al.: Immersive analytics: an introduction. Immersive Analytics. LNCS, vol. 11190, pp. 1–23. Springer, Cham (2018). https://doi.org/10.1007/978-3-030-01388-2_1
14. Fitch, W.T., Kramer, G.: Sonifying the body electric: superiority of an auditory over a visual display in a complex, multivariate system. In: Santa Fe Institute Studies on the Sciences of Complexity, Proceedings, vol. 18, pp. 307–307. Addison-Wesley Publishing Co (1994)
15. Fonnet, A., Prié, Y.: Survey of immersive analytics. IEEE Trans. Visual. Comput. Graphics 1–22 (2019)
16. Gračanin, D.: Immersion versus embodiment: embodied cognition for immersive analytics in mixed reality environments. In: Schmorrow, D.D., Fidopiastis, C.M. (eds.) AC 2018. LNCS (LNAI), vol. 10915, pp. 355–368. Springer, Cham (2018). https://doi.org/10.1007/978-3-319-91470-1_29
17. Hoppenstedt, B., et al.: Applicability of immersive analytics in mixed reality: usability study. IEEE Access **7**, 71921–71932 (2019)

18. Johansson, M.: VR for your ears: dynamic 3D audio is key to the immersive experience. IEEE Spectr. **56**(2), 24–29 (2019)
19. Levene, H.: Robust tests for equality of variances. Contributions to probability and statistics. Essays in honor of Harold Hotelling, pp. 279–292 (1961)
20. Marquardt, A., Trepkowski, C., Eibich, D., Maiero, J., Kruijff, E.: Non-visual cues for view management in narrow field of view augmented reality displays. In: Proceedings of the IEEE International Symposium for Mixed and Augmented Reality 2019 (2019)
21. Mauney, B.S., Walker, B.N.: Creating functional and livable soundscapes for peripheral monitoring of dynamic data. In: Proceedings of the Tenth International Conference on Auditory Display (2004)
22. Prouzeau, A., Lhuillier, A., Ens, B., Weiskopf, D., Dwyer, T.: Visual link routing in immersive visualisation. In: Proceedings of the 2019 International Conference on Interactive Surfaces and Spaces, pp. 241–253. ACM, New York (2019)
23. Shapiro, S.S., Wilk, M.B.: An analysis of variance test for normality (complete samples). Biometrika **52**(3–4), 591–611 (1965). https://doi.org/10.1093/biomet/52.3-4.591
24. Srinivasan, A., Stasko, J.: Orko: facilitating multimodal interaction for visual exploration and analysis of networks. IEEE Trans. Visual Comput. Graphics **24**(1), 511–521 (2017)
25. Su, S., Perry, V., Dasari, V.: Comparative study for multiple coordinated views across immersive and non-immersive visualization systems. In: Chen, J.Y.C., Fragomeni, G. (eds.) HCII 2019. LNCS, vol. 11574, pp. 321–332. Springer, Cham (2019). https://doi.org/10.1007/978-3-030-21607-8_25
26. Wagner Filho, J.A., Rey, M.F., Freitas, C.M., Nedel, L.: Immersive visualization of abstract information: an evaluation on dimensionally-reduced data scatterplots. In: Proceedings of the 2018 IEEE Conference on Virtual Reality and 3D User Interfaces (VR 2018), pp. 483–490. IEEE (2018)
27. Yang, Y., Dwyer, T., Jenny, B., Marriott, K., Cordeil, M., Chen, H.: Origin-destination flow maps in immersive environments. IEEE Trans. Visual Comput. Graphics **25**(1), 693–703 (2018)

Perception of Illusory Body Tilt Induced by Electrical Tendon Stimulation

Nozomi Takahashi[1][(✉)], Tomohiro Amemiya[1,2], Takuji Narumi[1], Hideaki Kuzuoka[1,2], Michitaka Hirose[3], and Kazuma Aoyama[1,2]

[1] The Graduate School of Information Science and Technology, The University of Tokyo, Tokyo, Japan
nozomi@cyber.t.u-tokyo.ac.jp
[2] Virtual Reality Educational Research Center, The University of Tokyo, Tokyo, Japan
[3] Research Center for Advanced Science and Technology, The University of Tokyo, Tokyo, Japan

Abstract. In virtual reality (VR) experiences, discordance between visual and somatic sensation reduces a user's immersive experience. To address this problem, we introduced somatic sensation using illusory force induced by tendon electrical stimulation (TES). The effect of TES applied to the tibialis anterior muscle tendon (TA) and Achilles' tendon (AC) on force sensation and center of pressure (CoP) in standing position was investigated. Result suggests that application of TES to TA and AC induces force sensation to evoke body sway in the backward and forward direction respectively, while effects on CoP appear to be contradicted to our hypothesis. According to the descriptions of induced sensation of body tilt obtained from questionnaires, this technique is a promising way to present information of ground in VR.

Keywords: New Haptic and Tactile Interaction · Electrical tendon stimulation · Galvanic stimulation · Proprioceptive sensation·

1 Introduction

Recently, Head-Mounted Display (HMD) are becoming less expensive, making them more accessible to general consumers and increase the popularity of Virtual Reality (VR) experiences. Although an HMD allows a user to explore virtual environment, it only provides visual experiences, and the user's body movement is strongly limited by the real environment where the user exists. One of such limitations is difference of terrain shape between virtual and real environments. While VR designers often present tilted surfaces in virtual environments, the user's actual environment is typically a flat floor. In the real environment, when the floor changes from horizontal to slope, it causes certain changes in one's somatic sensation. However, such phenomenon does not happen in VR. This gap between visual and somatic sensations can deteriorate the user's immersive experience. To resolve this problem, we attempted to find an effective way to induce

© Springer Nature Switzerland AG 2021
S. Yamamoto and H. Mori (Eds.): HCII 2021, LNCS 12765, pp. 357–368, 2021.
https://doi.org/10.1007/978-3-030-78321-1_27

perception of postural changes to make the user feel as if the surface changes between flat and tilted.

Methods to induce a sensation of posture change are mainly divided into mechanical and perceptual solutions. Mechanical solutions physically change the structure of the real world. For instance, Schmidt et al. [1] developed "Level-Ups", boot-shaped devices whose height from the floor can be controlled by actuators. With these devices, users can walk up and down in a virtual environment. This type of method is intuitive for the users, as the method is able to precisely control a presented sensation. However, mechanical systems tend to be complicated, expensive, and large. In addition, as users cannot see their step in the VR experience, the change in physical structure can be dangerous. Because of these reasons, mechanical solutions are not optimal for VR experiences. On the other hand, perceptual solutions use psychological characteristics of humans to induce illusory sensations. For example, Nagao et al. [2] used visuo-haptic interaction to simulate a sense of walking up and down in a virtual environment. They provided a user with haptic stimulation by placing small bumps on the floor. When stepping on the bumps, the use interpreted the sensation as edges of stairs in the virtual environment, making them feel as if they were ascending or descending the stairs. This type of solution does not need any complicated devices, and it can be used easily and safely compared with mechanical solutions. Thus, this study mainly focuses on using perceptual solutions to induce illusory posture sensations.

The control of human posture depends on visual, vestibular, and somatosensory inputs [3]. Therefore, presentation of these sensations is expected to be useful for making users perceive illusory postural changes. In particular, vestibular sensation is thought to have a direct relationship to perception of body tilt. As a simple way of presenting vestibular sensation, galvanic vestibular stimulation (GVS) is widely used. This technique applies direct current to electrodes placed on the bilateral mastoids and temples and induces virtual acceleration (and/or virtual head motion) to lateral, anteroposterior, and yaw rotational directions. However, the intensity of GVS to the anteroposterior direction is weaker than that of bilateral directions. To alleviate this disadvantage, more effective ways of inducing illusory body tilt in anteroposterior directions is needed.

In order to induce a sensation of anteroposterior body tilt, we focused on proprioceptive sensation of leg joints. Proprioceptive sensation transmits information about a joints angle, velocity, and tension. Since human legs are the only part that supports the whole body weight, generation of proprioceptive sensation to the legs is expected to effectively change one's perception of body tilt by presenting illusory joint angles. One technique to generate proprioceptive sensation is tendon vibration. Applying approximately 100 Hz vibration to muscle tendons stimulates muscle spindles, which detect changes in muscle length, and induces illusory movement of the joint [4]. Though this is a promising technique with which to generate illusory body tilt, vibrators need to be fixed at flexible joints and it may be difficult to obtain stable application of vibrations. Electrical tendon stimulation (TES) which targets to muscle tendons is also known to induce illusory movement [5]. Application of current to tendons stimulates Golgi tendon organs (GTO), which detect changes in muscle tension, and it induces illusory movement. Since only light weight and flexible gel electrodes need to be attached for electrical stimulation and current can be provided from a small stimulator, this method should be suitable for virtual reality purposes.

For these reasons, this paper uses TES as a way of generating anteroposterior illusory body tilt. The purpose of our study was to investigate if TES can be used to modify the terrain and posture perception in a VR experience. Here, we present the initial report of the effects of TES to the legs on body sway and subjective sensation.

2 Materials and Methods

Five healthy adult males participated in the experiment. The average of their age was 24.2 years old. The experiment was conducted in accordance with the safety standards approved by the local ethics research committee at the University of Tokyo, Japan. The experiment was explained to the participants prior to their participation, and they signed a letter of consent. The experiment protocol was performed in accordance with the ethical standards provided in the Declaration of Helsinki.

In our experiments, the effect of TES on anteroposterior body tilt was investigated. Participants' sensation during stimulation was also investigated using a questionnaire.

2.1 Stimulation

As a geometrical characteristic, the angle change of the ankle joint is thought to have the largest influence on the tilt of the upper body among all leg joints. For this reason, stimulating tendons around the ankle joint are appropriate for our purposes. It is considered that the tibialis anterior muscle tendon (TA) and Achilles' tendon (AC) are the main drivers of anteroposterior angle changes of the joint among these tendons. Therefore, these tendons were chosen as the targets for electrical stimulation. Considering that electrical stimulation is most prominent immediately beneath the electrodes, they were placed around the targeted tendons. Since stimulations with high current induce pain sensation at the stimulated position and its threshold is different among individuals, current values were set at the highest value within the pain threshold of each participant. The upper limit of current value was 2.5 mA. Based on the previous study conducted by Takahashi et al. [5], frequency was fixed at 80 Hz. The alternate current square wave was used as the current waveform.

2.2 Measurement of Body Tilt

The change of Center of Pressure (CoP) over time was measured using a Wii Balance Board (Nintendo Co., Ltd.). CoP data was obtained in two dimensions (x and y) in the coordinate system of Fig. 1. Frequency of record was set at 100 Hz. To unify posture during the measurement process, participants were told to follow instructions below.

1. Put both feet together (Ronberg's elect position) at the center of the balance board.
2. Put both hands on the sides of the body.
3. Look at the marker on the front wall in the eyes opened condition trial.

Height of the marker was set at the height of the eyes for each individual, and the balance board was placed 2 meters from the marker.

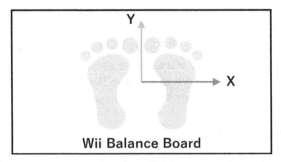

Fig. 1. Coordinate system of recorded CoP

2.3 Experimental Conditions

Since this is the first study to investigate the effect of TES to the ankle tendon on the body tilt sensation and body sway, it is difficult to estimate the optimal condition to induce body tilt sensation and to evoke body sway. Therefore, to investigate the effect of electrical stimulation in detail, we controlled four kinds of parameters. The first parameter was the state of the participant's eyes, opened or closed because the absence of visual input might affect postural changes induced by electrical stimulation. The second parameter is the stimulated tendon, TA or AC. Stimulation to these tendons is expected to have opposite effects on anteroposterior posture. As the third parameter, the effects of using a parallel or perpendicular direction of current were investigated because the appropriate current direction for TES is not clear. In the Parallel condition, a line connecting the two electrodes was parallel to the running of the stimulated tendon, whereas in the perpendicular condition, the electrodes were placed perpendicular to the tendon. The last parameter was whether we stimulate left leg, right leg, or both legs. We speculated that stimulating only one leg could induce a force sensation in lateral or diagonal directions. In total, we tested 26 experimental condition, with a combination of 4 parameters ($2 \times 2 \times 2 \times 3 = 24$) plus two conditions without electrical stimulation, eyes opened and closed.

2.4 Position of Electrodes

The position of the electrodes was determined by the combination of target tendon location and direction of the current path. Figure 2 shows four of the different electrode positions used in the experiment.

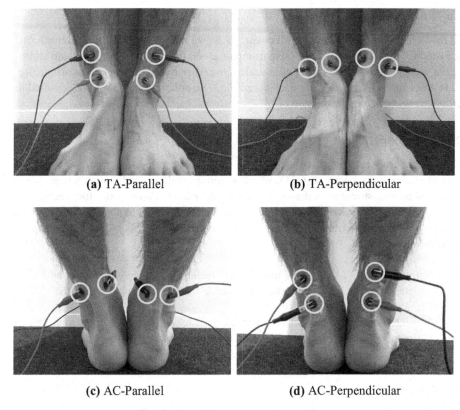

(a) TA-Parallel   (b) TA-Perpendicular

(c) AC-Parallel   (d) AC-Perpendicular

Fig. 2. Four different electrode positions

2.5 Questionnaire

After each trial, participants were asked the following questions.

Q1. Did you feel force sensation in the lateral direction?
Q2. Did you feel force sensation in the anteroposterior direction?
Q3. Did you feel any other sensations?

Q1 and Q2 were answered by choosing a number from 1 to 7. For Q1, a score of 1 meant "strong leftwards force sensation was felt", and 7 indicated "strong rightwards force sensation was felt". For Q2, 1 meant "strong forwards force sensation was felt", and 7 indicated "strong backwards force sensation was felt". In both of these questions, a score of 4 was used to indicate "no force sensations were felt". Q3 was answered by free writing to investigate any details of sensation. Previous study [5] have shown that force sensation can be induced by TES. Since this experiment was an initial study to determine whether TES to the ankle tendons was effective, we asked the "force sensation" rather than the sensation of body tilt to examine whether these tendons were affected by TES.

2.6 Procedure

At first, electrodes were placed at all positions shown in Fig. 2, and current strength was adjusted for each position. Next, the participant stood on Wii Balance Board and stimulation was applied for 30 s. After stimulation, the participant filled in the questionnaire. This procedure was repeated for each condition. The order of conditions was shuffled to eliminate sequence effects.

2.7 Hypothesis

GTO detects muscle tension [6]. Takahashi et al. [5] reported that application of current to tendon would activate GTO, which is interpreted as muscle contraction. Since electrical tendon stimulation does not elicit actual muscle contraction, force sensation to the opposite direction of muscle contraction is perceived. It was also reported that the intensity of force sensation can be controlled by current value. Accordingly, we propose following hypotheses.

H1. Stimulation of TA induces an illusory backwards force sensation, and CoP counterbalances forwards. The opposite effect is expected when the AC is stimulated.
H2. The intensity of a force sensation in the parallel condition is larger than that in the perpendicular condition since the amount of the current passing through the tendon in the parallel condition is larger than that in the perpendicular condition.

3 Results

3.1 Sensation During Stimulation

Figures 3 and 4 show the results for the participants' answers to Q1 and Q2 with their eyes opened and closed, respectively. The red line in these figures represents neutral answer that means "no force sensations were felt". In Q1, answered value under 4 means "leftwards force sensation was felt", and over 4 means "rightwards force sensation was felt". In Q2, under 4 means "forwards force sensation was felt", and over 4 means "backwards force sensation was felt".

Three-factor ANOVA was conducted on intensity of force sensation to each direction (lateral and anteroposterior) with eye state (opened or closed), position of stimulation (TA or AC), and direction of current path (parallel or perpendicular) used as the independent factors. In the lateral direction, a significant three-way interaction ($p = 0.0327$) was identified. In the anteroposterior direction, a significant main effect in the position of stimulation ($p = 0.0121$) and a significant two-way interaction between eye state and direction of current path ($p = 0.0299$) was identified.

In addition, Steel's multiple comparison test was conducted to clarify the effect of each stimulus condition on force perception compared to the no-stimulus condition. The result of Steel's multiple comparison test showed that TA-Parallel with eyes opened, TA-Perpendicular with eyes opened, and TA-Parallel with eyes closed induced significant amount of backward force sensation. Furthermore, it showed that AC-Parallel with eyes opened induced a significant amount of forwards force sensation.

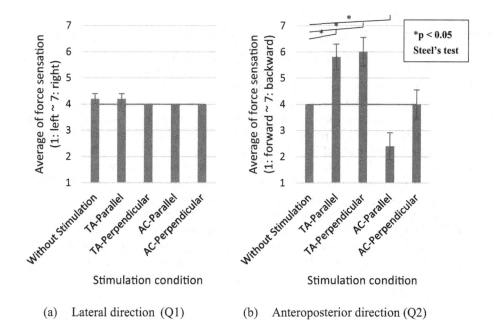

(a)   Lateral direction (Q1)   (b)   Anteroposterior direction (Q2)

Fig. 3. Average of perceived force direction by TES with eyes opened (Color figure online)

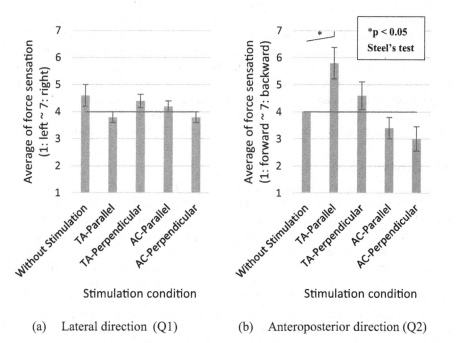

(a)   Lateral direction (Q1)   (b)   Anteroposterior direction (Q2)

Fig. 4. Average of perceived force direction by TES with eyes closed (Color figure online)

3.2 Change of CoP

Initial CoP position was regarded as the origin, and the largest displacement of CoP from the origin in four directions (rightward, leftward, forward, and backward) was calculated. The averages of these changes with eyes opened is shown in Fig. 5, and the averages with eyes closed are displayed in Fig. 6.

Three-factor ANOVA was conducted on the largest displacement of CoP in each direction. Eye state (opened or closed), position of stimulation (TA or AC), and direction of current path (parallel or perpendicular) were used as the factors. In the leftwards direction, significant main effects in the eye state ($p = 0.0329$), and the position of stimulation ($p = 0.0105$) were found. In both the forwards and backwards directions, significant main effects of the position of stimulation were found ($p = 0.0168$ and 0.0083, respectively). No significant interaction effect was found.

In addition, a multiple comparison test of Steel was conducted to clarify the effect of each stimulus condition on change of CoP compared to the no-stimulus condition. Steel's multiple comparison test showed that the largest forwards displacement of CoP was significantly reduced with TA-Perpendicular. In addition, it showed that the largest backwards displacement of CoP was significantly increased with TA-Perpendicular.

4 Discussion

4.1 Perception of Force Sensation

Figures 3 and 4 show that electrical stimulation to TA and AC induces force sensation in either a backward or forward direction, respectively. Although this was based on subjective responses, it is consistent with our hypothesis. However, whether the force sensation was induced by stimulating the GTO is unclear. We suspect that the force sensation was induced by stimulating the tendon because the direction of force sensation occurred in an opposite direction when muscles near the electrodes were contracted. For example, the tibialis anterior muscle exists at the front of the legs and could be stimulated when the electrodes were attached on the front part of the legs. If the tibialis anterior muscles were stimulated, the subjects should have felt front force sensation.

4.2 Effect of TES on CoP

Figures 5 and 6 shows that the application of perpendicular electrical stimulation to TA caused significant backwards changes in CoP. Conversely, perpendicular AC stimulation caused significant forward changes in CoP, especially when the eyes were closed. This is in complete opposition to our hypothesis. One explanation for this is the awareness and standing strategy of the participants during stimulation. In the experiment, we did not force participants to keep their body stable during stimulation. Because of this, participants might not correct their posture against a force sensation, and the position of CoP may have changed to the direction of force sensation. Alternatively, the initial posture of the participants may have influenced these findings. During the experiments, electrical stimulation and recording started simultaneously, and the initial CoP position was used as the origin. As a consequence, it is possible that the participants' posture is accidentally unstable at first and a biased position of CoP may have been used as the origin. To exclude these possibilities, the process of recording CoP should be reconsidered.

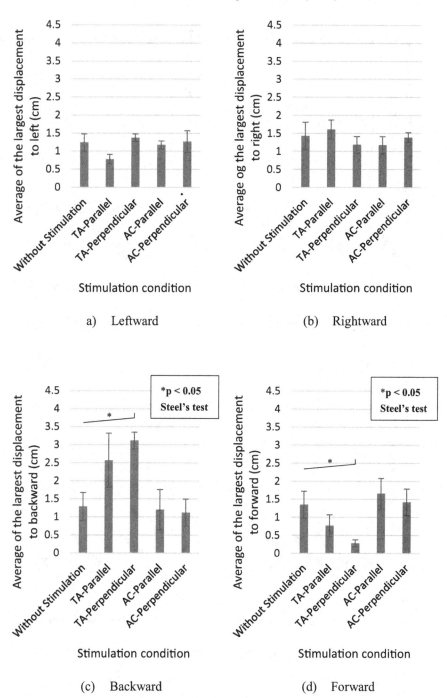

a)   Leftward

(b)   Rightward

(c)   Backward

(d)   Forward

Fig. 5. Average of the largest displacement to four directions with eyes opened

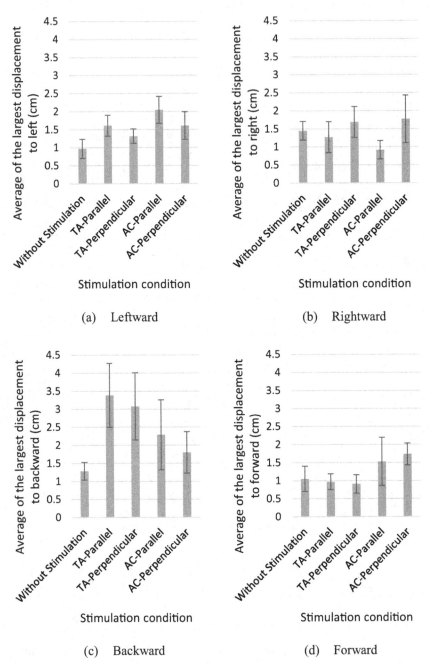

(a)   Leftward

(b)   Rightward

(c)   Backward

(d)   Forward

Fig. 6. Average of the largest displacement to four directions with eyes closed

4.3 Effect of the Direction of Current Path

Result of three-way ANOVA didn't show significant effect of current direction on antero-posterior force sensation. This result is opposition to our hypothesis that the intensity of the force sensation is larger in the parallel condition. One possible explanation to this fact is that the amount of current passing through tendons was large enough to induce strong force sensation both in the parallel and perpendicular condition.

According to the questionnaire, perpendicular condition would stimulate nerves which dominate haptic sensation. Since the vibration sensation induced by perpendicular stimulation may disturb the immersive experience, the parallel stimulation is more appropriate for VR experiences.

4.4 Use of TES to Present Terrain Information

The purpose of this study is to find an effective way to induce the perception of postural changes in order to present information of ground tilt in a virtual environment. In our study, two participants reported that they felt an unconscious change of posture. These comments suggest that application of TES to the tendons around the ankle joint can change the perception of body tilt. In addition, two participants reported that sensation of body tilt arose from the lower body. This is clearly different from the sensation of body tilt induced by GVS, and seems to be suitable for presenting information of ground tilt. Based on other comments, however, no one interpreted force sensation during stimulation as a change of ground tilt. This may be explained by the absence of visual stimulation during the experiment. Since visual input of the participants was not altered in the experiment, participants might simply interpret that an external force was applied to the stimulated position. To investigate the effect of TES on perception of ground tilt, further experiments simultaneously presenting force sensation and altered visual stimulation should be conducted.

5 Conclusions

In this paper, we proposed to apply TES to a user as a new method to induce the user's illusory sensation of body tilt to simulate he/she is standing on a tilted ground. We applied TES to TA and AC to investigate its effect on perception of body tilt and position of CoP. Our results suggest that application of TES to TA and AC induces force sensation in an anteroposterior direction. In addition, comments from participants indicate that the sensation of body tilt arises from lower body. These results show that our approach is effective in inducing sensation as if the participants are standing on a tilted ground. Our results also indicate that stimulation applied parallel to the targeted tendon is most suitable for minimizing undesirable sensations. However, CoP data recorded during stimulations appears to contradict the effect of TES. To investigate the detailed effect of TES on actual postural changes, further experiments are necessary.

In future research, we will apply the same stimulation while providing visual stimulation using an HMD.

References

1. Schmidt, D., et al.: Level-ups: motorized stilts that simulate stair steps in virtual reality. In: Proceedings of the 33rd Annual ACM Conference on Human Factors in Computing Systems, pp. 2157–2160 (2015)
2. Nagao, R., Matsumoto, K., Narumi, T., Tanikawa, T., Hirose, M.: Infinite stairs: simulating stairs in virtual reality based on visuo-haptic interaction. In: ACM SIGGRAPH 2017 Emerging Technologies, pp. 1–2 (2017)
3. Oie, K.S., Kiemel, T., Jeka, J.J.: Multisensory fusion: simultaneous re-weighting of vision and touch for the control of human posture. Cogn. Brain. Res. **14**(1), 164–176 (2002)
4. Ribot-Ciscar, E., Rossi-Durand, C., Roll, J.P.: Muscle spindle activity following muscle tendon vibration in man. Neurosci. Lett. **258**(3), 147–150 (1998)
5. Takahashi, A., Tanabe, K., Kajimoto, H.: Relationship between force sensation and stimulation parameters in tendon electrical stimulation. In: Hasegawa, S., Konyo, M., Kyung, K.-U., Nojima, T., Kajimoto, H. (eds.) AsiaHaptics 2016. LNEE, vol. 432, pp. 233–238. Springer, Singapore (2018). https://doi.org/10.1007/978-981-10-4157-0_40
6. Moore, J.C.: The Golgi tendon organ: a review and update. Am. J. Occup. Therapy **38**(4), 227–236 (1984)

Wearable Haptic Array of Flexible Electrostatic Transducers

Ian Trase[1], Hong Z. Tan[2], Zi Chen[1], and John X. J. Zhang[1]([envelope])

[1] Thayer School of Engineering, Dartmouth College, 14 Engineering Drive,
Hanover, NH 03755, USA
john.zhang@dartmouth.edu
[2] School of Electrical and Computer Engineering at Purdue University,
West Lafayette, IN 47907, USA

Abstract. We demonstrate a wearable flexible electrostatic transducer (FET) matrix that conforms to the skin and can be used to generate complex haptic signals. The transducers consist of a pair of flexible electrodes that are attracted based on electrostatic force. We designed and built a 2 by 2 matrix of transducers from a pair of flexible films, with the design optimized for indentation into the skin. A transducer with a 10 mm by 10 mm footprint and a height of 2 mm effectively generated perceptible stimulus on the skin when operated at 150 V–250 V. Three psychophysical experiments were carried out to evaluate the properties of the new device as a wearable haptic display. Experiment 1 showed that the matrix of transducers was able to linearly boost perceived intensity when compared to a single transducer. Experiment 2 indicated that it was possible to indistinguishably replicate multi-frequency signals delivered to one transducer using multiple transducers that each operated at one of the frequencies. In Experiment 3, four movement-based stimulation patterns were designed using tactile illusion, and the identification rates were significantly above chance level. Our findings demonstrate that the compact, flexible, and scalable transducer array is well suited as a new type of actuator for wearable haptics research and applications.

Keywords: Haptics · Transducer · Electrostatic · Flexible electrodes · Sensing · Wearable

1 Introduction

Wearable haptic technology is found in a variety of consumer devices, generally using linear resonant actuators (LRAs) or eccentric rotary motors (ERMs). These technologies generate vibrations at one or more frequencies that can be perceived on the skin. While small and efficient, these technologies are rigid and lack the ability to deliver complex signals over a wide frequency range. Other methods of haptic actuation include piezoelectric actuation and voice coils, but these have higher voltage or power consumption requirements. There is a need for flexible and compact haptic actuator arrays that can conform to the skin and deliver complex signals over a wide frequency and intensity ranges.

© Springer Nature Switzerland AG 2021
S. Yamamoto and H. Mori (Eds.): HCII 2021, LNCS 12765, pp. 369–385, 2021.
https://doi.org/10.1007/978-3-030-78321-1_28

Haptic arrays have been developed for a variety of applications. Tactile sensing "skins" for robots are a current area of research using both triboelectric [1] and piezoelectric films [2]. These skins seek to sense a broad range of frequencies while still retaining flexibility. As most tactile sensing technologies are reversible (especially piezoelectric systems), these designs provide insight into actuation array systems as well. Both research and commercial activities are being conducted with ultrasonic transduction arrays that can generate sensation remotely in air, away from the array itself. Most current research uses piezoelectric printed polymer membranes (PPTs) to achieve actuation [3, 4], but work has also been done with thin flexible electrostatic actuators [5]. Arrays have been shown integrated into a wirelessly charged and controlled device on the skin [6], and work on soft piezoelectrics has also been conducted [7].

Significant research has been conducted into the use of haptic actuator arrays to generate tactile sensations, either through direct skin contact or through remote ultrasonic transduction in air. Haptic arrays can take advantage of several important psychophysical properties of the skin to deliver rich sensations. A key parameter when designing haptic arrays is the two-point discrimination threshold [8], which describes the distance at which two stimuli on the skin are indistinguishable from a single stimulus. This threshold is often quite large, for example on the order of 30–40 mm on the forearm [9]. An array of haptic transducers needs to have an inter-actuator spacing above this threshold for the delivered stimuli to be perceived as separate events. Haptic arrays with spacing below the two-point threshold can deliver complex stimuli that are perceived as a single, richer stimulus. Most day-to-day haptic events fall into this second category. By carefully designing stimuli, the total amount of information transferred can be increased [10]. We can exploit this effect to generate several well-known haptic illusions, including the apparent motion effect [11]. In this phenomenon, a series of discrete stimuli are perceived as a single stimulus moving smoothly across the surface of the skin. Through effects like this, an array can be used to generate smooth signals from a relatively sparse array of discrete systems, though still below the two-point discrimination distance. Algorithms have been generated to optimize this sensation, including the Tactile Brush algorithm [12] and the algorithm developed by Park et al. [13].

We developed a wearable haptic array based on our previous research into flexible electrostatic transducers [14, 15]. The transducers consist of a flexible buckled electrode bonded to an unbuckled electrode with an air gap in between. When an alternating voltage is applied across the two electrodes, they move closer and farther apart to generate perceivable vibration. The flexible and buckled nature of the electrodes makes them resistant to pull-in and able to operate at large length scales compared to the electrostatic force distance [16, 17]. The transducers can operate at a range of frequencies while under a variety of strain conditions, making them well-suited for conformal operation on the skin. These transducers were integrated into a two by two array that allowed for independent control of each tactor (tactile stimulator). The shape and response of the transducer elements were verified using COMSOL.

From a perception point of view, a transducer array opens new opportunities for delivering stimuli that are more complex than those possible with a single transducer. For example, two or more tactors can generate signals that are perceived to be more intense than the maximum intensity possible with a single tactor, thereby extending the achievable intensity range [18]. In terms of the waveforms used to drive tactors, many haptic applications call for stimuli with rich spectral contents in order to achieve a large set of distinct stimuli (e.g., [19–21]). Since it is generally more difficult to develop broad bandwidth tactors than to develop resonant-type tactors with tunable resonant frequencies, it is desirable to use two nearby tactors each driven at a single frequency to emulate the sensation of a stimulus containing more than one frequency components. Finally, it has also been shown that movement-based stimuli are highly effective at increasing the information transmission achievable with haptic interfaces [22]. More than one tactor is needed in order to generate tactile movement illusions [11]. To assess the efficacy of using the transducer array to deliver complex stimuli, three psychophysical experiments were conducted to test (1) whether simultaneously activating multiple tactors can lead to a perceivable increase in stimulus intensity, (2) the perceptual difference between a single tactor driven by a signal with two frequency components and two tactors each driven at one frequency, and (3) the distinctiveness of four simulated movement patterns generated by the tactor array. The present research explores the development of a new, flexible haptic array for wearable applications. This array was characterized through theory and simulation, as well as through a psychophysical study. The results verify that the array can be used to deliver a rich range of tactile stimuli while still being comfortable and conformal to the skin.

2 Device Design and Characterization

2.1 Electromechanical Theory

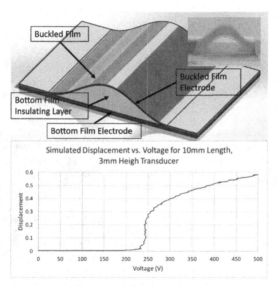

Fig. 1. A) Schematic of individual flexible electrostatic transducer. Inset: photograph of example transducer. B) Simulated voltage-displacement curve for a 10 mm × 10 mm transducer, with turn-on voltage at 240 V.

Each transducer is composed of a pair of thin films that have been coated with a gold electrode on one side. The film that goes in contact with the skin, shown in Fig. 1A, is buckled such that the point of maximum extension presses into the skin. The other film remains flat and undergoes only minor curvature when deformed against the skin. The buckled film is coated with the gold electrode on the side facing the flat film, while the electrode of the flat film is fully encased within that film's multiple layers. Thus, the buckled film begins in a prestressed state while the flat film begins in a relaxed state. When a voltage is applied between the two electrodes, a Coulombic attraction force is generated between the two films. For the voltages and geometry of the device in question, this electrostatic force is negligible for most of the length of the buckled film. However, the force becomes very high at points where the film separation is low. This includes both the region where the two films are in constant contact, as well as the point at which the buckled film begins to leave the flat film. The high force felt at this point is enough to move the buckled film closer to the flat film by an incremental amount. This motion further moves the point of film contact, allowing for increased force at a new location. This positive feedback loop continues until the downward pressure from the electrostatic force is balanced by restoring pressure from the bending force of the buckled film. The approximate governing equations for this relationship can be found in [14, 15], but the phenomena has no closed-form solution and is best studied through simulation. Figure 1B shows an example voltage-displacement curve. Here we can see the characteristic "turn-on" voltage of the transducer, as well as the relatively flat displacement response. These properties make the FET act as a pseudo-binary actuator, with an on state, an off state, and a narrow region where voltage has a very large effect on displacement.

2.2 Device Design

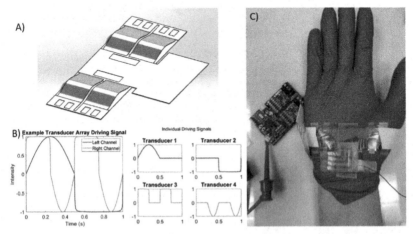

Fig. 2. A) Schematic of transducer array. B) Example waveform used to drive array, with components separated out for clarity. C) Photograph of electrical setup for transducer array, showing Boreas 1901 development kit with array on palm, ready for psychophysical experiment.

A basic schematic of the transducer array is shown in Fig. 2A. The array is composed of two composite films that have been laser-cut and bonded into place. The top buckled film consists of a piece of 25 μm thick Kapton film coated with a 40 nm thick layer of gold as an electrode. This film begins as a single piece of Kapton film that was then laser-cut by a Laser Pro Spirit GLS into the prescribed shape. An acrylic mask 1/16" thick was then placed over the cut Kapton, and 20 nm of gold was sputter-coated onto the film. The top film features gold traces that lead away from the center of the device and off to the side. These traces lead to wire interconnects, which need to be moved away from the transducers to maintain the low profile and conformal nature of the device. The bottom film begins as a lightly prestretched piece of 2.5 μm thick Mylar film taped on its edges to a glass slide. Another acrylic mask was used to help deposit a 20 nm thick layer of gold to the Mylar. A ~ 5 μm thick layer of 20:3 base to curing agent PDMS is then spin coated onto the Mylar, and the composite is then cured at 70 °C for 2 h. After 1 h, a laser-cut piece of Kapton film is adhered to the PDMS, and the Mylar-Gold-PDMS-Kapton composite is excised from the glass slide using a razorblade at the end of 2 h. The two films are then aligned such that the gold-coated side of the top film and the Mylar side of the bottom film are facing each other. They are bonded using tape at a series of prescribed locations, such that each of the four transducers in the top film buckles into a precise shape and the two films maintain good contact everywhere else. In this configuration, the two electrodes are separated only by the 2.5 μm Mylar film, which allows for large electric fields at low applied voltages. The ultimate product is a two by two array, in which each transducer has a footprint of 10 mm by 10 mm.

The top film has four separate gold electrodes, each connected to a separate wire. These wires are connected to the four terminals of a Boreas 1901 Development Kit. This development kit consists of a pair of BOS1901 piezoelectric drivers with a microchip allowing it to be used as a USB audio device. Each driver is capable of supplying anywhere from 0 V to 95 V to either of two channels, though only one channel may be active at a given time. Using this configuration, all four transducers on the flexible array can be controlled using a single audio signal from a computer or any device capable of providing USB audio. The left stereo channel controls one pair of transducers, while the right channel controls the other pair. Within each channel, the positive component of the audio signal controls one transducer while the negative component controls the other. An image of the Boreas Kit connected to the flexible array with an inset of the general waveform shape is shown in Fig. 2B. The bottom film electrode was connected through a wire to the powered terminal of a Trek Model 2210 high voltage amplifier controlled through a Keysight 33522B signal generator and isolated using a Triad Magnetics MD-250-E medical grade isolation transformer. The high voltage amplifier was used to provide a DC bias of −150 V to the bottom electrode, allowing the voltage differential across each transducer to range from 150 V to 245 V. As the displacement-vs-voltage response of the flexible transducers is highly nonlinear, applying a bias voltage increases the change in the displacement generated by the Boreas chip. This effect is shown through experiment and simulation in Fig. 3A and B. The relationship between displacement and voltage at −150 V is used to calibrate the decibel change of the signals during the psychophysical experiments.

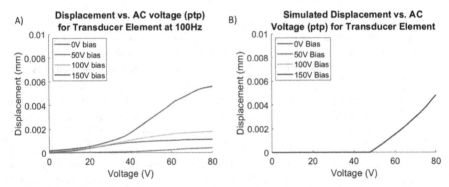

Fig. 3. A) Experimental displacement-voltage curves for an individual transducer on the array at different DC biases. B) Simulated displacement-voltage curves for the same set of parameters.

To experimentally determine the displacement of the transducer at different voltage and frequency conditions, a PDV-100 laser vibrometer was used. The vibrometer was set up such that the laser was illuminating the point of maximum extension on the buckled film. The analog vibrometer signal was then collected by a MATLAB program using a National Instruments USB-4431 DAQ, converted to a displacement signal, and correlated with the driving signal generated by the Boreas Kit. For the linear vibrometer test, a series of signals at frequencies from 1 Hz to 2000 Hz with linearly ramping amplitudes from 0 V to 80 V over 5 s were tested. Each signal was tested for 4 separate DC biases from the high voltage amplifier: 0 V, −50 V, −100 V, and −150 V. The voltage-displacement relationship of the transducers is highly nonlinear, and thus receives a significant benefit from the DC offset generated by the high voltage amplifier. This effect was also observed in user testing, in which participants could not feel vibration even at the maximum amplitude unless the bias was set to −150 V. Experiment and simulations were also conducted to understand the frequency response of the transducers. It was desired for the transducers to have large displacements from 0 to 500 Hz, as this is the frequency range of interest for most haptic stimulation. Figure 4A and B show that the transducer has a reasonably large response in the desired range, and a flat frequency dependence at lower frequencies. Lastly, we conducted simulations to explore the restoring force and actual displacement that the transducer exerted on the skin when actuating, summarized in Fig. 4C and D. For this data, the skin and the device were both approximated as connected springs with prescribed natural lengths and spring constants. The spring constant of the transducer is highly nonlinear, as the shape of the transducer changes significantly as it is compressed, so the force-displacement relationship was directly simulated in COMSOL. The results were used to build a function to describe the spring constant of the transducer at a given displacement. Human skin is a complicated mechanical system, and a wide range of elasticities have been reported for it. Skin has significant viscoelastic properties and responds differently to different displacement magnitudes and frequencies. For this analysis, we simplified elasticities in the reported range to get a sense for the potential response. As such, the results should be viewed as an approximate estimate of the interface at the skin and device.

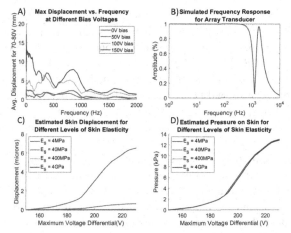

Fig. 4. A) Experimental displacement-frequency curves for a single transducer on the array at different bias voltages, showing a strong resonance around 800 Hz. B) Simulated frequency response for the transducer, showing a flat response until around 800 Hz. C) Estimated skin displacement by device for different skin elasticities. D) Estimated pressure on skin by device for different skin elasticities.

2.3 Characterization

A variety of experiments were conducted on the fabricated transducer array to determine relationships between displacement, voltage, and frequency, as well as the effects of crosstalk, noise, and complex signals. The results are shown in detail in the sections below.

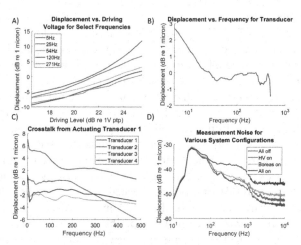

Fig. 5. Experimental displacement data for transducer array. A) Displacement vs. driving voltage for select frequencies. B) Example displacement-frequency curve at 330 V input level. C) Crosstalk levels as a function of frequency when transducer 1 is driven at 330 V. D) Measurement noise for different electrical configurations.

A laser vibrometer was used to measure the displacement of the four transducers in the array for different driving conditions. In a basic test, the displacement was measured for a parametric array of voltage and frequency conditions, and the results are summarized in Fig. 5A and B. The first figure shows how the displacement increases with driving level as expected, with higher frequencies providing lower levels of displacement at all driving levels. The second figure shows a characteristic displacement-frequency curve for a given voltage, displaying a characteristic drop-off in displacement intensity as the frequency increases, with a relatively flat region in the area of interest from ~20 Hz–300 Hz.

Figure 5C shows the results from a crosstalk test, where a single transducer (#1) was driven and the displacement of the undriven transducers was measured. We can see that the displacement of the undriven transducers was lower but still significant. The graph shows displacement differences ranging from a difference of around 8 dB at low frequencies to 4 dB at high frequencies. This significant crosstalk can be attributed to the reasonably high level of connectivity between the 4 transducers. All the buckled transducer films are fabricated from a single piece of Kapton film, and kirigami-style cuts are made to create the 4 buckled films. Since the transducers are not mechanically isolated from each other, small vibrations (on the order of 1 μm or less) can travel laterally through the film to generate vibrations in the other transducers. It is expected that at higher displacement levels, the dB difference between the driving transducer and the other transducers would be much lower. In addition, better mechanical isolation between the transducers should further reduce the crosstalk. For example, fully separating the buckled film during manufacturing will reduce crosstalk significantly.

Figure 5D summarizes results from tests conducted to determine the noise floor of the transducers under various electrical conditions. The transducer array is connected on one end to a high voltage amplifier controlled by a signal generator, and on the other end to a piezo driver driven by a desktop PC. Both of these components are a source of noise, as is the natural environment. Tests were conducted with both electrical pieces on, one on and the other off, and both off. The displacement was again measured with a laser vibrometer. The results are in line with what was expected, with all electronics on producing the most noise and all off producing the least. It was initially expected that the amplifier would be a source of more noise than the piezo driver due to the amplifier's higher power draw and voltages, but this was not the case. It may be that the noise from the PC passed through the piezo driver is larger than the amplifier noise. In addition, the sum of the independent noise from the amplifier and the driver was smaller than the noise when both were used together, indicating that there is potential interference between the two. In any case, these results indicate that the magnitude of the noise is much lower than even the lowest displacements achievable through driving the transducers.

Figure 6 explores the results of various frequency-based experiments. In Fig. 6A, the phase difference between the pre-amplified driving signal and the measured displacement is compared. It is very clear that the relationship is based on linear phase, which can be attributed to the effect of the high voltage amplifier. The transducer acts as a capacitor in-circuits, and thus one would expect its phase delay to be highly nonlinear. A linear phase indicates that the transducer suffers from very little phase delay, and in general the electronics used to drive the devices contributed more to the delay than the transducer itself. Similarly, Fig. 6B describes the group delay for various voltages, which is a low constant. This is again due to the high voltage amplifier, but the levels are low enough not to be a concern for the target frequency range.

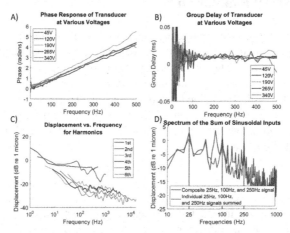

Fig. 6. Frequency analysis of transducer array. A) Phase difference between the pre-amplified driving signal and the measured displacement. B) Group delay of the transducer array for a selection of driving voltages. C) Analysis of higher harmonics for transducer array. D) Comparison of multispectral driving signal outputs.

Figure 6C explores the amplitude of higher harmonics when the transducer is driven at a frequency. The graph makes it clear that the higher-order harmonics (3rd and up) are greatly diminished compared to the fundamental and 2nd frequencies. For a few frequencies, the 2nd harmonic is even larger than the base frequency. This is expected behavior, as the transducer acts as a partial frequency-doubling system. The force experienced by the transducer is roughly proportional to the square of the voltage. Because of this, both the positive and negative portions of the driving sinusoids have the same effect, and the frequency is effectively doubled. Thus, we would expect to see similar levels for the first two frequencies, and lower levels thereafter.

Figure 6D examines the ability of the transducer array to reproduce signals composed of multiple base frequencies. The three frequencies used for the psychophysical evaluation were chosen as inputs (25 Hz, 100 Hz, and 250 Hz). Each frequency was played alone at a low amplitude to eliminate clipping, and the results were recorded. A signal composed of all three frequencies playing at once, each at the same amplitude as in the solo test, was then used to drive the transducer and the results recorded. The three solo frequency results were then summed and compared to the combined signal results, as shown in the figure. If the spectrum of the two signals matched exactly, it would indicate that the transducer was able to perfectly add frequencies together with no distortion. We can see that the results are quite similar, and the levels of the three frequencies in question are almost identical. The combined signal, however, suffers from increased amplitude at all the harmonics of the 25 Hz signal. This indicates that the summation of signals is nonlinear and may be because the relationship between voltage and displacement in the transducer is nonlinear to begin with.

3 Psychophysical Evaluation

We conducted three psychophysical experiments to determine the efficacy of the transducer array at delivering haptic stimuli to the skin. In the first experiment, the ability of the transducer array to generate higher perceived intensity levels than a single tactor was studied. In the second experiment, the participants discriminated between a stimulus produced by one tactor that contained two frequency components, and another stimulus produced by two separate tactors each driven at one of the two frequencies. The final experiment measured the ability of the transducer array to deliver four distinct haptic signals using a tactile movement illusion.

3.1 Methods

Participants. Ten participants (5 females, age 22–33 years old) took part in each of the three experiments. No participant suffered from any known sensorimotor deficiency. The participants gave their informed consent to the protocol approved by Dartmouth College's Institutional Review Board (IRB). They were compensated for their time.

Apparatus. A transducer array was mounted to the palm of the dominant hand, as shown in Fig. 2C. Each participant wore a nitrile glove sized such that the glove was slightly tight but not constricting by self-report. The glove served as an insulating failsafe in case the Kapton film suffered an electrical breakdown and it may reduce the perceptibility of vibrations on the skin. The participant rested their hand lightly on a table with the palm and device facing down. The array was mounted to the palm such that the buckled film was facing inwards, and the flexible film conformed to the skin. This caused each transducer to be lightly compressed. When a voltage is applied the buckled film is pulled away from the skin and the pressure on the skin is reduced. The bottom film of the array was connected to a high voltage amplifier controlled by a signal generator that provided a –150 V DC bias. Each of the four transducers in the buckled array was connected to a different terminal in the Boreas Kit driven by a PC, such that they could be independently controlled. A MATLAB program delivered audio signals to the Boreas chip, and the experimenter used the program to record each participant's responses. All participants wore headphones to prevent sound from the FET being used as audio cues.

Procedure. Prior to the three experiments, the perceived intensities of the four transducers were calibrated in two steps following the methods described by Reed et al. [19]. It has been shown that the equal-sensation contours for vibrotactile perception are mostly parallel in decibel scale for touch [23]. Therefore, the perceived stimulus intensity can be specified by calculating the ratio of the driving voltage over that at the human detection threshold in dB scale (called dB sensation level, or dB SL). The detection threshold was measured for each participant on a single transducer (the reference transducer) using a 3-interval, 2-alternative, 1-up 3-down adaptive forced-choice procedure [24]. On each trial, the participant received three stimulus intervals. One randomly selected interval contained a 500-ms signal at 100 Hz, while the other 2 intervals were empty. The timing of the 3 intervals were indicated visually through an LED on the Boreas kit and the inter-interval gap was kept at 500 ms. The participant's task was to indicate during which of the 3 intervals a signal was felt. The voltage was decreased after 3 consecutive correct responses and increased after 1 incorrect response. The threshold obtained with this transformed up-down method corresponds to the 79.4% point on the psychometric function [25]. The adaptive procedure continued until 8 reversals had been obtained. A reversal is defined as the voltage changing from decreasing to increasing, or vice versa. The voltage changed by 5 dB for the first 2 reversals for faster convergence, and by 2 dB for the remaining 6 reversals for better resolution of the estimated thresholds. The local maxima and minima at the last 6 reversals were averaged to get an estimate of the threshold. In the second step, the relative intensities of the four tactors were calibrated using a method of adjustment [24]. For each of the 3 non-reference tactors (target tactor), the participant felt 3 100-Hz vibrations: the first on the reference tactor, the second on the target tactor, and the third on the reference tactor again. The reference signal was always at 10 dB SL. The participant was asked to respond whether the target signal felt stronger or weaker than the reference signal. The target signal voltage was adjusted in 2-dB steps until the participant indicated that the reference and target signals were equally intense. The dB measurements for all tests were measured in terms of the equivalent zero-load displacement at that voltage. Thus, a 10 dB signal had $\sqrt{10}\times$ the displacement (as measured by a laser vibrometer) of a 0 dB SL signal, but not necessarily $\sqrt{10}\times$ the driving voltage amplitude. With the individual thresholds and adjustment values for the 3 target tactors, it was then possible to assign voltage multipliers for all four tactors to ensure that all participants felt all four tactors at desired intensities specified in dB SL during the subsequent experiments. These voltage multipliers were also used for the 25 Hz and 250 Hz signals, as the perceived intensity at different frequencies was assumed to be the same as dB SL [23].

Experiment 1: The difference in perceived intensities between single-transducer actuation and multi-transducer actuation was studied in a 1-interval, 4-alternative forced-choice identification experiment [24]. All signals were 500 ms in duration and 10 dB SL in intensity. Each participant completed 3 blocks (at 25 Hz, 100 Hz or 250 Hz, in a randomly assigned order) of 40 trials for a total of 120 trials. Before each block of 40 trials, the participant felt the 4 signals, one each with 1, 2, 3 and 4 transducers actuating simultaneously. For the signals with 1, 2, or 3 tactors, the actuated transducers were randomly chosen on each trial so as not to associate the number of actuated tactors with their locations. On each trial, one of the stimulus alternatives was chosen by randomization with replacement, and the participant indicated the perceived intensity by responding with an integer between 1 and 4. Trial-by-trial correct-answer feedback was provided. The first 10 trials in each 40-trial block were discarded to avoid training effect, and the remaining 30 trails were analyzed.

Experiment 2: The perceptual difference between two sinusoidal signals delivered via two tactors (each following a single-frequency input) and one tactor was studied in a 1-interval, 2-alternative forced-choice signal detection experiment. We hypothesized that the two signals would be indistinguishable due to the inability of the somatosensory system to assign a frequency to a particular spatial location when two tactors are located within the two-point discrimination threshold. There were three experimental conditions involving different combinations of frequencies: 25 Hz and 100 Hz, 25 Hz and 250 Hz, and 100 Hz and 250 Hz. All signals were 500 ms in duration. Each participant completed 3 blocks (in a randomly assigned order) of 40 trials for a total of 120 trials. Prior to each condition, the perceived intensities of the two stimulus alternatives were equalized using a method of adjustment. The participant first felt two tactors each being driven at one of the frequencies (Signal A in Fig. 7A), then a single tactor being driven by the sum of the two single-frequency signals (Signal B in Fig. 7B), and back to the two tactors. Once again, the tactors were randomly selected for each presentation. The voltage of the stronger signal (the two-tactor signal that was presented first and last) was set at 10 dB SL. The voltage for the weaker signal (the one-tactor signal in the middle) was adjusted until the single-tactor and two-tactor signals felt equally intense. During each 40-trial block, the two stimulus alternatives were labeled signal 1 and 2 and the participants were not informed of the nature of the two signals. On each trial, the participant felt one of the two stimuli that was selected by randomization with replacement and responded "1" or "2".

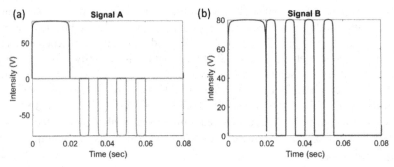

Fig. 7. A) Single period of signals sent to the two tactors for Signal A. B) Single period of signal sent to a single tactor for Signal B.

Experiment 3: The distinctiveness of movement patterns generated by the 2-by-2 tactor matrix was studied in a 1-interval, 4-alternative forced-choice identification experiment. The participant was shown a visual representation of four movement patterns: leftwards, rightwards, (clockwise) circular, and Z-shaped. The qualitative sketches of the perceived patterns are shown in Fig. 8. Prior to the experiment, the participants familiarized themselves with the patterns. The tactors were driven at 100 Hz at approximately 15 dB SL. On each trial, one of the four patterns was presented using randomization with replacement. The participant identified the pattern verbally which was recorded by the experimenter. Trial-by-trial correct-answer feedback was provided, and a total of 60 trials were collected per participant.

Data Analysis. The data from Experiment 1 were organized into 4-by-4 stimulus-response confusion matrices with the four rows corresponding to the number of tactors driven simultaneously, and the four columns the perceived intensity levels. There were a total of 30 confusion matrices for 10 participants and 3 frequencies. The data from the 10

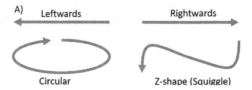

Fig. 8. A) qualitative perception of the four stimuli for experiment 3.

participants were then pooled at each frequency, resulting in three confusion matrices, one for each frequency. Percent-correct scores and information transfer were calculated from the three matrices (see [21]). For Experiment 2, the results were organized into 30 2-by-2 matrices with the rows corresponding to stimulus A and B and the columns the corresponding responses. The results were analyzed using signal detection theory to compute the sensitivity index d', a measure that is independent of response bias (see [24]). The d' scores from all 10 participants at the same frequency pair were then averaged, resulting in a total of three d' scores for the three experimental conditions. In general, a d' of above 1 indicates that the two signals are reliably distinguishable. The data from Experiment 3 were analyzed in the same manner as in Experiment 1.

3.2 Results and Discussion

Experiment 1: Figure 9 shows that the 4 levels of intensity are somewhat distinguishable, even with very limited training. The error patterns for all frequencies show that almost all errors are off by one, i.e. the participant was almost always close to the correct value. The overall percent-correct was 44% for the 25 Hz frequency, 50% for 100 Hz, 54% for 250 Hz, and 49% in aggregate. The overall error rate appeared frequency dependent, with the 25 Hz signals being more difficult to identify than the 100 Hz and 250 Hz signals. When information transfer was calculated, the values ranged from 0.47 bits at 25 Hz to 0.74 bits at 250 Hz. This means that even though the participants were able to discriminate the perceived intensities in a pairwise manner, they were unable to identify the four intensity levels. While it is theoretically possible for each transducer on the array to assume any intensity on a continuous distribution, it seems as though 4 levels are challenging for a user to identify. Indeed, 25 Hz signal is probably better suited to 2-3 intensity levels than to 4. Using multiple tactors to replicate different intensity levels can be useful when the tactor in question has a low dynamic range, or when the dynamic range is high, but the band of possible inputs is narrow and vulnerable to quantization error. In either of these cases, it may be better to use multiple transducers, each with only a few possible intensity values (or a single intensity value) to replicate the effect of a single transducer capable of taking on an infinite set of intensity values. While most transducers today fall into the latter category, that capability is unnecessary since a human cannot discriminate more than a few intensity values.

Experiment 2: The sensitivity indices for each configuration is summarized in Table 1, with the "All" row containing the average of the absolute values of the d's. the average standard deviation of d' was found to be $\sigma_{d'} = 0.41$.

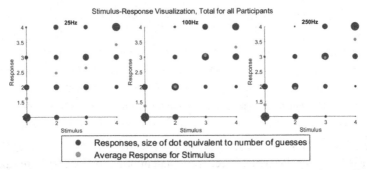

Fig. 9. Stimulus-response matrix visualization for intensity perception experiment. The red dots indicate the average response for each stimulus, while the blue dots indicate the total number of responses per stimulus, with larger dots indicating more responses. (Color figure online)

Table 1. Summary of d' values for combined frequency experiment.

Condition	d'	$\sigma_{d'}$	t-test
25 Hz & 100 Hz	0.40	0.41	1
25 Hz & 250 Hz	0.43	0.41	0
100 Hz & 250 Hz	0.47	0.41	0
All	0.44	0.41	1

By observing Table 1, it is clear that the d' values for all three configurations were below 1.0, indicating that the participants were unable to discriminate between two tactors each vibrating at a single frequency and one tactor being driven by the sum of two sinusoidal frequencies. This was expected since the distance between the two tactors were below the two-point discrimination threshold on the skin. A pairwise t-test was conducted, and the results show that the hypothesis that the 25 Hz and 100 Hz signal tests were indistinguishable passed a standard t-test, but the other two signal tests did not. The results from Experiment 2 further indicate that it may be possible to use this phenomenon to separate mechanically distinct parts of a transducer array while preserving the ability to generate complex stimuli. For example, the transducers in the current array are broadband, with a relatively flat response from 0–500 Hz, but an array could be constructed in which each transducer in the 2-by-2 array is tuned to a different resonant frequency. An array so constructed would still be able to produce results indistinguishable from a single transducer generating a complex waveform – even though each transducer in that theoretical array would be mechanically incapable of producing that signal. This would allow for transducers to be constructed with a stronger signal operating at a lower voltage, since each could be tuned for a specific frequency.

Experiment 3: The overall correct response rate was 64%. These are summarized in Fig. 10. Figure 10 explores how well participants were able to identify the four patterns, with some participants doing significantly better than others. An information transfer analysis was conducted on this data, showing that the average information transfer was 0.81 ± 0.33 bits (perfect transmission for this experiment would be 2 bits). More fine-tuned signals with a larger matrix should be able to increase the information transfer rate significantly. This study helps to confirm that the phantom tactile sensation [12] is functional even on a thin and flexible haptic matrix, and supports the idea that the other various tactile illusions will also function effectively on this device. Tactile illusions such as the phantom tactile sensation are important to be able to replicate on advanced haptic devices, as they provide increased design space for signal generation, as well as helping to create realistic sensations from relatively sparse arrays [21].

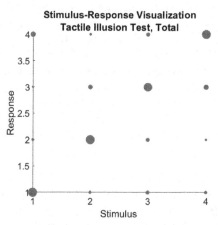

Fig. 10. Stimulus response visualization for the haptic illusion experiment. The blue dots indicate the total number of responses per stimulus, with larger dots indicating more responses. (Color figure online)

Qualitative sensation data were also collected through participant surveys at the end of each experiment. It was determined that no matter how many tactors were actuating, and no matter the signals being transduced, the device always presented to the wearer as a single stimulus point, rather than 1–4 discrete points. This is expected, as the transducer spacing is below the two-point discrimination threshold. When multiple tactors were actuated simultaneously, the phantom sensation illusion caused the single stimulus to appear to be in the middle of the two tactors as perceived by the participants, with the technical precise location governed by the relative intensity of the actuating tactors. Participants could tell when this phantom tactor moved in a certain direction but could not pinpoint its position on the skin. In other words, a phantom stimulus all the way to the right on the active array footprint followed by a stimulus in the middle felt qualitatively the same as a stimulus in the middle followed by a stimulus all the way to the left. The phantom sensation illusion also meant that certain experiments had trials that were not identical in sensation. For example, in the intensity perception experiment, a trial in which two tactors were actuated could in fact have any one of 6 tactor pairs. While this was unnoticed by the participants, this corresponds to 5 slightly different proximal stimuli (the two diagonal pairs both create a phantom tactor in the center of the device). It is likely that the location movement reduced overall participant success rates by introducing distracting cues.

4 Concluding Remarks

We have designed, fabricated, and tested a functional array of flexible electrostatic transducers. The transducers can deliver perceptible vibration to the skin at a variety of frequencies. We conducted a suite of experimental tests and simulations to determine the expected displacement and force of the transducers. An array was fabricated from Kapton, Mylar, and gold electrodes. Three psychophysical experiments were conducted with 10 participants. It was found that multiple transducers could be used to boost the perceived intensity versus a single transducer. When measuring the perceived difference between a single transducer actuating at a combined frequency and a pair of transducers each actuating at a single frequency, it was found that there was no perceivable difference, indicating that the participants could not use spatial cues to identify signals. Finally, we showed that the participants could use the transducer array to identify haptic symbols from a set of four transducers with reasonable accuracy without any training. This collection of experiments and user studies shows that this transducer array has the potential to be used in advanced haptic devices to provide increased communication capabilities.

Acknowledgements. The authors are grateful for the financial support of the National Science Foundation award (ECCS1509369), the National Institute of Health (NIH) Director's Transformative Research Award (R01HL137157), the Thayer School of Engineering PhD Innovation Program, and Facebook, Inc. under the SARA program. H. Z. Tan was partly supported by a Google Faculty Research Award 2019 and Grant No. 1954842-IIS from the National Science Foundation.

References

1. Zhou, K., et al.: Ultra-stretchable triboelectric nanogenerator as high-sensitive and self-powered electronic skins for energy harvesting and tactile sensing. Nano Energy **70**, 104546 (2020). https://doi.org/10.1016/j.nanoen.2020.104546
2. Seminara, L., et al.: Piezoelectric polymer transducer arrays for flexible tactile sensors. IEEE Sens. J. **13**(10), 4022–4029 (2013). https://doi.org/10.1109/JSEN.2013.2268690
3. van Neer, P.L.M.J., et al.: Feasibility of Using Printed Polymer Transducers for Mid-Air Haptic Feedback (2018)
4. van Neer, P., et al.: Development of a flexible large-area array based on printed polymer transducers for mid-air haptic feedback. 2019 Int. Congr. Ultrason. **38**, (2019). https://doi.org/10.1121/2.0001068
5. Kamigaki, T., Noda, A., Shinoda, H.: Thin and flexible airborne ultrasound phased array for tactile display. In: 2017 56th Annual Conference of the Society of Instrument and Control Engineers of Japan, SICE 2017, vol. 2017-Novem, pp. 736–739 (2017). https://doi.org/10.23919/sice.2017.8105623
6. Yu, X., et al.: Skin-integrated wireless haptic interfaces for virtual and augmented reality. Nature **575**(7783), 473–479 (2019). https://doi.org/10.1038/s41586-019-1687-0

7. Sorgini, F., et al.: Encapsulation of piezoelectric transducers for sensory augmentation and substitution with wearable haptic devices. Micromachines **8**(9) (2017). https://doi.org/10. 3390/mi8090270

8. Craig, J.C., Johnson, K.O.: The two-point threshold: not a measure of tactile spatial resolution. Curr. Dir. Psychol. Sci. **9**(1), 29–32 (2000). https://doi.org/10.1111/1467-8721.00054

9. Nolan, M.F.: Two-point discrimination assessment in the upper limb in young adult men and women. Phys. Ther. **62**(0031-9023) (Print), 965–969 (1982). https://doi.org/10.1093/ptj/62. 7.965

10. Novich, S.D., Eagleman, D.M.: Using space and time to encode vibrotactile information: toward an estimate of the skin's achievable throughput. Exp. Brain Res. **233**(10), 2777–2788 (2015). https://doi.org/10.1007/s00221-015-4346-1

11. Lederman, S.J., Jones, L.A.: Tactile and haptic illusions. IEEE Trans. Haptics 4(4), 273–294 (2011). https://doi.org/10.1109/TOH.2011.2

12. Israr, A., Poupyrev, I.: Tactile Brush: drawing on skin with a tactile grid display. Conference on Human Factors in Computing Systems- Proceedings, pp. 2019–2028 (2011). https://doi. org/10.1145/1978942.1979235

13. Song, J., Zhang, Y., Zhang, H., Wang, D.: Rendering Moving Tactile Stroke on the Palm Using a Sparse 2D Array, vol. 9774, no. 37, pp. 229–239 (2016). https://doi.org/10.1007/978- 3-319-42321-0

14. Trase, I., Xu, Z., Chen, Z., Tan, H., Zhang, J.X.J.: Thin-film bidirectional transducers for haptic wearables. Sens. Actuators A Phys. **303**, 111655 (2019). https://doi.org/10.1016/j.sna. 2019.111655

15. Trase, I.H., Xu, Z., Chen, Z., Tan, H.Z., Zhang, J.X.J.: Flexible electrostatic transducers for wearable haptic communication. In: Proceedings of World Haptics Conference 2019 (2019) (p. submitted)

16. Jebens, R., Trimmer, W., Walker, J.: Microactuators for aligning optical fibers. Sens. Actuators **20**(1–2), 65–73 (1989). https://doi.org/10.1016/0250-6874(89)87103-3

17. Legtenberg, R., Gilbert, J., Senturia, S.D., Elwenspoek, M.: Electrostatic curved elec-trode actuators. J. Microelectromech. Syst. **6**(3), 257–265 (1997). https://doi.org/10.1109/ 84.623115

18. Bensmaïa, S., Hollins, M., Yau, J., Carolina, N., Hill, C., Carolina, N.: Vibrotactile inten-sity and frequency information in the pacinian system: a psychophysical model. Percept. Psychophys. **67**(5), 828–841 (2005)

19. Reed, C.M., et al.: A Phonemic-Based Haptic Display for Speech communication. IEEE Trans. Haptics **12**(1), 2–17 (2019)

20. Tan, H.Z., Durlach, N.I., Reed, C.M., Rabinowitz, W.M.: Information transmission with a multifinger tactual display. Percept. Psychophys. **61**(6), 993–1008 (1999). https://doi.org/10. 3758/BF03207608

21. Tan, H.Z., Choi, S., Lau, F.W.Y., Abnousi, F.: Methodology for maximizing information transmission of haptic devices: a survey. Proc. IEEE **108**(6), 945–965 (2020). https://doi.org/ 10.1109/JPROC.2020.2992561

22. Park, G., Cha, H., Choi, S.: Transactions on haptics, and undefined 2018. Haptic Enchanters: Attachable and Detachable Vibrotactile Modules and Their Advantages (2018). https://ieeexp lore.ieee.org

23. Verrillo, R.T., Fraioli, A.J., Smith, R.L.: Sensation magnitude of vibrotactile stimuli. Percept. Psychophys. **6**(6), 366–372 (1969). https://doi.org/10.3758/BF03212793

24. Jones, L.A., Tan, H.Z.: Application of psychophysical techniques to haptic research. IEEE Trans. Haptics **6**(3), 268–284 (2013). https://doi.org/10.1109/TOH.2012.74

25. Levitt, H.: Transformed up-down methods in psychoacoustics. J. Acoust. Soc. Am. **49**(2B), 467–477 (1971)

Investigation of Sign Language Motion Classification by Feature Extraction Using Keypoints Position of OpenPose

Tsukasa Wakao[1], Yuusuke Kawakita[1], Hiromitsu Nishimura[2], and Hiroshi Tanaka[1(✉)]

[1] Department of Information and Computer Sciences, Kanagawa Institute of Technology, Atsugi, Kanagawa, Japan
s1721072@cco.kanagawa-it.ac.jp, {kwkt, h_tanaka}@ic.kanagawa-it.ac.jp
[2] Department of Information Media, Kanagawa Institute of Technology, Atsugi, Kanagawa, Japan
nisimura@ic.kanagawa-it.ac.jp

Abstract. So far, on the premise of using a monocular optical camera, sign language motion classification has been performed using a wristband and color gloves with different dyeing on each finger. In this method, the movement of sign language is detected by extracting the color region using color gloves. However, this method has problems such as the burden on the signer of wearing color gloves and the change in color extraction accuracy due to changes in ambient light, resulting in difficulty in ensuring stable classification accuracy.

Therefore, we used OpenPose, which can detect the movements of both hands without the need for colored gloves, to classify sign language movements. Feature element extraction was performed using the keypoint position obtained from OpenPose. Then, we proposed three methods as feature element for classifying each motion and compared their classification accuracy. In method 1, feature element is obtained directly from the keypoint positions of the neck, shoulder, elbow, and wrist. Method 2 is a scheme of obtaining from the relative distance from the target keypoint position around the neck. In method 3, the feature element was 30 elements, which is the sum of the 24 elements obtained in method 1 and the 6 elements obtained in method 2.

In the classification experiment, cross-validation was performed using the feature quantity obtained from the sign language motion videos of five people, and the accuracy of each method was investigated. In method 1, B (68.05%), A (62.56%), C (62.19%), D (61.49%), E (56.75%), average 62.21%, in order from the signer with the highest average classification accuracy. In method 2, B (75.31%), A (75.09%), D (73.28%), E (69.97%), C (69.81%), average 72.69%. Method 3 gave B (70.72%), A (69.65%), C (66.13%), D (64.27%), E (62.72%), and an average of 66.30% classification accuracy.

Keywords: Sign language · Motion · Classification · OpenPose · LSTM · Cross-validation

© Springer Nature Switzerland AG 2021
S. Yamamoto and H. Mori (Eds.): HCII 2021, LNCS 12765, pp. 386–399, 2021.
https://doi.org/10.1007/978-3-030-78321-1_29

1   Introduction

Sign language is widely used all over the world for communication between hearing-impaired people [1]. However, it is extremely difficult to learn any sign language, and the barriers to communication with hearing person are very large. The current situation is that written conversation and character input devices are usually used, but it takes more time than sign language. Similar to the speech translation [2] that has already been commercialized in some parts, if sign language translation becomes possible, the barrier between the hearing impaired and the hearing person can be made extremely small. Many sign language recognition studies have been conducted for this purpose [3–5].

Conventionally, as a means for detecting arms and fingers motions, the KINECT, which is a depth sensor, a gyro sensor, or a data glove, which has a built-in distortion sensor, is often used [6–8]. However, in recent years, many studies have been conducted using an optical camera. The framework for acquiring the position information of body nodes called keypoint can be provided, based on the skeletal information of the body such as OpenPose [4, 9]. By using an optical camera, it is possible to detect facial expressions and the shape of the lips, and there is a possibility that sign language classification can be realized only with the camera, and the merit of using it is great.

The authors have detected the movement with an optical camera by extracting a predetermined color region from the image of the sign language movement wearing a wristband and color gloves, and have classification the sign language movement by combining methods such as SVM and kNN [10]. Figure 1 shows the usage pattern we are assuming. Sign language movements are classification by using the built-in camera of the smartphone. Then, the result is output as a display or voice on a smartphone. However, there are problems that it is necessary to wear a wristband and gloves, and the detection accuracy of the color region changes depending on the ambient light conditions.

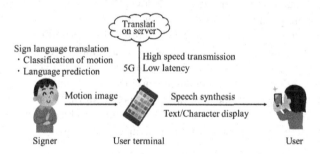

Fig. 1. Final usage image

In order to solve this problem, the authors decided to obtain the feature element from the coordinate positions of the key points that can be obtained from OpenPose. Then, in this paper, we describe the result of sign language motion classification by LSTM [11], which has been applied to many classification problems in recent years, using the feature quantity. Section 2 describes how to create sign language motion data from OpenPose, and Sect. 3 describes the features of sign language motion used in learning model creation and classification experiments from motion data. Specifically, we propose a method 1: a method of directly obtaining from the coordinate position, a method 2: a method of using the distance between key points, and a method 3: a method of using a feature element that combines method 1 and method 2. Section 4 shows LSTM and training process. Then, Sect. 5 shows the experimental results and consideration for the results.

2   Data Creation Using OpenPose

Figure 2 shows the data creation method used in this experiment. The keypoint position in obtained from the sign language motion video using OpenPose. Since there are some missing parts, the feature element is extracted after performing linear interpolation. Details will be described in Sect. 2.3.

Fig. 2. Data creation method

2.1   Sign Language Motion Video Data

The motions to be classified are included in Smart Deaf [12], which is a video dictionary for sign language learning developed and sold by KCC Co., Ltd. It is a sign language word in the health/illness category that is often used. Our laboratory has a sign language motion video database that we have acquired so far, and we use it. The followings are the conditions of capturing the motion data below [10].

The sign language motions of a signer wearing colored gloves were acquired at a 30 fps (frames per second) time interval. The resolution was 600×800 pixels. The distance between the camera and the signer was 1 m, assuming a realistic conversational situation. Video data was obtained under the supervision of the actual sign language user, who confirmed that each sign language motion was correct.

Among them, since the purpose of this study is to confirm the method, we targeted words that can be expressed by one-handed movement, which can shorten the learning model creation time. Table 1 shows the sign language words used.

Table 1. Target sign language words

Asthma	Atopic	Bald	Blushing	Breath
Cancer	Catheter	Cecum	Contact lens	Diabetes
Fever	Headache	Nap	Nausea	No smoking
Otolaryngology	Physical condition	Remove	Runny nose	Sickness
Smoking	Stroke	Take medicine	Tooth decay	Urinary

2.2   Keypoints Acquisition

With OpenPose, a total of 25 keypoint positions can be acquired when the entire body is captured by the camera. The keypoint position is acquired by inputting the sign language motion video into OpenPose as input data. After that, only the required keypoint positions are extracted.

Figure 3 shows the keypoints selected in this experiment. In this experiment, we selected the keypoints of the neck, shoulder, elbow, and wrist, which are expected to be effective when used as feature element for sign language motion classification.

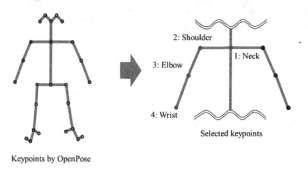

Fig. 3. Keypoints by OpenPose and selected keypoints

2.3   Preprocessing by Linear Interpolation

Since the keypoint position may not be obtained by OpenPose, linear interpolation is performed. In this experiment, feature element extraction is performed using the keypoint position after this interpolation. Figure 4 shows an example of keypoint (wrist) positions motion "diabetes" after linear interpolation.

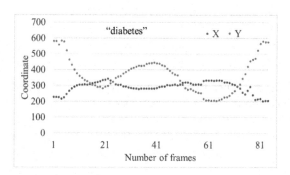

Fig. 4. Keypoints position change after linear interpolation

2.4 Dataset Configuration

Table 2 shows the data set of the sign language motion video used in the investigation. There are five signers, A, B, C, D, and E, there are 25-word types, and the number of samples per word is 25/person, so the number of total samples is $5 \times 25 \times 25 = 3125$.

Table 2. Data set prepared for experiment

	Person	Number of samples for each word	Sample total for each word
Sign language data	A, B, C, D, E	25	125 (5×25)

Word: 25 (kinds), Total: 3125 (125×25)

3 Feature Elements Creation

3.1 Method 1

From the keypoint position (x, y), the feature element of the motion of each word is created as representing the position, speed, and locus of the sign language motion [10]. The position was calculated by normalizing the keypoint position with a video resolution of 800 × 600 [pixel]. This feature element indicates the motion position in the view range of the camera. The elements are obtained by expression 1.

$$\left. \begin{array}{l} Px_{ij} = \frac{x_{ij}}{800} \\[4pt] Py_{ij} = \frac{y_{ij}}{600} \end{array} \right\} \tag{1}$$

Here,

- i: i^{th} frame of motion data
- j: keypoint (1: neck, 2: shoulder, 3: elbow, 4: wrist)

The velocity was calculated from the difference between the frames at the keypoint position. The valued are calculated from expression 2.

$$\left. \begin{array}{l} Vx_{ij} = x_{ij} - x_{i-1j} \\ Vy_{ij} = y_{ij} - y_{i-1j} \end{array} \right\} \tag{2}$$

Here,

- i: i^{th} frame of motion data
- j: keypoint (1: neck, 2: shoulder, 3: elbow, 4: wrist)

The locus was calculated by normalizing the magnitude and position of the movement using the mean value and its standard deviation of the keypoint position. Its expression is as follows.

$$\left.\begin{array}{l} Tx_{ij} = \frac{x_{ij}-\bar{x}}{A} \\ Ty_{ij} = \frac{y_{ij}-\bar{y}}{A} \end{array}\right\} \tag{3}$$

Here,

- i: i^{th} frame of motion data
- j: keypoint (1: neck, 2: shoulder, 3: elbow, 4: wrist)
- A: standard deviation
- n: the total number of frames of sign language motion data

$$A = \sqrt{\frac{1}{n}\sum_{i=1}^{n}\left((x_{ij}-\bar{x})^2 + (y_{ij}-\bar{y})^2\right)}$$

$$\left.\begin{array}{l} \bar{x} = \frac{1}{n}\sum_{i=1}^{n}x_{ij} \\ \bar{y} = \frac{1}{n}\sum_{i=1}^{n}y_{ij} \end{array}\right\}$$

These above operations were applied to the keypoint positions of the neck, shoulder, elbow, and wrist, and a total of 24 elements of $4 \times 3 \times 2$ were obtained as feature element.

3.2 Method 2

The two-dimensional vectors showing the positional relationship from the neck to the shoulder, the neck to the elbow, and the neck to the wrist were obtained with reference to the neck at the keypoint position. Each element is normalized by the distance from the neck to the shoulder, and their calculated value were used as the feature element [13]. Calculation formula 4 is shown below.

$$\left.\begin{array}{l} x_{ij} = \frac{x_{ij}-x_{i1}}{l_i} \\ y_{ij} = \frac{y_{ij}-y_{i1}}{l_i} \end{array}\right\} \tag{4}$$

Here,

- i: i^{th} frame of motion data
- j: keypoint (1: neck, 2: shoulder, 3: elbow, 4: wrist)

$$l_i = \sqrt{x_{i1}^2 + y_{i1}^2}$$

By the above operation, a total of 6 elements of 3×2 were acquired as feature element.

3.3 Method 3

The feature quantity was 30 elements, which is the sum of the 24 elements feature element calculated by method 1 and the 6 elements feature element calculated by method 2.

4 Classifier Creation

4.1 LSTM Structure and Parameters

The structure of the LSTM network and the setting parameters were in Table 3, those are the same as in our past investigation [14].

Table 3. LSTM network and setting parameters

Structure	
Hidden units	100
Fully connected layer	25
Parameters	
Max epoch	600
Solver	Adam
Mini batch size	128

4.2 Training Model

When creating the training model, we used a total of 2500 (=625 × 4) sign language motion videos of B, C, D, E/A, C, D, E/A, B, D, E/A, B, C, E/A, B, C, D out of a total of 3125 on the premise of cross-validation. We decided to create 3 training models for each of these 5 data set, that is, a total of $5 \times 3 = 15$ for each method for confirmation of result stability. The operating environment was OS: Windows 10 Education, CPU: Intel Core i5-8400, GPU: GeForce GTX 1070Ti and memory: 16 GB.

When creating a training model, training ends when the maximum EPOCH is reached, and the training model is saved. Therefore, the training loss converged halfway, but there are cases where it deviates from the convergence due to recalculation. A training model that deviates from convergence has low classification accuracy and does not perform optimal training. In order to solve this problem, we set the training end condition to end training when the average of the difference (absolute value) of the training losses of the last 6 times is less than $1e^{-4}$. Figure 5 shows an example of a training curve that actually converges according to the end condition.

The training time required at this case was 42 min. and 13 s. In this experiment, all training satisfied the end condition and converges, the required times were as follows.

- Method 1: 6 min 38 s to 28 min 7 s (average: 15 min 29 s standard deviation: 349 s)

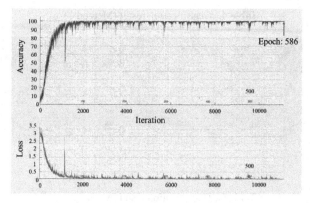

Fig. 5. Example of training history

- Method 2: 11 min 15 s to 42 min 13 s (average: 24 min 33 s standard deviation: 460 s)
- Method 3: 5 min 52 s to 27 min 5 s (average: 12 min 52 s standard deviation: 290 s)

The training time for method 2 (6 elements) with the least feature element was the longest, and the training time for method 3 (30 elements) with the largest feature element was the shortest. In addition, there was no significant change in the training time between method 1 (24 elements) and method 3 (30 elements).

5 Experimental Results and Discussion

5.1 Experimental Results

Cross-validation was performed to evaluate the stability of classification accuracy. Before that, in order to confirm each of the created training models, the data (5 patterns) of 4 people used in the training were used as training data, and the data was input to each created training model to examine the validity of the training model. After that, the data (5 patterns) of one person other than the person used in the training was input as evaluation data, and the classification accuracy was evaluated.

Result by Method 1. The classification results are shown in Table 4. In method 1, the classification accuracy of 95.24 to 99.36% was obtained for the data used in the training, and it is considered that the training was sufficiently performed. When the evaluation data was used, the classification accuracy of 53.92 to 69.92% was obtained. The average classification accuracy of each signer was B (68.05%), A (62.56%), C (62.19%), D (61.49%), E (56.75%) in order from the highest signer.

Result by Method 2. The classification results are shown in Table 5. In method 2, the classification accuracy of 91.92 to 98.76% was obtained for the training data, and it is considered that the training was sufficiently performed even in this case. When the evaluation data was used, the classification accuracy of 61.12 to 76.96% was obtained. The average classification accuracy of each signer was B (75.31%), A (75.09%), D (73.28%), E (69.97%), C (69.81%) in order from the highest signer.

Table 4. Classification result (Method 1)

Signer	Sex	Physique (shoulder width)	1st time		2nd time		3rd time		Average	
			Training	Evaluation	Training	Evaluation	Training	Evaluation	Training	Evaluation
A	M	217	97.52	69.92	96.24	60.00	95.24	57.76	96.33	62.56
B	M	215	98.72	69.12	99.36	68.32	97.52	66.72	98.53	68.05
C	M	203	98.16	65.60	98.56	62.40	98.40	58.56	98.37	62.19
D	M	204	98.72	60.00	98.92	64.32	98.32	60.16	98.65	61.49
E	M	194	97.80	60.00	98.76	53.92	98.12	56.32	98.23	56.75
Average			98.18	64.93	98.37	61.79	97.52	59.90	98.02	62.21

[%]

Table 5. Classification result (Method 2)

Signer	Sex	Physique (shoulder width)	1st time		2nd time		3rd time		Average	
			Training	Evaluation	Training	Evaluation	Training	Evaluation	Training	Evaluation
A	M	217	97.72	76.32	97.16	76.32	97.72	72.64	97.53	75.09
B	M	215	96.64	75.04	95.56	73.92	95.84	76.96	96.01	75.31
C	M	203	96.84	61.12	96.20	72.80	97.92	75.52	96.99	69.81
D	M	204	95.32	71.04	98.76	75.84	97.88	72.96	97.32	73.28
E	M	194	91.92	74.56	95.00	68.64	96.84	66.72	94.59	69.97
Average			95.69	71.62	96.54	73.50	97.24	72.96	96.49	72.69

[%]

Results by Method 3. The classification results are shown in Table 6. Method 3 obtained a classification accuracy of 96.32 to 99.60% for the training data, which is considered to be the best training process among the three methods. However, the classification accuracy of 59.04 to 73.92% was obtained. When the evaluation data was used, the average classification accuracy of each signer was B (70.72%), A (69.65%), C (66.13%), D (64.27%), E (62.72%) in order from the highest signer.

Table 6. Classification result (Method 3)

Signer	Sex	Physique (shoulder width)	1st time		2nd time		3rd time		Average	
			Training	Evaluation	Training	Evaluation	Training	Evaluation	Training	Evaluation
A	M	217	98.60	72.32	99.24	70.08	97.20	66.56	98.35	69.65
B	M	215	98.24	65.28	99.28	73.92	98.60	72.96	98.71	70.72
C	M	203	99.60	69.76	98.12	63.20	98.84	65.44	98.85	66.13
D	M	204	97.64	59.04	98.88	69.12	97.16	64.64	97.89	64.27
E	M	194	97.44	64.16	96.32	61.12	97.52	62.88	97.09	62.72
Average			98.30	66.11	98.37	67.49	97.86	66.50	98.18	66.70

[%]

5.2 Discussion

Table 7 shows the average, standard deviation, maximum value, minimum value, person with the highest average, and person with the lowest average from aspect of classification accuracy for each method.

Table 7. Summarized classification result

	Method 1	Method 2	Method 3
Average	62.21	72.69	66.70
Standard deviation	4.7	4.1	4.3
Max. value	69.92	76.96	73.92
Min. value	53.92	61.12	59.04
Signer with highest (average)	B (68.05)	B (75.31)	B (70.72)
Signer with lowest (average)	E (56.75)	E (69.81)	E (62.72)

[%]

Looking at the average value of the classification accuracy, the method 2 gave the highest result even though the feature element was the smallest. Next is the method 3. The worst result was method 1. From at the standard deviation, the variability of the results was similar for all methods. The maximum and minimum values were the same as the average value. From the results of this experiment, it is considered that the feature element extraction method of the method 2 is superior to the method 1. It was also found that the classification accuracy of the method 3 was an intermediate result between the method 2 and the method 1, and that the method of simply increasing the feature element cannot improve the classification accuracy.

Next, when comparing the signer with the highest average classification accuracy, the classification accuracy with the evaluation data set to the signer B was the highest in all methods. Comparing the signer with the lowest average classification accuracy, the results were obtained as the signer E for the method 1 and the method 3, and the signer C for the method 2. In the method 1, the feature element calculation formula (1) shows at which position within the view range of the camera the sign language motion was performed. It is possible that signer E physique is the smallest in size, and the position data is different from the others.

On the other hand, in the method 2, the feature element calculation formula (4) is normalized so as to absorb the influence of the difference in physique, so it seems that the effect is appaired. In the method 3, the feature element is the sum of the 24 elements of the method 1 and the 6 elements of the method 2, so the result was influenced by the method 1.

Table 8 shows the top 3 sign language words with the highest average classification accuracy among the signer with the highest average classification accuracy.

Table 8. Top 3 sign language words with high accuracy

	1ˢᵗ rank	2ⁿᵈ rank	3ʳᵈ rank
Method 1 (B)	atopic, breath, cecum, fever	diabetes	smoking
Method 2 (B)	cecum, fever	atopic, nausea	catheter
Method 3 (B)	cecum, fever, tooth-decay	smoking	diabetes

As can be seen from Table 8, "cecum" and "fever" are common in the first word. Therefore, we focused on these words and confirmed the sign language motion video. Figure 6 shows the sign language motion of the two words.

 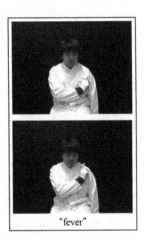

"cecum"   "fever"

Fig. 6. Sign language motion with high accuracy

Both of these sign language words were simple motions and short sign language motions, and the motions themselves were clearly different. This seems to be an easy-to-classify motion.

Table 9 shows the sign language word worst3, which has the lowest classification accuracy among the signer with the lowest average classification accuracy.

As can be seen from Table 9, the sign language word "sickness" is common. Therefore, we focused on the sign language word "sickness" here. Figure 7 shows an example of the confusion matrix.

Table 9. Worst 3 sign language words with low accuracy

	1st rank	2nd rank	3rd rank
Method 1 (F)	sickness	contact-lens	blushing
Method 2 (C)	sickness	tooth-decay	remove
Method 3 (F)	smoking	sickness	blushing, no-smoking, tooth-decay

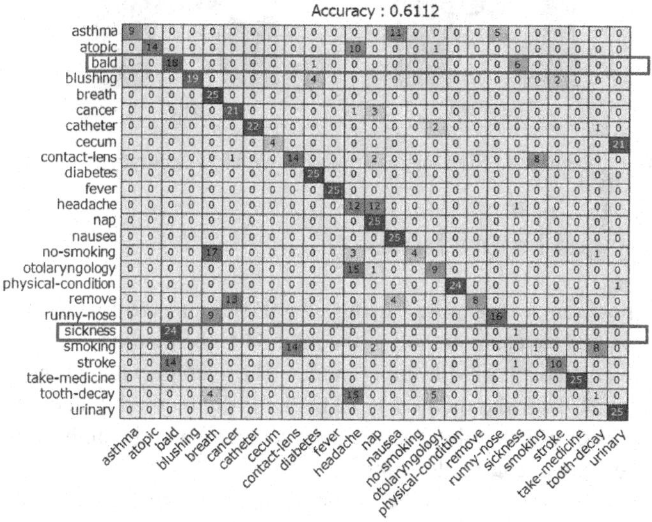

Accuracy : 0.6112

	asthma	atopic	bald	blushing	breath	cancer	catheter	cecum	contact-lens	diabetes	fever	headache	nap	nausea	no-smoking	otolaryngology	physical-condition	remove	runny-nose	sickness	smoking	stroke	take-medicine	tooth-decay	urinary
asthma	9	0	0	0	0	0	0	0	0	0	0	0	11	0	0	0	0	5	0	0	0	0	0	0	0
atopic	0	14	0	0	0	0	0	0	0	0	10	0	0	0	1	0	0	0	0	0	0	0	0	0	0
bald	0	0	18	0	0	0	0	0	0	1	0	0	0	0	0	0	0	0	6	0	0	0	0	0	0
blushing	0	0	0	19	0	0	0	0	0	0	0	4	0	0	0	0	0	0	0	0	2	0	0	0	0
breath	0	0	0	0	25	0	0	0	0	0	0	0	0	0	0	0	0	0	0	0	0	0	0	0	0
cancer	0	0	0	0	0	21	0	0	0	0	1	3	0	0	0	0	0	0	0	0	0	0	0	0	0
catheter	0	0	0	0	0	0	22	0	0	0	0	0	0	0	0	0	0	2	0	0	0	0	0	1	0
cecum	0	0	0	0	0	0	0	4	0	0	0	0	0	0	0	0	0	0	0	0	0	0	0	0	21
contact-lens	0	0	0	0	0	0	1	0	14	0	0	2	0	0	0	0	0	0	0	0	8	0	0	0	0
diabetes	0	0	0	0	0	0	0	0	0	25	0	0	0	0	0	0	0	0	0	0	0	0	0	0	0
fever	0	0	0	0	0	0	0	0	0	0	25	0	0	0	0	0	0	0	0	0	0	0	0	0	0
headache	0	0	0	0	0	0	0	0	0	0	12	12	0	0	0	0	0	1	0	0	0	0	0	0	0
nap	0	0	0	0	0	0	0	0	0	0	0	0	25	0	0	0	0	0	0	0	0	0	0	0	0
nausea	0	0	0	0	0	0	0	0	0	0	0	0	0	25	0	0	0	0	0	0	0	0	0	0	0
no-smoking	0	0	0	0	17	0	0	0	0	0	0	3	0	0	4	0	0	0	0	0	0	0	0	1	0
otolaryngology	0	0	0	0	0	0	0	0	0	0	15	1	0	0	9	0	0	0	0	0	0	0	0	0	0
physical-condition	0	0	0	0	0	0	0	0	0	0	0	0	0	0	0	24	0	0	0	0	0	0	0	0	1
remove	0	0	0	0	13	0	0	0	0	0	0	0	0	0	4	0	0	8	0	0	0	0	0	0	0
runny-nose	0	0	0	9	0	0	0	0	0	0	0	0	0	0	0	0	0	16	0	0	0	0	0	0	0
sickness	0	0	24	0	0	0	0	0	0	0	0	0	0	0	0	0	0	0	0	0	0	0	0	0	0
smoking	0	0	0	0	0	0	14	0	0	0	2	0	0	0	0	0	0	0	1	0	0	8	0	0	0
stroke	0	0	14	0	0	0	0	0	0	0	0	0	0	0	0	0	0	1	0	10	0	0	0	0	0
take-medicine	0	0	0	0	0	0	0	0	0	0	0	0	0	0	0	0	0	0	0	0	0	25	0	0	0
tooth-decay	0	0	0	0	4	0	0	0	0	0	15	0	0	0	5	0	0	0	0	0	0	0	1	0	0
urinary	0	0	0	0	0	0	0	0	0	0	0	0	0	0	0	0	0	0	0	0	0	0	0	0	25

Fig. 7. Example of confusion matrix with low accuracy

Looking at Fig. 7, the sign language word "sickness" has been classified as "bald". Therefore, we confirmed the sign language motion videos of "sickness" and "bald". Figure 8 shows the sign language motion of the two words.

Both words are motions that bring the hand over the head, and it can be judged that it is difficult to classify them only by the keypoint positions of the neck, shoulder, elbow, and wrist that are currently used. Therefore, the classification accuracy may be improved by adding a new finger keypoint position.

"sickness"   "bald"

Fig. 8. Sign language motion with low accuracy

6 Summary and Future Issues

The keypoint position was acquired using OpenPose, and the feature elements were devised and obtained for each method of the method 1, the method 2, and the method 3. Using them, we created a LSTM training model. The average value of the classification accuracy was 62.21% for the method 1, 72.69% for the method 2, and 66.70% for the method 3. The result was that the classification accuracy was the highest in the Method 2, which has the fewest feature elements.

The keypoint positions used in the feature element extraction are the four elements of neck, shoulder, elbow, and wrist, but the expected results could not be obtained for sign language words that have similar keypoint position changes and different finger motions. Therefore, in the future, it will be necessary to use the keypoint positions of the finger of the hand that can be obtained by OpenPose.

In this experiment, LSTM was used as a classifier, but if the optimum structure can be found for the increase in feature elements, the classification accuracy may be improved. Furthermore, it is necessary to apply conventional machine training methods such as SVM, and DT and compare the classification accuracy.

References

1. Pfau, R., Steinbach, M., Woll, B.: Sign Language: An International Handbook, p. 1138. Walterde Gruyter, Berlin (2012)
2. Sourcenext Corporation, Pockettalk. https://pocketalk.jp/. (in Japanese)
3. Tornay, S., Razavi, M., Camgoz, C.N., Bowden, R., Doss, M.M.: HMM-based approaches to model multichannel information in sign language inspired from articulatory features-based speech processing. In: IEEE International Conference on Acoustics, Speech and Signal Processing (ICASSP), pp. 2817–2821 (2019). https://doi.org/10.1109/ICASSP.2019.8683167

4. Park, C.-I., Sohn, C.-B.: Data augmentation for human keypoint estimation deep learning based sign language translation. Electronics **9**(8), 1257 (2020). https://doi.org/10.3390/electronics9081257
5. Mukushev, M., Sabyrov, A., Imashev, A., Koishybay, K., Kimmelmany, V., Sandygulova, A.: Evaluation of manual and non-manual components for sign language recognition. In: Proceedings of the 12th Conference on Language Resources and Evaluation (LREC 2020), pp. 6073–6078 (2020)
6. Zafrulla, Z., Brashear, H., Starner, T., Hamilton, H., Presti, P.: American sign language recognition with the Kinect. In: Proceedings of the 13th International Conference on Multimodal Interfaces, pp. 276–286 (2011)
7. Jitcharoenpory, R., Senechakr, P., Dahlan, M., Suchato, A., Chuangsuwanich, E., Punyabukkana, P.: Recognizing words in Thai Sign Language using flex sensors and gyroscopes. In: i-CREATe2017, 4 p. (2017)
8. Oz, C., Leu, C.M.: American sign language word recognition with a sensory glove using artificial neural networks. Eng. Appl. Artif. Intell. **24**(7), 1204–1213 (2011)
9. Ko, K.S., Kim, J.C., Jung, H., Cho, C.: Neural sign language translation based on human keypoint estimation. Appl. Sci. **9**, 2683 (2019). https://doi.org/10.3390/app9132683
10. Ozawa, T., Okayasu, Y., Dahlan, M., Nishimura, H., Tanaka, H.: Investigation of sign language recognition performance by integration of multiple feature elements and classifiers. In: Yamamoto, S., Mori, H. (eds.) HIMI 2018. LNCS, vol. 10904, pp. 291–305. Springer, Cham (2018). https://doi.org/10.1007/978-3-319-92043-6_25
11. Liu, G., Guo, J.: Bidirectional LSTM with attention mechanism and convolutional layer for text classification. Neurocomputing **337**, 325–338 (2019). https://doi.org/10.1016/j.neucom.2019.01.078
12. KCC Corporation, Smart Deaf. http://www.smartdeaf.com/. (in Japanese)
13. Yanagisawa, T., Ishikawa, T., Watanabe, H.: Sign language analysis with monocular RGB camera. In: IEICE General Conference, D-12-34, p. 70 (2019). (in Japanese)
14. Kawaguchi, K., Wang, W., Ohta, E., Nishimura, H., Tanaka, H.: Basic investigation of sign language motion classification by feature extraction using pre-trained network models. In: IEEE Pacific Rim Conference on Communications, Computers and Signal Processing (PacRim2019), 4 p. (2019)

Author Index

Printed in the United States
by Baker & Taylor Publisher Services